D1349534

Mims' Pathogenesis
of
Infectious Disease

Mims' Pathogenesis
of
Infectious Disease

Fifth Edition

Cedric A. Mims
Formerly
Department of Microbiology
Guy's Hospital Medical School, UMDS
London, UK

Anthony Nash
Department of Veterinary Pathology
University of Edinburgh
Edinburgh, UK

John Stephen
School of Biological Sciences
University of Birmingham
Edgbaston, UK

ACADEMIC PRESS

An imprint of Elsevier Science

Amsterdam Boston London New York Oxford Paris
San Diego San Francisco Singapore Sydney Tokyo

Academic Press
An imprint of Elsevier Science
84 Theobald's Road, London WC1X 8RR, UK
http://www.academicpress.com

Academic Press
An imprint of Elsevier Science
525 B Street, Suite 1900, San Diego, California 92101-4495, USA
http://www.academicpress.com

ISBN 0–12–498264–6 (HB)
0–12–498265–4 (PB)

Library of Congress Card Number: 00–105900

A catalogue record for this book is available from the British Library

Typeset by Phoenix Photosetting, Chatham, Kent
Printed and bound in Great Britain by MPG Books Ltd, Cornwall, UK

02 03 04 05 06 MP 9 8 7 6 5 4 3

Preface to the Fifth Edition

The past five years have seen the extraordinary development of gene sequencing of microorganisms. Indeed, earlier days have been referred to as 'pregenomic'. At the time of writing, about 30 different microbes have been completely sequenced, including *M. tuberculosis*, *T. pallidum* and *H. pylori*, and many more are expected in the coming years. In principle, this gives us the ability to probe the inner depths of microbial pathogenicity, but it is not easy and so far the big spinoff for our understanding of pathogenesis is promised rather than delivered. We have to identify the key genes and then find out how the gene products operate. It has been said that the gene sequence of a microbe is like the Rosetta stone – impressive to see, but to have value it must be translated. It will be an immense help if we can become better at predicting protein function from sequence.

Infectious diseases are thriving. In the case of smallpox (and soon polio), the disease can be eliminated from the earth before details of its pathogenesis have been unravelled. Nevertheless, we need to keep studying pathogenesis, because understanding it lends a helping hand to therapy, control of transmission, vaccine development and to the science of immunology. It is no accident that the recent Nobel laureates, Peter Doherty and Rolf Zinkernagel, made their discovery of the MHC restriction of cytotoxic T-cells in the course of studies on the pathogenesis of a virus infection of mice.

The book has been updated, while retaining the structure of earlier editions. Pathogenetic principles remain much the same, but we now see the details slowly filled in.

C. A. Mims

Preface to the Fourth Edition

Further advances in immunology and in the molecular analysis of pathogenesis make this new edition overdue. Microbial toxins, in particular, are well-represented. This time the original author has had the good fortune to be assisted by Professor Nigel Dimmock, Professor Tony Nash and Dr John Stephen. We hope that the original flavour of the book has not been lost.

The chapter sequence is unchanged. It is becoming more fashionable to think of infection as a *conflict* between parasite and host as a series of host defences and the parasite's 'answer' to them, as set out in the First Edition. Infectious diseases continue to threaten us, and the study of pathogenesis prospers. Indeed, it seems likely that more basic research on pathogenesis is needed if we are to develop good vaccines or therapies for AIDS, or if we are to understand how new infections (from vertebrates, arthropods or the environment) may learn the crucial trick of transmission from person to person.

C. A. Mims

Preface to the Third Edition

I have once again updated the text, but the general layout of the book still seems appropriate and has not been altered. I continue to look at things from the point of view of the infectious agent, which is perhaps becoming a more respectable thing to do.

The first edition was written in 1975–6 and the last line in the text contained my ultimate justification for the study of pathogenesis. It refers to the need for greater knowledge of disease processes and pathogenicity because it helps with 'our ability to deal with any strange new pestilences that arise and threaten us', and it is still included on page 386. The emergence of AIDS has provided an immense stimulus to pathogenesis studies, particularly those dealing with the interaction of viruses with the immune system.

Studies of microbial pathogenesis are flourishing these days, and the final analysis of virulence at the molecular level has begun. Molecular biology, pathology and immunology will come together to explain just how a given gene product contributes to disease, giving not only intellectual satisfaction to scientists but also a rich fallout for human and veterinary medicine.

December, 1986   C. A. Mims

Preface to the
Second Edition

I have brought things up to date, especially in the fast moving fields of phagocytes and immunology. Important subjects such as diarrhoea and persistent infections are given greater attention, and there is a brief look at infectious agents in human diseases of unknown aetiology, a confusing area for which an overview seemed timely. Otherwise the general layout of the book is unaltered, and it is still quite short. There are more references, but not too many, and they remain at the end of the chapters so as not to weigh down the text or inhibit the generalisations!

At times I have been taken to task for calling viruses 'microorganisms'. I do so because there is no collective term embracing viruses, chlamydias, rickettsias, mycoplasmas, bacteria, fungi and protozoa other than 'infectious agents', and for this there is no equivalent adjective. I prefer to take this particular liberty with microbiological language rather than stay tied to definitions (which can end up with tautologies such as 'viruses are viruses').

March, 1982   C. A. Mims

Preface to the First Edition

For the physician or veterinarian of course, the important thing about microorganisms is that they infect and cause diseases. Most textbooks of medical microbiology deal with the subject either microbe by microbe, or disease by disease. There are usually a few general chapters on the properties of microorganisms, natural and acquired resistance to infection etc., and then the student reads separately about each microbe and each infectious disease. It is my conviction that the centrally significant aspect of the subject is the mechanism of microbial infection and pathogenicity, and that the principles are the same, whatever the infectious agent. When we consider the entry of microorganisms into the body, their spread through tissues, the role of immune responses, toxins and phagocytes, the general features are the same for viruses, rickettsiae, bacteria, fungi and protozoa. This book deals with infection and pathogenicity from this point of view. All microorganisms are considered together as each part of the subject is dealt with. There are no systematic accounts of individual diseases, their diagnosis or their treatment, but the principal microorganisms and diseases are included in a series of tables and a figure at the end of the book.

Just as the virologist has needed to study not only the virus itself but also the cell and its responses to infection, so the student of infectious diseases must understand the body's response to infection as well as the properties of the infecting microorganism. It is hoped that this approach will give the reader an attitude towards infection and pathogenicity that will be relevant whatever the nature of the infectious agent and whatever the type of infectious disease. Most of the examples concern infections of man, but because the principles apply to all infections, the book may also prove of value for the student of veterinary or general science.

C. A. Mims

For creatures your size I offer
    a free choice of habitat,
so settle yourselves in the zone
    that suits you best, in the pools
of my pores or the tropical
    forests of arm-pit and crotch,
in the deserts of my fore-arms,
    or the cool woods of my scalp

Build colonies: I will supply
    adequate warmth and moisture,
the sebum and lipids you need,
    on condition you never
do me annoy with your presence,
    but behave as good guests should
not rioting into acne
    or athlete's-foot or a boil.

From: 'A New Year Greeting' by W. H. Auden. (Epistle to a Godson and other poems. Published by Faber and Faber (UK) and Random House, Inc. (USA).)

Contents

1

General Principles

In general biological terms, the type of association between two different organisms can be classified as parasitic, where one benefits at the expense of the other, or symbiotic (mutualistic), where both benefit. There is an intermediate category called commensalism, where only one organism derives benefit, living near the other organism or on its surface without doing any damage. It is often difficult to use this category with confidence, because an apparently commensal association often proves on closer examination to be really parasitic or symbiotic, or it may at times become parasitic or symbiotic.

The same classification can be applied to the association between microorganisms and vertebrates. Generalised infections such as measles, tuberculosis or typhoid are clearly examples of parasitism. On the other hand, the microflora inhabiting the rumen of cows or the caecum of rabbits, enjoying food and shelter and at the same time supplying the host with food derived from the utilisation of cellulose, are clearly symbiotic. Symbiotic associations perhaps also occur between humans and their microbes, but they are less obvious. For instance, the bacteria that inhabit the human intestinal tract might theoretically be useful by supplying certain vitamins, but there is no evidence that they are important under normal circumstances. In malnourished individuals, however, vitamins derived from intestinal bacteria may be significant, and it has been recorded that in individuals with subclinical vitamin B_1 (thiamine) deficiency, clinical beri-beri can be precipitated after treatment with oral antibiotics. Presumably the antibiotics act on the intestinal bacteria that synthesise thiamine.

The bacteria that live on human skin and are specifically adapted to this habitat might at first sight be considered as commensals. They enjoy shelter and food (sebum, sweat, etc.) but are normally harmless. If the skin surface is examined by the scanning electron microscope, the bacteria, such as *Staphylococcus epidermidis* and *Proprionibacterium acnes*, are seen in small colonies scattered over a moon-like

1

landscape. The colonies contain several hundred individuals* and tend to get smeared over the surface. Skin bacteria adhere to the epithelial squames that form the cornified skin surface, and extend between the squames and down the mouths of the hair follicles and glands onto the skin surface. They can be reduced in numbers, but never eliminated, by scrubbing and washing, and are most numerous in moister regions such as the armpit, groin and perineum. The dryness of the stratum corneum makes the skin an unsuitable environment for most bacteria, and merely occluding and thus hydrating an area with polythene sheeting leads to a large increase in the number of bacteria. The secretions of apocrine sweat glands are metabolised by skin bacteria, and odoriferous amines and other substances such as 16-androstene steroids are produced, giving the body a smell that modern man, at least, finds offensive.† Deodorants, containing aluminium salts to inhibit sweating, and often antiseptics to inhibit bacterial growth, are therefore applied to the apocrine gland areas in the axillae. But for other mammals, and perhaps primitive man, body smells have been of great significance in social and sexual life. Not all body smells are produced by bacteria, and skin glands may secrete substances that are themselves odoriferous. Skin bacteria nevertheless contribute to body smells and could for this reason be classified as symbiotic rather than parasitic. There is also evidence that the harmless skin bacteria, by their very presence, inhibit the growth of more pathogenic bacteria, again indicating benefit to the host and a symbiotic classification for these bacteria.

A microbe's ability to multiply is obviously of paramount importance; indeed, we call a microbe dead or nonviable if it cannot replicate.‡ The ability to spread from host to host is of equal importance. Spread can be horizontal in a species, one individual infecting another by contact, via insect vectors and so on (Fig. 1.1). Alternatively spread can be 'vertical' in a species, parents infecting offspring via sperm, ovum, the placenta, the milk, or by contact. Clearly if a microbe does not spread from individual to individual it will die with the individual, and cannot persist in nature. The crucial significance of the ability of a microbe to spread can be illustrated by comparing the horizontal spread of respiratory and venereal infections. An infected individual can transmit influenza or the common cold to a score of others in the course of an

---

* The average size of these colonies is determined by counting the total number of bacteria recovered by scrubbing and comparing this with the number of foci of bacterial growth obtained from velvet pad replicas. The sterile pad is applied firmly to the skin, then removed and applied to the bacterial growth plate.

† The smell of feet encased in shoes and socks is characteristic, and in many European languages it is referred to as cheese-like. Between the toes lives *Brevibacterium epidermidis*, which converts L-methionine to methane thiol, a gas that contributes to the smell. A very similar bacterium is added to cheeses such as Brie to enhance odour and flavour.

‡ Sterilisation is the killing of all forms of microbial life, and appropriately the word means making barren, or devoid of offspring.

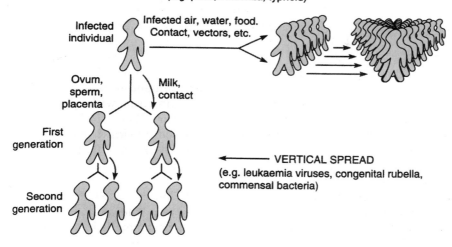

HORIZONTAL SPREAD
(e.g. polio, influenza, typhoid)

Infected individual

Infected air, water, food. Contact, vectors, etc.

Ovum, sperm, placenta

Milk, contact

First generation

Second generation

VERTICAL SPREAD
(e.g. leukaemia viruses, congenital rubella, commensal bacteria)

Fig. 1.1 Vertical and horizontal transmission of infection.

innocent hour in a crowded room. A venereal infection also must spread progressively from person to person if it is to maintain itself in nature, but even the most energetic lover could not transmit a venereal infection on such a scale. A chain of horizontal infection in this case, however, requires a chain of venery (sexual relations) between individuals. If those infected at a given time never had sexual relations with more than one member of the opposite sex, the total incidence could double in a lifetime, and when the infected people died the causative microbe would be eliminated. In other words, venereal infections must be transmitted to more than one member of the opposite sex if they are to persist and flourish. The greater the degree of sexual promiscuity, the greater the number of sex partners, the more successful such infections can be. Further discussion of sexually transmitted infection is included in the next chapter.

Only a small proportion of the microorganisms associated with humans give rise to pathological changes or cause disease. Vast numbers of bacteria live harmlessly in the mouth and intestines, on the teeth and skin, and most of the 150 or so viruses that infect humans cause no detectable illness in most infected individuals, in spite of cell and tissue invasion. This is to be expected because, from an evolutionary point of view, successful microbes must avoid extinction, persist in the world, multiply, and leave descendants. A successful parasitic microbe lives on or in the individual host, multiplies, spreads to fresh individuals, and thus maintains itself in nature (Table 1.1).

A successful parasitic microbe, like all successful parasites, tends to get what it can from the infected host without causing too much damage. If an infection is too often crippling or lethal, there will be a

Table 1.1.   Obligatory steps for infectious microorganisms

Step	Phenomenon	Requirement	Chapter
1. Attachment ± entry into body	Infection (entry)	Evade host's natural protective and cleansing mechanisms	2
2. Local or general spread in the body	Local events, spread	Evade immediate local defences, and the natural barriers to spread	3, 5
3. Multiplication	Multiplication	Multiply; many off-spring will die in host or *en route* to fresh host	
4. Evasion of host defences	Microbial answer to host defences	Evade phagocytic and immune defences long enough for full cycle in host to be completed	4, 6, 7
5. Shedding (exit) from body	Transmission	Leave body at site and on a scale that ensures spread to fresh host	2
6. Cause damage in host	Pathology, disease	Not strictly necessary but often occurs[a]	8

[a] Some damage may be inevitable if efficient shedding is to occur (e.g. common cold, diarrhoea, skin vesicles).

reduction in numbers of the host species and thus in the numbers of the microorganism. Thus, although a few microorganisms cause disease in a majority of those infected, most are comparatively harmless, causing either no disease, or disease in only a small proportion of those infected. Polioviruses, for instance, are transmitted by the faecal–oral route, and cause a subclinical intestinal infection under normal circumstances. But in an occasional host the virus invades the central nervous system, and causes meningitis, sometimes paralysis, and very occasionally death. This particular site of multiplication is irrelevant from the virus point of view, because growth in the central nervous system is quite unnecessary for transmission to the next host. If it occurred too frequently, in fact, the host species would be reduced in numbers and the success of the virus jeopardised. Well-established infectious agents have therefore generally reached a state of balanced pathogenicity in the host, and cause the smallest amount of damage compatible with the need to enter, multiply and be discharged from the body.

The importance of balanced pathogenicity is strikingly illustrated in the case of the natural evolution of myxomatosis in the Australian rabbit. After the first successful introduction of the virus in 1950 more than 99% of infected rabbits died, but subsequently new strains of virus appeared that were less lethal. Fewer infected rabbits died, so that the host species was less severely depleted. Also, because even those that died now survived longer, there were greater opportunities for the transmission of virus to uninfected individuals. The less lethal strains of virus were therefore selected out during the evolution of the virus in the rabbit population, and replaced the original highly lethal strains because they were more successful parasites. The rabbit population also changed its character, because those that were genetically more susceptible to the infection were eliminated. Rabies, a virus infection of the central nervous system, seems to contradict, but in fact exemplifies, this principle. Infection is classically acquired from the bite of a rabid animal and the disease in man is almost always fatal, but the virus has shown no signs of becoming less virulent. Man, however, is an unnatural host for rabies virus, and it is maintained in a less pathogenic fashion in animals such as vampire bats and skunks. In these animals there is a relatively harmless infection and virus is shed for long periods in the saliva, which is the vehicle of transmission from individual to individual. Rabies is thus maintained in the natural host species without serious consequences. But bites can infect the individuals of other species, 'accidentally' from the virus point of view, and the infection is a serious and lethal one in these unnatural hosts.

Although successful parasites cannot afford to become too pathogenic, some degree of tissue damage may be necessary for the effective shedding of microorganisms to the exterior, as for instance in the flow of infected fluids from the nose in the common cold or from the alimentary canal in infectious diarrhoea. Otherwise there is ideally very little tissue damage, a minimal inflammatory or immune response, and a few microbial parasites achieve the supreme success of causing zero damage and failing to be recognised as parasites by the host (see Ch. 7). Different microbes show varying degrees of attainment of this ideal state of parasitism.

The concept of balanced pathogenicity is helpful in understanding infectious diseases, but many infections have not yet had time to reach this ideal state. In the first place, as each microorganism evolves, occasional virulent variants emerge and cause extensive disease and death before disappearing after all susceptible individuals have been infected, or before settling down to a more balanced pathogenicity. Secondly, a microbe recently introduced into a host (e.g. human immunodeficiency virus (HIV) in humans) may not have had time to settle down into this ideal state. Thirdly, some of the microbes responsible for serious human diseases had appeared originally in one part of the world, where there had been a weeding out of genetically susceptible individuals and a move in the direction of a more balanced pathogenicity. Subsequent spread of the microorganism to a new continent has resulted in the infection of a different human population in whom

the disease is much more severe because of greater genetic susceptibility. Examples include tuberculosis spreading from resistant
Europeans to susceptible Africans or North American Indians, and
yellow fever spreading from Africans to Europeans (see pp. 368–9).
Finally, there are a number of microorganisms that have not evolved
towards a less pathogenic form in man because the human host is
clearly irrelevant for the survival of the microorganism. Microorganisms of this sort, such as those causing rabies (see above), scrub
typhus, plague, leptospirosis and psittacosis, have some other regular
host species which is responsible, often together with an arthropod
vector, for their maintenance in nature.* The pathogenicity for man is
of no consequence to the microorganism. Several human infections
that are spillovers from animals domesticated by man also come into
this category, including brucellosis, Q fever and anthrax. As humans
colonise every corner of the earth, they encounter an occasional
microbe from an exotic animal that causes, quite 'accidentally' from the
point of view of the microorganisms, a serious or lethal human disease.
Examples include Lassa fever and Marburg disease from African
rodents and monkeys, respectively.†

On the other hand, a microorganism from one animal can adapt to a
new species. Every infectious agent has an origin, and studies of
nucleic acid sequence homologies are removing these things from the
realm of speculation. Measles, which could not have existed and maintained itself in humans in the Palaeolithic era, probably arose at a
later stage from the closely related rinderpest virus that infects cattle.
New human influenza viruses continue to arise from birds, and the
virus of the acquired immunodeficiency syndrome (AIDS), the modern
pestilence (see p. 191), seems to have arisen from a very similar virus
infecting monkeys in Africa.

Microorganisms multiply exceedingly rapidly in comparison with
their vertebrate hosts. The generation time of an average bacterium is
an hour or less, as compared with about 20 years for the human host.
Consequently, microorganisms evolve with extraordinary speed in
comparison with their vertebrate hosts. Vertebrates, throughout their
hundreds of millions of years of evolution, have been continuously
exposed to microbial infections. They have developed highly efficient

---

* These infections are called *zoonoses* (see p. 53).

† Lassa fever is a sometimes lethal infection of man caused by an arenavirus (see Table
A.5, p. 423). The virus is maintained in certain rodents in West Africa as a harmless
persistent infection, and man is only occasionally infected. Another serious infectious
disease occurred in 1967 in a small number of laboratory workers in Marburg, Germany,
who had handled tissues from vervet monkeys recently imported from Africa. The
Marburg agent is a virus and has since reappeared to cause fatal infections in Zaire and
the Sudan, but nothing is known of its natural history. Monkeys are not natural hosts
and are probably accidentally infected, like man. Since 1976, Ebola virus, related to
Marburg, has caused dramatic local outbreaks in Zaire and Sudan. Like Lassa fever, it
can spread from person to person via infected blood, but its natural host is unknown.

recognition (early warning) systems for foreign invaders, and effective inflammatory and immune responses to restrain their growth and spread, and to eliminate them from the body (see Ch. 9). If these responses were completely effective, microbial infections would be few in number and all would be terminated rapidly; microorganisms would not be allowed to persist in the body for long periods. But microorganisms, faced with the antimicrobial defences of the host species, have evolved and developed a variety of characteristics that enable them to by-pass or overcome these defences. The defences are not infallible, and the rapid rate of evolution of microorganisms ensures that they are always many steps ahead. If there are possible ways round the established defences, microorganisms are likely to have discovered and taken advantage of them. Successful microorganisms, indeed, owe their success to this ability to adapt and evolve, exploiting weak points in the host defences. The ways in which the phagocytic and immune defences are overcome are described in Chs 4 and 7.

It is the virulence and pathogenicity of microorganisms, their ability to kill and damage the host, that makes them important to the physician or veterinarian. If none of the microorganisms associated with man did any damage, and none was notably beneficial, they would be interesting but relatively unimportant objects. In fact, they have been responsible for the great pestilences of history, have at times determined the course of history, and continue today, in spite of vaccines and antibiotics, as important causes of disease (see Table A.1). Also, because of their rapid rate of evolution and the constantly changing circumstances of human life, they continue to present threats of future pestilences. It is the purpose of this book to describe and discuss the mechanisms of infection and the things that make microorganisms pathogenic. This is the central significant core of microbiology as applied to medicine.

As molecular biological and immunological techniques are brought to bear on these problems, our understanding is steadily increasing beyond that of descriptive pathology to a more detailed understanding of host pathogen interactions at the cellular, genetic and biochemical levels. The power of the new technology stems from an ability to (1) mutate genes, (2) mobilise and transfect genes, or (3) in the case of multi-segmented viral genomes, reassort genes into progeny derived from different parents. Since methods now exist which readily allow the identification of the mutated/acquired/assorted genes, the acquisition or loss of a gene may then be correlated with the newly acquired phenotype, the gene isolated (i.e. *cloned*), sequenced, the corresponding amino acid sequence predicted and pre-existing data bases searched for comparisons with genes and their products already identified. By such means a great deal of biochemical information can be obtained about the microbial determinants involved in mediating different aspects of the complex infection process.

But perhaps the most exciting advances have yet to come. We are now moving into an area where it will be theoretically possible to look at the expression of several genes at once, rather than individually. This will be possible because of two new developments – genome sequencing and microarrays (or DNA chips). Complete genome sequences are now available, or soon will be, for the following bacterial pathogens: *Bordetella bronchiseptica, Bordetella parapertussis, Bordetella pertussis, Campylobacter jejuni, Clostridium difficile, Corynebacterium diphtheriae, Mycobacterium bovis, Mycobacterium leprae, Mycobacterium tuberculosis, Neisseria meningitidis, Salmonella typhi, Staphylococcus aureus, Streptococcus pyogenes, Streptococcus coelicolor, Yersinia pestis.* At the time of writing, at least nine other bacterial pathogens are being considered. Sequencing of the chromosomes of *Plasmodium falciparum, Leishmania major, Trypanosoma brucei* and *Dictyostelium discoideum* has also been completed or is in progress.

Microarrays consist of very large numbers of spots of DNA. Each spot is a unique DNA fragment. A chip the size of a microscope slide can contain tens of thousands of spots and hence the entire genome of some bacteria. By extracting mRNAs from bacteria grown in culture and from the same organism from an infection site (or grown in conditions which mimic infection conditions), and some neat colour chemistry, it will be possible to identify which gene(e) are expressed or repressed in the two situations. The main challenge is likely to be in reproducing conditions reflecting the *in vivo* situation when no animal models exist for a particular infection.

The book is largely based on the events listed in Table 1.1. To make sense of infectious diseases you need to know about the host's phagocytic and immune defences, and these are briefly set out in Chs 4, 6 and 9. There are additional chapters on resistance and recovery from infection, persistent infection, and the prevention of infection by vaccines.

References

Burnet, F. M. and White, D. O. (1972). 'The Natural History of Infectious Disease', 4th edn. Cambridge University Press.

Christie, A. H. (1987). 'Infectious Diseases, Epidemiology and Clinical Practice', 4th edn. Churchill Livingstone, Edinburgh.

Ewald, P. W. (1994). 'Evolution of Infectious Disease'. Oxford University Press, New York.

Fenner, F. (1959). Myxomatosis in Australian wild rabbits – evolutionary changes in an infectious disease. Harvey lectures 1957–8, Royal Society, London; 25–55.

Fenner, F. and Ratcliffe, F. N. (1965). 'Myxomatosis'. Cambridge University Press.

Mims, C. A. (1991). The origin of major human infections and the crucial role of person-to-person spread. *Epidemiol. Infect.* **106**, 423–433.

Mims, C. A., Playfair, J. H. L., Roitt, I. M., Wakelin, D. and Williams, R. (1998). 'Medical Microbiology', 2nd edn. Mosby, London.

Noble, W. C. (1981) 'Microbiology of the Human Skin'. Lloyd-Luke, London.

The interested student can quickly find information on which genomes have been completed or are in progress, and individual sequences by accessing web sites http://www.sanger.ac.uk/Projects/Microbes/ or http://www.sanger.ac.uk/Projects/Protozoa/

The information listed above was that available in November 1999.

2

Attachment to and Entry of Microorganisms into the Body

Introduction

Figure 2.1 shows a simplified diagram of the mammalian host. In essence, the body is traversed by a tube, the alimentary canal, with the respiratory and urinogenital tracts as blind diverticula from the alimentary canal or from the region near the anus. The body surface is covered by skin, with a relatively impermeable dry, horny outer layer, and usually fur. This gives a degree of insulation from the outside world, and the structure of skin illustrates the compromise between the need to protect the body, yet at the same time maintain sensory communication with the outside world, give mechanical mobility, and, especially in man, act as an important thermoregulatory organ. It is the largest 'organ' in the body, with a weight of 5 kg in humans.

The dry, protective skin cannot cover all body surfaces. At the site of the eye it must be replaced by a transparent layer of living cells, the conjunctiva. Food must be digested and the products of digestion absorbed, and in the alimentary canal therefore, where contact with the outside world must be facilitated, the lining consists of one or more layers of living cells. Also in the lungs the gaseous exchanges that take place require contact with the outside world across a layer of living cells. There must be yet another discontinuity in the insulating outer

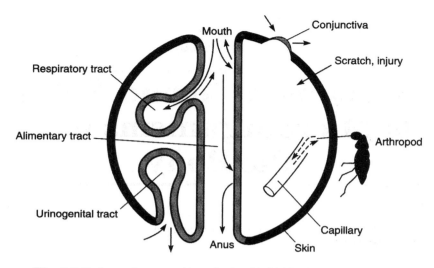

Fig. 2.1 Body surfaces as sites of microbial infection and shedding.

layer of skin in the urinogenital tract, where urine and sexual products are secreted and released to the exterior. The cells on all these surfaces are covered by a fluid film containing mucin, a complex hydrated gel that waterproofs and lubricates. In the alimentary canal the lining cells are inevitably exposed to mechanical damage by food and they are continuously shed and replaced. Shedding and replacement is less pronounced in respiratory and urinogenital tracts, but it is an important phenomenon in the skin, the average person shedding about 5 × 10^8 skin squames per day.

The conjunctiva and the alimentary, respiratory and urinogenital tracts offer pathways for infection by microorganisms. Penetration of these surfaces is more easily accomplished than in the case of the intact outer skin. A number of antimicrobial devices have been developed in evolution to deal with this danger, and also special cleansing systems to keep the conjunctiva and respiratory tract clean enough to carry out their particular function. In order to colonise or penetrate these bodily surfaces, microorganisms must first become attached, and there are many examples of specific attachments that will be referred to (see Table 2.1 where they are listed in some detail, it being an area of intense research activity). One striking feature of acute infectious illnesses all over the world is that most of them are either respiratory or dysentery-like in nature. They are not necessarily severe infections, but for sheer numbers they are the type that matter. In other words, infectious agents are for much of the time restricted to the respiratory and intestinal tracts.

It is of some interest to divide all infections into four groups (Fig. 2.2). First, those in which the microorganisms have specific mecha-

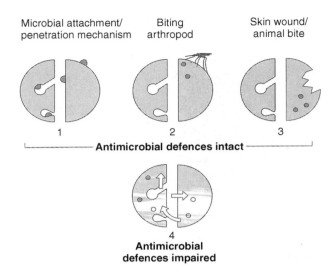

Fig. 2.2 Four types of microbial infection can be distinguished. Reprinted from Mims *et al.* (1998). Medical Microbiology, 2nd edn, Mosby, London.

nisms for attaching to and sometimes penetrating the body surfaces of the normal, healthy host. This includes the infections listed in Fig. 2.3. In the second group, the microorganism is introduced into the body of the normal healthy host by a biting arthropod, as with malaria, plague, typhus or yellow fever. Here the microorganism possesses specific mechanisms for infection of the arthropod, and depends on the arthropod for introduction into the body of the normal healthy host. The third group includes infections in which the microorganism is not by itself capable of infecting the normal healthy host. There must be some preliminary damage and impairment of defences at the body surface, such as a skin wound, damage to the respiratory tract initiated by a microbe from the first group, or an abnormality of the urinary tract interfering with the flushing, cleansing action of urine (see below). In the fourth group, there is a local or general defect in body defences. The opportunistic infections described later in this chapter come into this fourth group, and further examples are given in Ch. 11. A large proportion of the infections seen in hospitals comes into this category.

Adhesion/Entry: Some General Considerations

Adhesins are found in almost any class of surface structure present on microorganisms (Table 2.1). Adhesins are more than simply the deter-

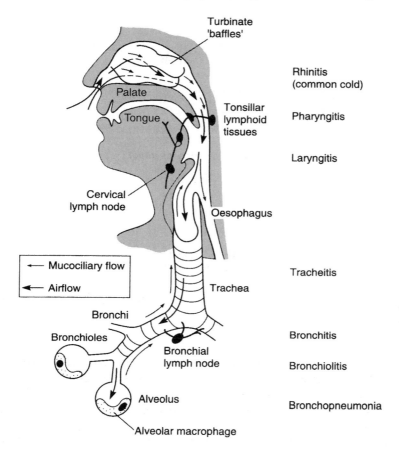

Fig. 2.3 Mechanisms of infection in the respiratory tract.

minants of pathogen location: they are effectors of important aspects of the biology of infection. The receptors on the eukaryotic cell surface which confer specificity to the initial binding comprise a relatively small number of oligosaccharides of transmembrane glycoproteins (see Table 2.1 for examples), which mediate cell–cell and cell–extracellular matrix interactions. They also play a key role in some cell signalling processes, particularly those involving actin rearrangements, by virtue of their contact with the cytoskeleton.

Since most pathogens possess more than one adhesin system, the fate of the interaction between the pathogen and the host will be determined by which receptor or sequential combination of receptors is activated. For example, pertussigen is an important toxin produced by *Bordetella pertussis* (see Ch. 8). The S2 and S3 subunits of the toxin B oligomer (Fig. 8.5), bind to the surface of macrophages, resulting in the upregulation of integrin CR3. The activated CR3 in turn binds with the filamentous haemagglutinin (FHA) adhesin of *B. pertussis*, leading to

Table 2.1. Some examples of specific attachments of microorganisms to host cell or body surface. In some cases the information on ligand receptor system is derived from *in vitro* studies on cultured cells[a]

Microorganism/disease	Target site or cell	Microbial ligand(s)	Receptor[b]	Strategy[c]
Viruses				
Influenza virus/flu	Respiratory epithelium	Viral haemagglutinin	Neuraminic acid	
Rhinovirus/common cold	Respiratory epithelium	Viral capsid protein	Intercellular adhesion molecules (ICAM-1)	
HIV-1/AIDS	CD4[+] T-cell	Viral envelope gp120 proteins	CD4 proteins	
Epstein–Barr virus/ glandular fever	B-cell	Viral envelope protein	CD21	
Herpes simplex virus/ cold sore/genital herpes	Most cells	g/C glycoprotein	Heparan sulphate	
Measles virus/measles	Most primate cells	Viral haemagglutinin	CD46 (membrane cofactor protein)	
Foot and mouth disease virus	Tissue culture cell	VP1	Vitronectin integrin receptor	
Coxsackie virus A9	Tissue culture cell	VP1	Integrins	
Bacteria				
Chlamydia/conjunctivitis/ urethritis	Conjunctivial/urethral epithelia	GAG[d], MOMP (major outer membrane protein; nonspecific, and specific attachment)	GAG receptors	
Mycoplasma pneumoniae/ atypical pneumonia	Respiratory epithelium	'Foot' on *Mycoplasma* surface	Neuraminic acid	
Neisseria meningitidis/ carrier state	Nasopharyngeal epithelium	Type IV Pili;[e] Opa (opacity associated) proteins	Heparin sulphate proteoglycan. Opa proteins also bind to vitronectin/integrins in HeLa and HEp-2 cells, and CD66 in neutrophils	
Neisseria gonorrhoeae/ gonorrhoea	Urethral epithelium			
Vibrio cholerae/cholera	Intestinal epithelium	Tcp (demonstrably important in humans); others[f]		

Organism/disease	Target tissue/cell	Adhesin	Receptor	
Escherichia coli				
ETEC/diarrhoea	Intestinal epithelium	K88 (pigs); K99 (calves, lambs)	[Neu5Glc(α2-3)Gal(β1-4)Glc(β1-1) ceramide	
		Colonization factors[g] (humans)		
EPEC/diarrhoea	Intestinal epithelium	Bfp,[h] Intimin (an OMP)	Tir (a bacterial protein; translocated intimin receptor), host cell co-factor	
	Colonic epithelium	Intimin	Tir	
EHEC/haemorrhagic colitis; haemolytic uraemic syndrome	Colonic epithelium	Intimin	Tir	
UPEC/pyelonephritis	Urinary tract	P fimbriae[i]	Gal(α1-4)Gal	
NMEC/neonatal meningitis	Endothelial and epithelial cells	S fimbriae[j]	α-Sialyl-(2-3)-β-galactose-containing receptor molecules	
S. typhimurium/gastroenteritis	Intestinal epithelium	Unclear		
S. enteritidis/gastroenteritis	Intestinal epithelium	Unclear		
S. typhi/enteric fever	Intestinal epithelium	Unclear		
Shigella spp./dysentery	Tissue culture cell	Ipa (invasion plasimd antigens) BCD	Integrin	
Streptococcus mutans/caries	Teeth	Glycosyl transferase, glucan ('glue')		
Streptococcus pyogenes/throat infections; other more serious infections	Pharyngeal epithelium[k]			
Listeria monocytogenes	Range of clinical disease	Internalins A, B	E-cahedrin (A)	
Legionella pneumophila/Legionnaires' disease	Macrophage	Adsorbed C3bi	Integrin (CR (complement receptor) 3	Masking
Mycobacteria tuberculosis	Macrophage	Adsorbed C3bi	CR3	Masking
Mycobacteria leprae	Schwann cells	?	α-Dystroglycan[l]	Masking
Treponema pallidum/syphilis	Tissue culture cell	Adsorbed fibronectin	Fibronectin receptor	Masking

Microorganism/disease	Target site or cell	Microbial ligand(s)	Receptor[b]	Strategy[c]
Bordetella pertussis / whooping cough	Respiratory epithelium, macrophage	Several adhesins (fimD, pertussis toxin, filamentous haemagglutinin, pertactin; others)	Several integrins	Multiple complex mimicry
Yersinia enterocolitica / diarrhoea	Intestinal epithelium	Invasin (OMP)	Integrins	
Protozoa				
Leishmania mexicana	Macrophage	Surface glycoprotein (Gp63)	CR3	Ancillary ligand recognition
Leishmania donovani	Macrophage	?	CR3	Ancillary ligand recognition
Leishmania major	Macrophage	Adsorbed C3bi	CR3	Masking
Histoplasma capsulatum	Macrophage	?	CD18	Ancillary ligand recognition
Plasmodium vivax / malaria	Erythrocyte of susceptible human	Merozoite (non-complement-mediated attachment) 'Duffy' antigen		
Plasmodium falciparum / malaria	Erythrocyte of susceptible human	Merozoite	Glycophorin A, B	
Trypanosoma cruzi	Tissue culture cell	Adsorbed fibronectin	Fibronectin receptor	Masking
Babesia / babesiasis in cattle	Erythrocyte	Complement-mediated attachment	CR3	
Giardia lamblia / diarrhoea	Duodenal, jejunal epithelia	Taglin[m] GLAM-1 on disc	Manose-6-phosphate	
Entamoeba histolytica / amoebic dysentery	Colonic epithelium	170 kDa Gal/GalNAcLectin	?	
Trypanosome cruzi / trypanosomiasis		Penetrin[n]	?	

a The table is not exhaustive; see Hoepelman and Tuomanen (1992), Virji (1997) and Kerr (1999).

b Receptors: cell adhesion molecules (CAMs), are transmembrane glycoproteins with extracellular, transmembrane and intracellular domains, and are the means whereby cells communicate with other cells and the extracellular environment (ECM). They play a key role in some cell signalling processes, particularly those involving actin rearrangements, by virtue of their contact with the cytoskeleton. There are six families of CAMs: the immunoglobulin-like superfamily, the cadherins, the receptor protein tyrosine phosphatases, the selectins, the hyaluronate receptors and the integrins. Integrins are heterodimers each consisting of an α chain (of which there are 14 known) and a β chain (of which there are eight known); there are at least 22 known integrins comprising different noncovalently-linked αβ combinations. In some cases the α and β chains comprise CD (cluster differentiation: based on computerised analysis of monoclonal antibody studies) marker proteins present on cell surfaces. For example, one subfamily (β$_2$ integrins) has CD18 as β component which when combined with CD11b constitutes CR3 (complement receptor type 3) which will bind C3bi.

c Some integrins bind to proteins which express a triplet motif RGD (Arg-Gly-Asp). Three different ways in which pathogens subvert normal cellular processes for their own ends: *mimicry*, expression of ligands with the RGD motif; *masking*, adsorption of the natural ligand on to the surface of the organism; *ancillary ligand recognition*, in which the pathogen interacts with domains on the integrin other than the RGD triplet. Some examples are given.

d GAG: a heparan sulphate-like glycosaminoglycan.

e Type I (common) fimbriae are mannose sensitive, i.e. their agglutinability with erythrocytes is blocked by mannose; Type II are resistant to blocking with mannose. Type III seems to have fallen into disuse as it was originally used to describe a type of thick hollow structure in soil bacteria including *Agrobacterium*. Type IV are characterized by structural subunits which share extensive N-terminal amino acid homology and all, except the fimbriae of *Vibrio cholerae*, contain the modified amino acid *N*-methylphenylalanine as the first residue of the mature protein. They are mainly polar in distribution, and confer on organisms a primitive form of surface translocation known as twitching. They are present on a number of Gram-negative bacteria including *Pseudomonas aeruginosa, Neisseria gonorrhoeae, Neisseria meningitidis, Moraxella bovis* (causative agent of infectious bovine keratoconjunctivitis) and *Dichelobacter nodosus* (*Bacteroides nodosus*). A *D. nodosus* pilus-based vaccine protects sheep against foot rot; this is the only example of an effective vaccine base on Type IV pili. In the absence of good animal models, the evidence that such fimbriae are involved in virulence is often epidemiological – a correlation with piliation and disease causation. Successful vaccines based on pili in general have proved elusive since there are many antigenic variants of pilus types and usually more than one important adhesin involved in virulence.

f Tcp: toxin co-regulated pili. Other adhesins implicated: mannose-fucose resistant haemagglutinin, O antigens of lipopolysaccharide, at least four haemagglutinins of different sugar specificity and three different fimbrial types.

g Adhesion is mediated by 'colonization factors' (CFs) the nomenclature of which is very confusing. An attempt has been made (Gaastra and Svennerholm, 1996, supported by Nataro and Kaper, 1998) to rationalise the situation for human enterotoxigenic *E. coli* (ETECs) by renaming them as coli surface antigens (CS, followed by a numeral, CS1, CS2, etc.). *E. coli* CFs are encoded on plasmids (which may also contain the genes for both types of enterotoxins described in Ch. 8) and determine both host and tissue specificity of infection. For example, important CFs produced by animal ETEC are not found on human strains, and CF K88 is found only in strains which infect pigs whereas K99-expressing strains will infect calves, lambs and pigs. CSs of human strains have been subdivided into four groups on a morphological basis – rigid rods, bundle forming, fibrillar and nonfibrillar (Nataro and Kaper, 1998) – the structures of which are schematised in Gaastra and Svennerholm (1996).

h Bfp: bundle-forming pili encoded on a large plasmid EAF (enteroadherent factor) responsible for local adhesion (LA) seen on HeLa cells; plasmid important for full virulence. Bfp not present in EHEC, ETEC, UPEC or NMEC, but has homologues in *S. typhimurium, S. dublin* and *S. choleraesuis*.

[i] P fimbriae. The majority of E. coli isolates associated with acute infantile pyelonephritis express mannose resistant adhesins which react with the Gal(α1–4)Gal moiety of the glycolipid part of blood group substance P expressed on uroepithelial cells. These adhesins are designated P fimbriae/pili of which there are several serological variants. The acronym Pap (pyelonephritis associated pili) is widely used in the genetic literature in this area. These pili are complex structures made up of several different proteins. The adhesin is present on the tip of the pilus and interacts with the Gal(α1–4)Gal disaccharide.

[j] S fimbriae. α-Sialyl-(2–3)-β-galactose-containing molecules – the receptor for S fimbriae – have been found in many tissues including epithelial elements of the human kidney, the choroid plexuses and ventricles of the baby rat, and human vascular endothelia. S fimbriae have been identified on some UPEC strains but are not as important as P fimbriae in pyelonephritis, perhaps due to the presence of soluble ligand-blocking receptors. S fimbriae are mainly found associated with strains which cause septicaemic and meningitic infections. The lack of organ specificity in the distribution of the S receptor must mean that there are additional factors involved in the invasion of the cerebrospinal fluid from the circulation.

[k] The mechanism of Streptococcus pyogenes adhesion is not well understood. It may be a multifactorial, two-step process. Step 1 involves complexes of surface proteins and lipoteichoic acid released from the membrane mediating initial weak reversible binding to many cell types. Step 2: greater specificity and strength of binding of the receptor ligand kind; M proteins and distinct fibronectin-binding proteins have been implicated by different laboratories. Probably strain/tissue specific.

[l] Exactly the same receptor is used by Lassa fever and LCM viruses.

[m] Taglin: trypsin-activated Giardia lamblia lectin. GLAM-1: Giardia lamblia adherence molecule-1.

[n] Penetrin, a 60 kDa surface protein of T. cruzi, the causative agent of Chagas' disease. It promotes selective adhesion to three extracellular matrix components (heparin, heparan sulphate and collagen). Purified penetrin also binds to host fibroblasts and confers on recombinant E. coli expressing penetrin, the ability to adhere to and invade nonphagocytic Vero cells. The adhesion of penetrin to fibroblasts and invasiveness of E. coli were inhibitable in a saturable manner by glycosaminoglycan and collagen.

the uptake of the organism. Viruses may also bind to more than one receptor. These may be used in invading different types of cell, or one receptor is for binding to the cell and another for penetration. HIV gp 120 binds to CD4 on susceptible cells (Table 2.1) and also to a chemokine receptor. People without the latter receptor recover from HIV infection!

The Skin

The skin is a natural barrier to microorganisms and is penetrated at the site of breaks in its continuity, whether macroscopic or microscopic (Table 2.2).

Microorganisms other than commensals (residents) are soon inactivated, probably by fatty acids (skin pH is about 5.5) and other materials produced from sebum by the commensals. In the perianal region, for instance, where billions of faecal bacteria are not only deposited daily, but then, in man at least, rubbed into the area, there is evidently an astonishing resistance to infection. Faecal bacteria are rapidly inactivated here, but the exact mechanism, and the possible role of perianal gland secretions, is unknown.

Table 2.2.   Microorganisms that infect the skin or enter the body via the skin

Microorganisms	Disease	Comments
Arthropod-borne viruses	Various fevers	150 distinct viruses, transmitted by infected arthropod bite
Rabies virus	Rabies	Bite from infected animals
Wart viruses	Warts	Infection restricted to epidermis
Staphylococci	Boils, etc.	Commonest skin invaders
Rickettsia	Typhus, spotted fevers	Infestation with infected arthropod
Leptospira	Leptospirosis	Contact with water containing infected animals' urine
Streptococci	Impetigo, erysipelas	
Bacillus anthracis	Cutaneous anthrax	Systemic disease following local lesion at inoculation site
Treponema pallidum and *pertenue*	Syphilis, yaws	Warm, moist skin is more susceptible
Yersinia pestis	Plague	Bite from infected rodent flea
Plasmodia	Malaria	Bite from infected mosquito
Trichophyton spp. and other fungi	Ringworm, athlete's foot	Infection restricted to skin, nails, hairs

Bacteria on the skin, as well as entering hair follicles and causing lesions (boils, styes), can also cause trouble after entering other orifices. Staphylococcal mastitis occurs in many mammals, but is of major importance in the dairy industry, and is thought to arise when the bacteria are carried up and past the teat canal of the cow as a result of vacuum fluctuations during milking.

Large or small breaks in the skin due to wounds are obvious routes for infection. The virus of hepatitis B or C can be introduced into the body if the needle of the doctor, tattooist, drug addict, acupuncturist or ear-piercer is contaminated with infected blood. Shaving upsets the antimicrobial defences in the skin and can lead to staphylococcal infection of the shaved area on the male face (sycosis barbae) or female axilla. Pre-operative shaving, although a well-established ritual, seems to enhance rather than prevent infection in surgical wounds. Various sports in which there is rough skin-to-skin contact can result in infections (streptococci, staphylococci, skin fungi) being transmitted at the site of minor breaks in the skin. It is called scrumpox, but is seen in judo and in wrestling as well as in rugby football.

Bites are also important sites for the entry of microorganisms.

Small bites

Biting arthropods such as mosquitoes, mites, ticks, fleas and sandflies penetrate the skin during feeding and can thus introduce pathogenic agents into the body. Some infections are transmitted mechanically, the mouthparts of the arthropod being contaminated with the infectious agent, and there is no multiplication in the arthropod. This is what happens in the case of myxomatosis. Fleas or mosquitoes carry myxoma virus on their contaminated mouthparts from one rabbit to another. When transmission is said to be biological, as in yellow fever or malaria, this means that the infectious agent multiplies in the arthropod, and, after an incubation period, appears in the saliva and is transmitted to the susceptible host during a blood feed. Mosquitoes or ticks, in the act of feeding, probe in the dermal tissues, emitting puffs of saliva as they do so. The mosquito proboscis may enter a blood capillary and is then threaded along the vessel, further injections of saliva occurring during the ingestion of blood. Infected saliva is thus introduced directly into the dermis and often into the vascular system, the counterpart of a minute intradermal or intravenous injection of microorganisms. Other diseases transmitted biologically by arthropods include typhus and plague, and in these cases the microorganisms multiply in the alimentary canal of the arthropod. Plague bacteria from the infected flea are regurgitated into the skin during feeding, and the human body louse infected with typhus rickettsiae defecates during feeding, the rickettsiae subsequently entering the body through the bite-wound.

Large bites

The classical infectious disease transmitted by a biting mammal is rabies. Virus is shed in the saliva of infected foxes, dogs, wolves, vampire bats, etc. and thus introduced into bite wounds. Human bites are not common, most people having neither the temperament nor the teeth for it. When they do occur, human bites can cause troublesome sepsis because of the fusiform and spirochaetal bacteria normally present in the mouth that are introduced into the wound. Teeth often make an involuntary inoculation of bacteria into skin during fist fights. The hero's decisive punch can then bring him knuckle sepsis as well as victory. Most cats carry *Pasteurella multocida* in their mouths, and cat bites, although less common than dog bites, are likely to cause infection. Bites from tigers or cougars can lead to *P. multocida* infection as well as bad dreams.

Respiratory Tract

Air contains a variety of suspended particles, and the total quantity seems large if one says that there are more than 1000 million tonnes of suspended particulate matter in the earth's atmosphere. Most of this is smoke, soot and dust, but microorganisms are inevitably present. Inside buildings there are 400–900 microorganisms per cubic metre, nearly all of them nonpathogenic bacteria or moulds. Therefore with a ventilation rate of 6 litres min^{-1} at rest, the average man would inhale at least eight microorganisms per minute or about 10 000 per day. Efficient cleansing mechanisms remove inhaled particles and keep the respiratory tract clean, and infection of the respiratory tract has to be thought of in relation to these mechanisms, which are designed to remove and dispose of inhaled particles, whatever their nature.

A mucociliary blanket covers most of the surface of the lower respiratory tract. It consists of ciliated cells together with single mucus-secreting cells (goblet cells) and subepithelial mucus-secreting glands. Foreign particles deposited on this surface are entrapped in mucus and borne upwards from the lungs to the back of the throat by ciliary action (Fig. 2.3). This has been called the mucociliary escalator. The nasal cavity (upper respiratory tract) has a similar mucociliary lining, and particles deposited here are also carried to the back of the throat and swallowed.* The average person produces 10–100 ml mucus from the nasal cavity each day and a similar amount from the lung. The terminal air spaces of the lower respiratory tract are the alveoli, and

---

* If human serum albumin aggregates labelled with ^{131}I are introduced into the nose of a volunteer, their movement can be followed with a crystal scintillation detector. The speed of movement is variable, but averages 0.5–1.0 cm min^{-1}.

these have no cilia or mucus but are lined by macrophages. IgG and secretory IgA are the predominant antibodies in the lower and upper respiratory tracts respectively, and afford specific defence once the immune system has been stimulated.

A great deal of experimental work has been carried out on the fate of inhaled particles, and particle size is of paramount importance. The larger the particle, the less likely it is to reach the terminal portions of the lung. All particles, whether viral, bacterial, fungal or inert, are dealt with in the same way. Larger visible particles are filtered off by the hairs lining the nostrils, and particles 10 mm or so in diameter tend to be deposited on the 'baffle plates' in the nasal cavity, consisting of the turbinate bones covered by nasal mucosa. Smaller particles are likely to reach the lungs, those 5 mm or less in diameter reaching the alveoli. Nearly all observations have been made with animals, but human subjects have been used on a few occasions. If a person inhales 5 mm particles of polystyrene tagged with ^{51}Cr and the fate of the particles is determined by external gamma measurements, about half of the labelled material is removed from the lungs within hours, after being deposited on the mucociliary escalator and carried up to the back of the throat. The rest is removed very slowly indeed with a half-life of more than 150 days, having been phagocytosed by alveolar macrophages after settling on alveolar walls. The marker particles in this experiment are nondegradable, and nonpathogenic microorganisms would have been disposed of more rapidly. Inhaled particles of soot are taken up by alveolar macrophages, some of which later migrate to the pulmonary lymph nodes. Town dwellers can be recognised in the postmortem room because of the grey colour of their pulmonary lymph nodes.*

If a microorganism is to initiate infection in the respiratory tract, the initial requirements are simple. First, the microorganism must avoid being caught up in mucus, carried to the back of the throat and swallowed. Second, if it is deposited in alveoli it must either resist phagocytosis by the alveolar macrophage, or if it is phagocytosed it must survive or multiply rather than be killed and digested.

It would seem inevitable that a microorganism has little chance of avoiding the first fate unless the mucociliary mechanisms are defective, or unless it has some special device for attaching firmly if it is lucky enough to encounter an epithelial cell. The highly successful myxoviruses, for instance, of which influenza is an example, have an attachment protein (the haemagglutinin) on their surface which specifically attaches to a receptor molecule (neuraminic acid of a glycoprotein) on the epithelial cells. A firm union is established (Fig. 2.4) and

---

* There is also a movement of macrophages from the lower respiratory tract up to the back of the throat on the mucociliary escalator. At least 10^7 macrophages a day are recoverable in normal rats or cats, a similar quantity in normal people, and more than this in patients with chronic bronchitis. This is a route to the exterior for macrophages laden with indigestible materials.

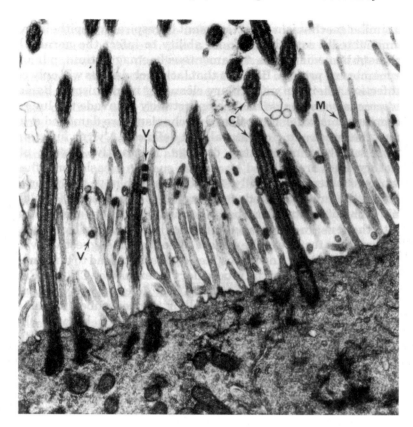

Fig. 2.4 Portion of ciliated epithelial cell from organ culture of guinea-pig trachea after incubation with influenza virus for 1 h at 4°C. Electron micrograph of thin section showing virus particles (V) attached to cilia (C) and to microvilli (M). The fluid between the cilia is watery, the viscous mucoid layer lying above the cilia. (Electron micrograph very kindly supplied by Dr R. Dourmashkin.)

the virus now has an opportunity to infect the cell. The common cold rhinoviruses also have their own receptors.† *Mycoplasma pneumoniae* has a special projection on its surface by which it attaches to neuraminic acid receptors on the epithelial cell surface. The bacterium responsible for whooping cough (*Bordetella pertussis*) has a similar

---

† Although made use of by invading microorganisms, receptors are clearly not there for this purpose, and serve other functions such as hormone binding, cell–cell recognition, etc. Sometimes virus receptors are present only on certain types of cell, which can account for cell tropisms and other features of the disease. For instance, the receptor for Epstein–Barr virus is the C3d receptor on B-cells, which are thus infected and undergo polyclonal activation (see p. 199), and the main receptor for HIV is the CD4 molecule on T helper cells, whose infection and depletion contributes to the serious immune deficit in AIDS.

mechanism for attachment to respiratory epithelium, and this undoubtedly contributes to its ability to infect the normal lung; attachment is mediated via a filamentous haemagglutinin, pili and an outer membrane protein. Bacteria that lack such devices will only establish infection when the mucociliary cleansing mechanism is damaged. *Streptococcus pneumoniae* has the opportunity to invade the lungs and cause pneumonia when mucociliary mechanisms are damaged or there is some other weakening of natural host defences. A virus infection is a common source of mucociliary damage. Destructive lesions of the respiratory tract are induced by viruses such as measles or influenza, and various bacteria, especially streptococci, then have the opportunity to grow in the lung and produce a secondary pneumonia. People with chronic bronchitis show disturbed mucociliary function, and this contributes to the low-grade bacterial infection in the lung which may be a semi-permanent feature of the disease. Also there is suggestive evidence that cigarette smoking and atmospheric pollutants lead to temporary or permanent impairment of the mucociliary defences (see Ch. 11). Finally, there are many ways in which natural host defences are weakened in hospital patients. Patients with indwelling tracheal tubes, for instance, are particularly susceptible to respiratory infection because the air entering the tracheal tube has been neither filtered nor humidified in the nose. Dry air impairs ciliary activity and the indwelling tube causes further epithelial damage. General anaesthesia decreases lung resistance in a similar way, and in addition depresses the cough reflex.

Certain microorganisms that infect the respiratory tract directly depress ciliary activity, thus inhibiting their removal from the lung and promoting infection. *Bordetella pertussis* attaches to respiratory epithelial cells and in some way interferes with ciliary activity. *Haemophilus influenzae* produces a factor that could be important *in vivo*. The factor slows the ciliary beat, interferes with its coordination and finally causes loss of cilia. At least seven ciliostatic substances are produced by *Pseudomonas aeruginosa*, which causes a devastating respiratory infection in those with cystic fibrosis (see p. 51). Ciliary activity is also inhibited by *Mycoplasma pneumoniae*. The mycoplasma multiply while attached to the surface of respiratory epithelial cells, and the ciliostatic effect is possibly due to hydrogen peroxide produced locally by the mycoplasma. Cilia are defective in certain inherited conditions. In Kartagener's syndrome, for instance, impaired ciliary movement leads to chronic infections in lung and sinuses. Spermatozoa are also affected, and males with this condition are infertile.

The question of survival of airborne microorganisms after phagocytosis by alveolar macrophages is part of the general problem of microbial survival in phagocytic cells, and this is dealt with more fully in Ch. 4. Tubercle bacilli tend to survive in the alveolar macrophages of the susceptible host, and respiratory tuberculosis (a disease of inspiration!) is thought to be initiated in this way. The common cold viruses, in contrast, which are very commonly phagocytosed by these cells, fail to survive and multiply, and therefore cause no perceptible infection in the lower respiratory tract. Growth of many of these viruses is in any case restricted at 37°C, being optimal at about 33°C, the temperature of nasal mucosa. Under certain circumstances the antimicrobial activity of alveolar macrophages is depressed. This occurs, for instance, following the inhalation of toxic asbestos particles and their phagocytosis by alveolar macrophages. Patients with asbestosis have increased susceptibility to respiratory tuberculosis. Alveolar macrophages infected by respiratory viruses sometimes show decreased ability to deal with inhaled bacteria, even those that are normally nonpathogenic, and this can be a factor in secondary bacterial pneumonias (see Ch. 8).

Normally the lungs are almost sterile, because the microorganisms that are continually being inhaled are also continually being phagocytosed and destroyed or removed by mucociliary action.

Gastro-intestinal Tract

The intestinal tract must take what is given during eating and drinking, and also various other swallowed materials originating from the mouth, nasopharynx and lungs. Apart from the general flow of intestinal contents, there are no particular cleansing mechanisms, unless diarrhoea and vomiting are included in this category. The lower intestinal tract is a seething cauldron of microbial activity, as can readily be appreciated from the microscopic examination of fresh faeces. Multiplication of bacteria is counter-balanced by their continuous passage to the exterior with the rest of the intestinal contents. A single *E. coli*, multiplying under favourable conditions, might well increase its numbers to about 10^8 within 12–18 h, the normal intestinal transit time. The faster the rate of flow of intestinal contents, the less the opportunity for microbial growth, so that there is a much smaller total number of bacteria in diarrhoea than in normal faeces. On the other hand, a reduced flow rate leads to increased growth of intestinal bacteria. This is not known to be harmful in individuals on a low-fibre diet, but is a more serious matter in the blind-loop syndrome. Here, surgical excision of a piece of intestine results in a blind length in which the flow rate is greatly reduced. The resulting bacterial overgrowth, especially in the small intestine, is associated with symptoms of malabsorption, because the excess bacteria metabolise bile acids needed for absorption of fats and also compete for vitamin B_{12} and other nutrients.

The commensal intestinal bacteria are often associated with the intestinal wall, either in the layers of mucus or attached to the epithelium itself. If a mouse's stomach or intestine is frozen with the contents intact and sections are then cut and stained, the various commensal bacteria can be seen in large numbers, intimately associated with the epithelial cells. This makes it easier for them to maintain themselves as permanent residents.

Helicobacter pylori are Gram-negative microaerophilic spiral bacteria which reside in the stomachs of humans and other primates. They can persist for years and possibly for life. They live in the mucus overlay of the gastric epithelium. They do not appear to invade the tissue, but the underlying mucosa is invariably inflamed, a condition termed chronic superficial gastritis, for which the organism is almost certainly responsible. To prove the association, two intrepid Australian research workers infected themselves by ingesting *H. pylori* and both developed gastritis, one lasting 14 days and the other nearly 3 years. Yet most infected individuals are asymptomatic. Nevertheless, the chronic inflammatory process is linked with peptic ulceration and gastric cancer, two of the most important diseases of the upper gastro-intestinal tract. *H. pylori* infection precedes ulceration, is nearly always present, and eradication of the organism by antibiotic therapy results in healing of the ulcer and a very low rate of ulcer recurrence. However, since many more people carry the organism than have ulcers, there must be other as yet unidentified predisposing factors which play a role in disease. There is also an association between *H. pylori* infection and gastric cancer. Urease is produced in abundance by this organism. Presumably it acts on urea, present in low concentrations, to form ammonia which locally neutralises acid and thus enables the bacteria to survive in this hostile environment. The bacterial cytotoxin, sheathed flagella and adhesins may also play a role in the survival of this organism in its very peculiar niche. The role of these products in the causation of ulcers and maybe cancer remains to be elucidated.

Pathogenic intestinal bacteria must establish infection and increase in numbers, and they too often have mechanisms for attachment to the epithelial lining so that they can avoid being carried straight down the alimentary canal with the rest of the intestinal contents. Indeed, their pathogenicity is likely to depend on this capacity for attachment or penetration. The pathogenicity of cholera, for instance, depends on the adhesion of bacteria to specific receptors on the surface of intestinal epithelial cells, and other examples are included in Table 2.1. Clearly, the concentration and thus the adsorption of bacterial toxins will also be affected by the balance between production and removal of bacteria in the intestine. Certain protozoa cause intestinal infections without invading tissues, and they too depend on adherence to the epithelial surface. *Giardia lamblia* attaches to the upper small intestine of man by means of a sucking disc, assisted by more specific binding (Table 2.1).

The likelihood of infection via the intestinal tract is certainly affected by the presence of mucus, acid, enzymes and bile. Mucus protects epithelial cells, perhaps acting as a mechanical barrier to infection, and contains secretory IgA antibodies that protect the immune individual against infection. Motile microorganisms (*Vibrio cholerae*, certain strains of *E. coli*) can propel themselves through the mucus layer and are thus more likely to reach epithelial cells to make specific attachments.* *Vibrio cholerae* also produces a mucinase that probably helps its passage through the mucus. Microorganisms infecting by the intestinal route are often capable of surviving in the presence of acid, proteolytic enzymes and bile. This also applies to microorganisms shed from the body by this route. The streptococci that are normal human intestinal inhabitants (*Streptococcus faecalis*) grow in the presence of bile, unlike other streptococci. This is also true of other normal (*E. coli, Proteus, Pseudomonas*) and pathogenic (*Salmonella, Shigella*) intestinal bacteria. It is noteworthy that the gut picornaviruses (hepatitis A, coxsackie-, echo- and polioviruses) are resistant to bile salts and to acid. The fact that tubercle bacilli resist acid conditions in the stomach favours the establishment of intestinal tuberculosis. Most bacteria, however, are acid sensitive and prefer slightly alkaline conditions.† Intestinal pathogens such as *Salmonella* or *Vibrio cholerae* are more likely to establish infection when they are sheltered inside food particles or when acid production in the host is impaired (achlorhydria). Volunteers who drank different doses of *Vibrio cholerae* contained in 60 ml saline showed a 10^4-fold increase in susceptibility to cholera when 2 g of sodium bicarbonate were given with the bacteria. Classical strains of cholera were used, and the minimal disease-producing dose without bicarbonate was 10^8 bacteria. Similar experiments have been done in volunteers with *Salmonella typhi*. The minimal oral infectious dose was 10^3–10^4 bacteria, and this was significantly reduced by the ingestion of sodium bicarbonate.

The intestinal tract differs from the respiratory tract in that it is always in motion, with constantly changing surface contours. The surface is made up of villi, crypts and other irregularities, and the villi themselves contract and expand. Particles in the lumen are moved about a great deal and have good opportunities for encounters with living cells; this is what the alimentary canal is designed for, if food is to be mixed, digested and absorbed. Viruses, by definition, multiply only in living cells; thus enteric viruses must make the most of what are primarily chance encounters with epithelial cells. Polio-, coxsackie- and echoviruses and the human diarrhoea viruses (rotaviruses, certain adenoviruses, etc.) form firm unions with receptor substances on the

---

* Nonmotile microorganisms, in contrast, rely on random and passive transport in the mucus layer. How important is mucus as a physical barrier? Gonococci and *Chlamydia* are known to attach to spermatozoa, and they could be carried through the cervical mucus as 'hitch-hikers' so that spermatozoa could help transmit gonorrhoea and non-specific urethritis.

† The standard Sabouraud's medium for the isolation of yeasts and moulds has an acid pH (5.4) in order that bacterial growth should be generally inhibited.

surface of intestinal epithelial cells, thus giving time for the penetration of virus into the cell. On the other hand, enteric bacteria that enter the mucosa are able to increase their numbers by growth in the lumen before entry, but it is not surprising that there are also mechanisms for bacterial attachment to epithelial cells (Table 2.1). Penetration of viruses into cells is discussed at a later stage, but it takes place either by endocytosis (phagocytosis into virus-sized vesicles) of the virus particle, or by fusion of the membrane of enveloped viruses with the cell membrane so that the contents of the virus particle enter the cell. These alternatives are not so distinct, because the virus particle in an endocytic vesicle is still in a sense outside the cell and still has to penetrate the cell membrane for its contents to be released into the cytoplasm, and enveloped viruses achieve this by fusion.

Most epithelial cells, whether epidermal, respiratory or intestinal, are capable of phagocytosis, but this is on a small scale compared with those specialist phagocytes, the macrophages and polymorphonuclear leukocytes (see Ch. 4). Certain pathogenic bacteria in the alimentary canal are taken into intestinal epithelial cells by a process that looks like phagocytosis. As seen by electron microscopy in experimental animals, *Salmonella typhimurium* (Fig. 2.5) attach to microvilli forming the brush border of intestinal epithelial cells. The microvilli degenerate locally at the site of attachment, enabling the bacterium to enter the cell, and the breach in the cell surface is then repaired. A zone of degeneration precedes the bacterium as it advances into the apical cytoplasm. In general, commensal intestinal bacteria do not appear to be taken up when they are attached to intestinal epithelium. We are beginning to understand intestinal invasion at the molecular level. The sequelae to penetration of the epithelium will depend on bacterial multiplication and spread (see Ch. 3), on toxin production, cell damage and inflammatory responses (see Ch. 8).

Microbial toxins, endotoxins and proteins can certainly be absorbed from the intestine on a small scale, and immune responses may be induced. Antibodies to materials such as milk, eggs and black beans can be detected when they form a significant part of the diet, and insulin is absorbed after ingestion as shown by the occurrence of hypoglycaemia. Diarrhoea promotes the uptake of proteins, and absorption of protein also takes place more readily in the infant, especially in species such as the pig or horse that need to absorb maternal antibodies from milk. As well as large molecules, particles the size of viruses can be taken up from the intestinal lumen,* and this occurs in Peyer's patches. Peyer's patches are isolated collections of lymphoid tissue lying immediately below the intestinal epithelium. The epithelial cells here are highly specialised (so-called M (microfold) cells) and

---

* When rats drink water containing very large amounts of bacteriophage T7 (diameter 30 nm), intact infectious phages are recoverable from thoracic duct lymph within 20 min.

Fig. 2.5 Attachment and entry of *S. typhimurium* into the enterocytes of rabbit distal ileum. Note the attachment structures, the elongation and swelling of microvilli. The organisms enter via the brush border and not apparently via the tight junctions. (Reproduced with permission from Figure 11, Worton *et al.* (1989). *J. Med. Microbiol.* **29**, 283–294.)

take up particles and foreign proteins, delivering them to underlying immune cells with which they are intimately associated by means of cytoplasmic processes. When large amounts of a reovirus for instance (see Table A.5) are introduced into the intestine of a mouse, the uptake of virus particles by M cells and delivery to immune cells, from whence they reach local lymph nodes, can be followed by electron microscopy. It seems appropriate that microorganisms in the intestine are sometimes 'focused' into immune defence strongholds.

The normal intestinal microorganisms of man are specifically adapted to life in this situation, and most of them are anaerobes of the

Bacteroides group, although *E. coli*, enterococci, lactobacilli and diphtheroids are common. Because of the acid pH, the stomach harbours
small numbers of organisms, but the total numbers increase as the
intestinal contents move from the small to the large intestine, and
there are about 10^8–10^{10} bacteria g^{-1} in the terminal ileum, increasing
to $10^{11} g^{-1}$ in the colon and rectum. Bacteria normally compose about a
quarter of the total faecal mass. The normal flora are in a balanced
state, and tend to resist colonisation with other bacteria. Possible
mechanisms include killing other bacteria by bacteriocins (see
Glossary), competition for food substances or attachment sites, and the
production of bacterial inhibitors. For instance, in mice the resident
coliforms and *Bacteroides* produce acetic and propionic acids which are
inhibitory for *Shigella* (dysentery) bacteria. Patients treated with
broad-spectrum antibiotics show changes in normal intestinal flora
and this may allow an abnormal overgrowth of microorganisms, such
as the fungus *Candida albicans* and *Clostridium difficile*. In breast-fed
infants, the predominant bacteria in the large bowel are lactobacilli
and their metabolic activity produces acid and other factors that
inhibit other microorganisms. As a result of this, and perhaps also
because of antibacterial components present in human milk, breast-fed
infants resist colonisation with other bacteria, such as the pathogenic
strains of *E. coli*. Bottle-fed infants, on the other hand, lacking the
protective lactobacilli, are susceptible to pathogenic strains of *E. coli*,
and these may cause serious gastroenteritis.

Intestinal microorganisms that utilise ingested cellulose serve as
important sources of food in herbivorous animals. In the rabbit, for
instance, volatile fatty acids produced by microorganisms in the
caecum yield 20% of the daily energy requirements of the animal. The
rumen of a 500 kg cow is a complex fermentation chamber whose
contents amount to 70 litres. In this vast vat 17 species of bacteria
multiply continuously, utilising cellulose and other plant materials,
and protozoa (seven genera) live on the bacteria. As the microbial mass
increases, the surplus passes into the intestine to be killed, digested
and absorbed. Large volumes of CO_2 and methane are formed and
expelled from both ends of the cow. The passage of methane represents
a loss of about 10% of the total energy derived from food. In man,
intestinal bacteria do not normally have a nutritive function; they
break down and recycle the components of desquamated epithelial
cells, and perhaps synthesise vitamins, but this is unimportant under
normal circumstances.

Mechanisms of attachment to and invasion of the gastro-intestinal tract

Pursuit of the mechanisms and the microbial determinants responsible
for invasion and replication within intestinal mucosae by enteric

pathogens is one of the most actively researched areas in bacterial pathogenicity. Diarrhoeal disease is caused by invasive species (e.g. Salmonellae, Shigellae and enteroinvasive *E. coli*) as well as noninvasive species (e.g. *V. cholerae,* enteropathogenic and enterotoxigenic *E. coli*) and is still responsible for a huge proportion of the total morbidity and mortality in developing countries; a separate section is devoted to this topic in Ch. 8. Also, in recent years the incidence in the UK and other developed countries of *Salmonella* and *Campylobacter* infections has risen dramatically and remains high. Intense (and highly competitive!) research is being conducted to elucidate the mechanisms of invasion. Recently there has been a veritable explosion of new molecular genetic information and what follows is an overview of the biological significance of this fast-moving field.

General considerations

Before dealing with the molecular mechanistic work, it is important to record a note of caution regarding experimental systems. Several attachment/invasin systems, operative in cultured cells, have been described in *Yersinia* spp., yet only the chromosomally encoded 'invasin' appears to be important in interaction with M cells in Peyer's patches through which *Yersinia* penetrate the gut. Moreover, once they have negotiated the M cell barrier, *Yersinia* are essentially extracellular pathogens, yet for years they have been studied as paradigms of intracellular pathogens! The most widely used system for modelling human typhoid-like infections caused by *S. typhi* is infection of the mouse with *S. typhimurium* in which the ratios of oral to intraperitoneal LD_{50} values obtained for parent and mutant strains are compared. By this means one can deduce whether a mutation has affected the ability of the pathogen to negotiate the gut mucosa or to survive some later stage in the complex host pathogen interaction as, for example, an encounter with macrophages. However, the gut mucosa is a complex highly organised tissue (see Figs 8.19 and 8.20) and increases in oral LD_{50} values do not *per se* indicate whether this is due to failure to negotiate the epithelial layer of enterocytes, or to handle the hypertonic conditions at the tips of villi, or to spread across the deeper submucosal layers. In any case, if one is interested in the mechanisms of diarrhoeal disease induced by *S. typhimurium*, one cannot use mice since *S. typhimurium* in mice (as its name implies) is equivalent to *S. typhi* in man and causes a systemic infection rather than a localised mucosal infection. For the latter the best small laboratory model is the rabbit.

Two major molecular biological developments have occurred which dominate current research into bacterial pathogenicity. The first is the recognition of 'pathogenicity islands' (PAIs, PaIs, PIs; the acronyms still vary) and other mobile virulence elements (see Kaper and Hacker 1999). PAIs carry one or more virulence factor (e.g. adhesins, invasins, iron uptake systems, toxins and secretion systems, and doubtless others), and are present in the genome (chromosome or plasmid) of pathogenic bacteria but absent from the genome of related nonpathogens. They range in size from 10 to 200 kb and often have different G+C content, suggesting their acquisition by horizontal transfer of DNA into new hosts. PAIs are often flanked by direct repeat sequences with tRNA genes, a common target for insertion into the chromosome. The *Shigella* virulence plasmid has been called an 'archipelago' of PAIs and smaller elements (1–10 kb) 'islets'. The second development concerns the recognition of at least four secretion systems in Gram-negative bacteria. Type I is sec (secretory system) independent and exemplified by the secretion of α-haemolysin of *E. coli* ; it secretes proteins from the cytoplasm across the inner and outer membranes in one step facilitated by a small number of ancillary genes. Type II secretion is sec-dependent and secretes effector proteins using the general secretory pathway (GSP). Type III systems are complex, sec-independent systems closely related to the flagella assembly system; they are also activated on contact with host cells. They are known to be extremely important in a growing number of pathogens (as we shall see) and facilitate translocation of bacterial effector molecules directly into the host cell membrane or cytoplasm. Type IV involves secretion of proteins where all the necessary information for transmembrane negotiation inheres in the secreted protein itself. Now for specific examples.

Enteropathogenic *E. coli* (EPEC)

EPEC was the first serotype of *E. coli* to be incriminated as a pathogen. Its designation as EPEC is unfortunate as all pathogenic *E. coli* are in a real sense enteropathogenic, but the nomenclature is rigidly embedded. It is essentially a noninvasive pathogen with only rare reports of its presence inside human gut epithelial cells; it can, however, be internalised by cultured cells. The pathognomonic lesion of EPEC is the pedestal type 'attaching and effacing' (A/E) lesion induced on microvilli-bearing enterocytes resulting in 'intimate' type of adherence (Fig. 2.6). The adherence is different to the nonintimate adherence exhibited by *V. cholerae* and enterotoxigenic *E. coli*, which both attach via adhesins that project from the organism to the host cell. There are two main genetic elements which confer virulence on EPEC: (1) *bfp* genes (bundle-forming pili (BFP) encoded in the EPEC adherence factor (EAF) plasmid); and (2) the genes encoding the determinants of A/E encoded in the chromosomally located LEE (locus of enterocyte effacement) pathogenicity island. BFP have been shown in human volunteer studies to be important, although not absolutely necessary, in the colonisation of EPEC; their expression is regulated by the *per* (plasmid-encoded regulator) genes located in the EAF plasmid. The *per* regulator also controls expression of other membrane proteins

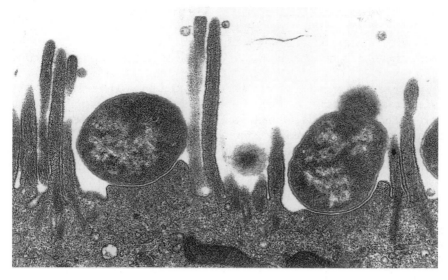

Fig. 2.6 Classical EPEC-induced A/E pedestal. (Reproduced from Knutton *et al*. (1987). *Infect. Immun.* **55**, 69–77. We thank Dr Stuart Knutton for the figure and publishers for permission to use this figure.)

and thus acts as a global regulator, a feature now increasingly recognised in pathogenic bacteria. Upon contact with epithelial cells, expression of LEE is triggered, the sequelae to which is summarised in Fig. 2.7. The incredible fact is that the organism expresses and inserts its own receptor Tir (translocated intimin receptor), which after phosphorylation and interaction with intimin triggers a signalling cascade which results (by as yet unknown mechanisms) in diarrhoea.

Enterohaemorrhagic *E. coli* (EHEC)

Unlike EPEC which colonises predominantly the small intestine, EHEC colonises the colon. The mechanism of initial attachment is not clear, but sequential attachment is via a Tir–intimin interaction. The question as to how the tissue tropism is determined is an intriguing and as yet unanswered one. After initial attachment the clinical outcome of infection is quite different from that induced by EPEC and this is discussed in Ch. 8.

Shigella

Initial entry of *S. dysenteriae* into the colonic mucosa is via M cells in follicle-associated epithelia (FAE) through which they migrate without killing the M cell. Shigellae are then able to infect intestinal epithelial cells via their basolateral membranes: they do not penetrate via brush borders. The genetic system encoding this invasive phenotype is a PAI

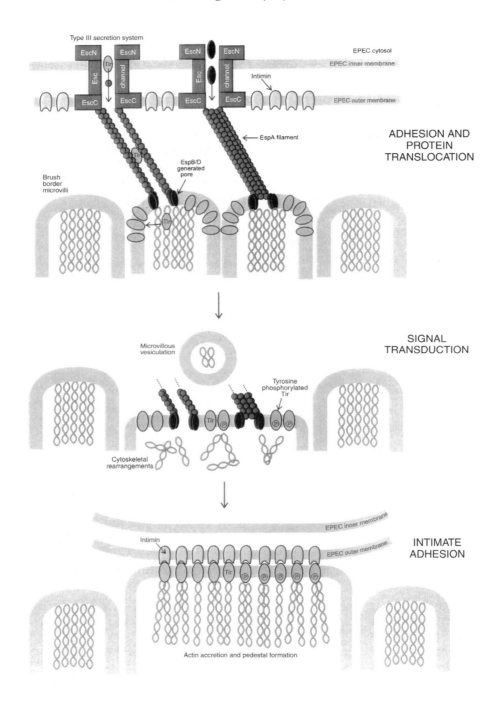

in the virulence plasmid, which includes the Mxi/Spa secretory apparatus and the Ipa (invasion plasmid antigen B) proteins which are the
major effectors of entry (Fig. 2.8). Infected epithelial cells are induced
to release the inflammatory cytokines interleukin-8 (IL-8) and tumour
necrosis factor-α (TNFα). In addition, *Shigella* infect macrophages
inducing IpaB-mediated apoptosis and IpaB-mediated release from
those macrophages of IL-1β, another potent inflammatory cytokine.
The inflammatory response destabilises epithelial integrity and
permeability by the extrusion of polymorphonuclear (PMN) cells
thereby allowing direct access of more *Shigella* to the basolateral
membranes of epithelial cells. Interaction with PMN cells results in
killing of organisms with concomitant release of tissue damaging granules. Within minutes after entry, *S. flexneri* escapes from its vacuole by
virtue of IpaB which, in addition to triggering entry, is responsible for
lysis of the vacuolar membrane and escape of the organisms into the
cytoplasm. *Shigella* then express the *olm* (organelle-like movement)
phenotype, which allows the spread of organisms throughout the cytoplasm along actin stress cables which run between anchorage sites of
cells adhering to substrata. The details of the mechanism responsible
for this movement are not known but cytoskeletal actin is involved in
one of several ways, possibly by myosin-like proteins giving rise to an
ATP-powered movement. In addition they express the *ics* (intra-inter-
cellular spread) phenotype which allows colonisation of the adjacent
cells. This second type of movement, also seen with intracellular
Listeria monocytogenes, is accompanied by the appearance of 'comet'-
like rearrangements of actin in the cell. Rapid polymerisation of actin
filaments occurs localised at one end of the bacterium, resulting in a

Fig. 2.7 Mechanism of formation of EPEC-induced A/E pedestal. Cell contact
stimulates the expression of LEE-encoded proteins and the assembly of a
protein translocation apparatus (translocon). The translocon consists of pores
in the bacterial envelope (EscC-generated pore) and in the host membrane
(EspB/D-generated pore) with the pores connected by a hollow EspA filament,
thereby providing a continuous channel from the bacterial to the host cell
cytosol. Energy is thought to be provided by EscN protein. The translocon is
used to translocate Tir into the host cell where it becomes inserted into the
host cell membrane; EPEC Tir (but not EHEC O157:H7 Tir) becomes phosphorylated on tyrosine residues following translocation. Translocated Tir and/or
other as yet unidentified effector proteins transduce signals that induce breakdown of the brush-border microvillous actin cytoskeleton with consequent
vesiculation of the microvillous membrane. Although the mechanisms are
unknown, localised translocation of effector proteins results in localised
cytoskeletal changes. Intimate adhesion and pedestal formation results from
the interaction of intimin and Tir and the accumulation of actin (and other
cytoskeletal proteins) beneath intimately attached bacteria following microvillous effacement. (Adapted from Frankel *et al.* (1998). *Mol. Microbiol.* **30**,
911–921. We thank Dr Stuart Knutton for the figure and publishers for permission to use this figure.)

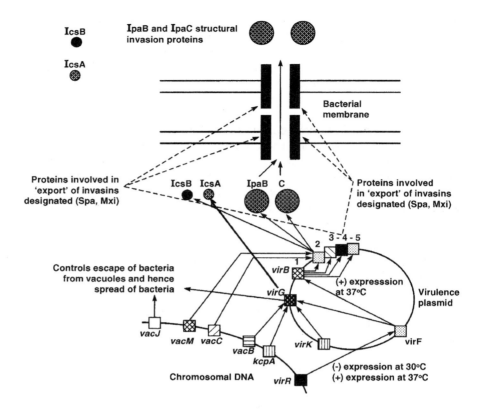

Fig. 2.8 A highly *schematic* simplified representation of the genetic systems involved in the invasiveness of *Shigella*. This is a highly complex process involving both plasmid- and chromosomal-borne genes which encode: the structural proteins that actually mediate internalisation (*ipaB, C*), random internal movement (*icsA*), vacuolar escape and intercellular spread processes (*ipaB, virG, icsB, vacJ*); proteins (*mxi / spa* components of the Type III secretion system) that mediate the export or surface positioning of several polypeptides including the actual invasins; and others (*virR, virF*) involved in regulating the expression of the biologically active determinants. *Note*: (a) Plasmid regions 1–5 represent the PAI encoding the Type III system. (b) The expression of some genes is temperature dependent, probably reflecting the need to regulate expression outside and inside the body. (c) *vir*, virulence; *vac*, virulence-associated chromosomal virulence gene; *ipa*, invasion plasmid antigen; *inv*, invasion; *spa*, surface presentation antigen; *mxi*, membrane expression of invasion plasmid antigens; *ics*, intra–inter-cellular spread; *kcp*, keratoconjunctivitis provocation.

forward movement of the bacterium. Nucleation and polymerisation of actin is mediated by *virG* (*icsA*). The organism is now propelled towards the cell membrane and into a protrusion of the adjacent cell membrane now surrounded by a double membrane. The latter is ruptured by the product of *icsB* gene thereby releasing the organism into adjacent cells where rapid intracytoplasmic multiplication can again take place. The intragastric inoculation of macaque monkeys with *icsA* mutants shows a dramatic loss of virulence proving that the mechanisms described above were not mere cell artefacts. The role of Shiga toxin is discussed in Ch. 8.

Salmonella

Much of the existing information on *Salmonella* invasion relates to studies in mice from which the near dogma has developed that entry via M cells in the FAE is a *sine qua non* for intestinal invasion. However, this is clearly not the case in rabbits, calves and pigs where concurrent entry of *Salmonella* into M cells and enterocytes can be seen. There are at least two invasive biotypes of *Salmonella* which cross conventional serotypic boundaries.

Histotoxic *Salmonella*
The main feature of the early damage to epithelia caused by histotoxic strains of *Salmonella* serotype Typhimurium is a toxin-mediated detachment of enterocytes from rabbit terminal ileum which is preceded by cleavage of tight junctions (Fig. 2.9C). This leads to the release of microvilli-bearing cells which degenerate rapidly into spherical highly vacuolated entities. Similar lesions can be produced in rabbit tissues challenged *in vivo* and *in vitro* with live histotoxic Dublin strains. Sterile supernates from rabbit gut challenged *in vitro* with a histotoxic Typhimurium strain induce an almost identical picture of epithelial disintegration when added to fresh tissue from the same animal. In calves and pigs, histotoxic strains of Dublin cause extensive tissue damage to both absorptive epithelium (AE) and to follicle-associated epithelium. *Salmonella* serotype Choleraesuis is not histotoxic. These observations are of crucial importance in attempting to understand the pathogenesis of *Salmonella* infection. By virtue of their ability to denude epithelia, these organisms open up new routes of invasion and tissue transmigration. It is important to point out that if one uses Caco-2 cells (which are gut-derived, tight junction-forming, microvilli-expressing cells) as model epithelia, one cannot demonstrate the tight junction cleavage by *Salmonella* serotype Dublin as was shown *in vivo*.

Nonhistotoxic *Salmonella*
Nonhistotoxic *Salmonella* cause shortening of whole villi. Here the picture, as observed in the rabbit ileal loop model, is totally different from the one described for histotoxic *Salmonella*. Bacteria enter via brush borders (Fig. 2.9A, B) and bacteria-laden cells are shed. There is no evidence of a rapid initial cleavage of tight junctions. The time scale of events is quite different with maximum cell shedding leading to truncation of villi occurring at 12–14 h post-challenge. Behind the extrusion of bacteria-laden cells the epithelium is resealed. The significance of these observations is discussed in Ch. 8. Virulent strains induce a massive influx of PMN cells.

Fig. 2.9 (A) Non-histotoxic *Salmonella*. The scanning electron microscopic picture shows many rabbit ileal villi whose tips are being extruded some 12–14 h after infection with a nonhistotoxic strain of *S. typhimurium*. (B) Transmission electron micrograph of the same tissue as in (A) showing the extruded cells to be laden with internalised bacteria. (C) Histotoxic *Salmonella* cause the rapid detachment of enterocytes with little initial infection of cells via brush borders; note that the bacteria are in the lumen surrounding the detached cell. (A, B) are reproduced with permission from Wallis *et al.* (1986). *J. Med. Microbiol.* **22**, 39–49; and (C) from Lodge *et al.* (1999). *J. Med. Microbiol.* **48**, 811–818.

In recent years, there has been a huge effort, still actively ongoing, to discover the molecular genetic basis of virulence of *Salmonella*. As a result, at least five 'pathogenicity islands' (PIs) have been recognised in the *Salmonella* chromosome. It is premature to give a detailed coverage of this complex field in relation to *Salmonella*, as the field is rapidly developing, but a few emerging points are summarised. First, the genes recognised in these *Salmonella* PI (SPI) clusters are mainly to do with invasion of eukaryotic cells, intracellular survival and systemic infection. Second, at least SPI-1 and SPI-2 are known to encode Type III secretion systems. Some of the secreted proteins have been recognised and are involved in the translocation of effector molecules into eukaryotic target cells, thereby promoting invasion. Appendage structures have been observed whereby Typhimurium attaches to gut epithelia (see Fig. 2.5) which are remarkably similar to those described for EPEC but have not yet been fully characterised, but that is only a matter of time.

Yersinia

As indicated above, *Yersinia* has long been regarded as an intracellular pathogen, but histopathological examination shows clearly that after penetration of the intestinal epithelium through M cells, and destruction of Peyer's patches, *Y. enterocolitica* is found in lymphoid follicles where it is potently anti-phagocytic by virtue of its virulence plasmid PAI (see Ch. 4).

Campylobacter jejuni

No detailed definitive molecular biological information exists as yet comparable to that described for EPEC and *Shigella*. However, *C. jejuni* is one of those pathogens for which the genomic sequence is now known. Projects are now underway to combine this knowledge with the microarray technique referred to in Ch. 1. The future is exciting as study of the genome has revealed very little in common with the other well-studied enteric pathogens. Keep watching this space!

Giardia lamblia

Giardia lamblia colonises the human small bowel and causes diarrhoea. There are two candidate adhesins: taglin (trypsin-activated *Giardia lamblia* lectin) and GLAM-1 (*Giardia lamblia* adherence molecule-1). The current perception is that initial contact of the parasite with the gut wall is via taglin, which is distributed round the surface of the parasite, and that the disc-specific GLAM-1 (there may be more such adhesins) present on this organelle is responsible for the avid attachment of the disc to the target cell surface.

Entamoeba histolytica

The trophozoite form of *Entamoeba histolytica* lives in the lumen of the large bowel, the only known reservoir for this parasite. The trigger mechanisms which convert this organism into the pathogen causing serious invasive amoebiasis are not known. The ability to adhere to colonocytes *in vivo* seems to be the exception rather than the rule. More is known about the putative determinants of gut damage and the factors important in the spread of the organism formation of lesions in the liver (Ch. 8).

Oropharynx

The throat (including tonsils, fauces, etc.) is a common site of residence of microorganisms as well as of their entry into the body.

The microbial inhabitants of the normal mouth and throat are varied, exceedingly numerous, and are specifically adapted to life in this environment. Bacteria are the most numerous, but yeasts (*Candida albicans*) and protozoa (*Entamoeba gingivalis, Trichomonas* *tenax*) occur in many individuals. Oral bacteria include streptococci, micrococci and diphtheroids, together with *Actinomyces israeli* and other anaerobic bacteria. Some of these are able to make very firm attachments to mucosal surfaces, and others to teeth which provide a long-term, nondesquamating surface. *Streptococcus mutans*, for instance, uses the enzyme glycosyl transferase to synthesise glucan (a high molecular weight polysaccharide) from sucrose. The glucan forms an adhesive layer, attaching bacteria to the surface of teeth and to other bacteria (Table 2.1). If there are no teeth, as in the very young or the very old, *Streptococcus mutans* has nothing to 'hold on to' and cannot maintain itself in the mouth. The dextran-containing secretions constitute a matrix in which various other bacteria are present, many of them anaerobic. It forms a thin film attached to the surface of the tooth which is called dental plaque, and is visible as a red layer when a dye such as erythrosine is taken into the mouth. Dental plaque is a complex microbial mass containing about 10^9 bacteria g^{-1}. Certain areas of the tooth are readily colonised, especially surface fissures and pits, areas next to the gum, and contact points between neighbouring teeth. The film is largely removed by thorough brushing, but re-establishes itself within a few hours. When teeth are not cleaned for several days the plaque becomes quite thick, a tangled forest of microorganisms (Fig. 2.10). Dietary sugar is utilised by bacteria in the plaque and the acid that is formed decalcifies the tooth and is responsible for dental caries. The pH in an active caries lesion may be as low as 4.0. Unless the bacteria, the sugar (and the teeth) are present, dental caries does not develop. When monkeys are fed on a caries-producing diet, the extent of the disease can be greatly reduced by vaccination against *Streptococcus mutans*,* and vaccines are being developed for use against caries in man. Caries is already becoming less common, but if the vaccines are effective, caries (and many dentists) could one day be eliminated. Western individuals with their tightly packed, bacteria-coated teeth and their sugary, often fluoride-deficient diet, have been badly affected, and it is legitimate to regard dental caries as one of their most prevalent infectious diseases.

Periodontal disease is another important dental condition that affects nearly everyone (and most animals) to a greater or lesser extent. The space between the tooth and gum margin has no natural cleansing mechanism and it readily becomes infected. This results in inflammation, with accumulation of polymorphs and a serum exudate.

---

* Antibodies are protective, as shown when orally administered monoclonal antibody to outer components of *S. mutans* prevented colonization. It is noteworthy that in one study of 11 agammaglobulinaemic patients, all were badly affected by caries, and four lost all their teeth quite rapidly between the age of 20 and 30 years. Local antibody to the relevant bacteria would be protective, and additional antimicrobial forces are present in crevicular spaces. The crevicular space is a small fluid-filled cleft between the edge of the gum and the tooth, containing antibodies (IgG, IgM), complement and phagocytic cells derived from plasma.

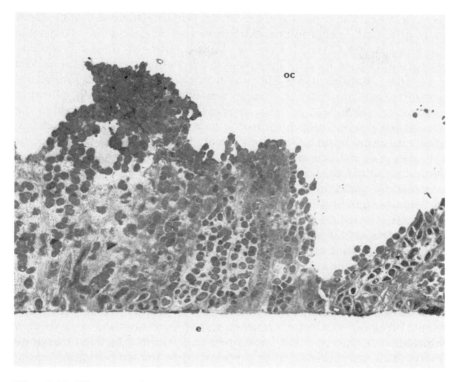

Fig. 2.10 Electron microscope section through dental plaque at gingival margin of a child's tooth, showing microcolonies of cocci. Thickness of plaque from enamel (e) to free border (oral cavity, oc) above is variable. Magnification, ×4000. (Photograph kindly supplied by Dr H. N. Newman, Institute of Dental Surgery, Gray's Inn Road, London.)

The inflamed gum bleeds readily and later recedes, while the multiplying bacteria can cause halitosis. Eventually the structures that support the teeth are affected and teeth become loose as bone is resorbed and ligaments weakened. Bacteria such as *Actinomyces viscosis*, *Actinobacillus actinomycetemcomitans*, and especially *Porphyromonas gingivalis* are commonly associated with periodontal disease.

Certain strains of streptococci adhere strongly to the tongue and cheek of man but not to teeth, and can be shown to adhere to the epithelial cells in cheek scrapings. Pharyngeal cells can be obtained by wiping the posterior pharyngeal wall with a wooden applicator stick, and experiments show that virulent strains of *Streptococcus pyogenes* make firm and specific attachments to these cells by means of lipoteichoic acid on threads (pili) protruding from the bacterial surface. Presumably the corynebacteria responsible for diphtheria also have surface structures that attach them to epithelium in the throat. As in the intestines, the presence of the regular microbial residents makes it more difficult for other microorganisms to become established. Possible mechanisms for this interference were mentioned in the preceding section. Changes in oral flora upset the balance. For instance, the yeast-like fungus *Candida albicans* is normally a harmless inhabitant of the mouth, but after prolonged administration of broad-spectrum antibiotics, changes in the normal bacteria flora enable the pseudomycelia of *Candida albicans* to penetrate the oral epithelium, grow and cause thrush.

Saliva is secreted in volumes of a litre or so a day, and has a flushing action in the mouth, mechanically removing microorganisms as well as providing antimicrobial materials such as lysozyme (see Glossary) and secretory antibodies. It contains leucocytes, desquamated mucosal cells and bacteria from sites of growth on the cheek, tongue, gingiva, etc. When salivary flow is decreased for 3–4 h, as between meals,* there is a four-fold increase in the number of bacteria in saliva. Disturbances in oral antimicrobial and cleansing mechanisms may upset the normal balance. In dehydrated patients, or those ill with typhus, typhoid, pneumonia, etc., the salivary flow is greatly reduced, and the mouth becomes foul as a result of microbial overgrowth, often with some tissue invasion. Vitamin C deficiency reduces mucosal resistance and allows the normal resident bacteria to cause gum infections. As on all bodily surfaces, there is a shifting boundary between harmless coexistence of the resident microbes and invasion of host tissues, according to changes in host resistance.

During mouth breathing the throat acts as a baffle on which larger inhaled particles can be deposited, and microorganisms in saliva and nasal secretions are borne backwards to the pharynx. Microorganisms in the mouth and throat need to be attached to the squamous epithelial surface or find their way into crevices if they are to avoid being washed away and are to have an opportunity to establish infection. The efficiency of infection may be increased by the act of swallowing. As material from the nasal cavity, mouth and lung is brought to the pharynx, that great muscular organ the tongue pushes backwards with a vigorous thrust and firmly wipes this material against the pharyngeal walls. One of the earliest and most regular symptoms of upper respiratory virus infections is a sore throat, suggesting early viral growth in this area, with an inflammatory response in the underlying tissues. It may also signify inflammation of submucosal lymphoid tissues in the tonsils, back of the tongue, and throat, which form a defensive ring guarding the entrance to alimentary and respiratory tracts.

---

* Salivary flow continues between meals, the average person swallowing about 30 times an hour.

Urinogenital Tract

Urine is normally sterile, and since the urinary tract is flushed with urine every hour or two, invading microorganisms have problems in gaining access and becoming established. The urethra in the male is sterile, except for the terminal third of its length, and microorganisms that progress above this point must first and foremost avoid being washed out during urination. That highly successful urethral parasite, the gonococcus, owes much of its success to its special ability to attach very firmly to the surface of urethral epithelial cells, partly by means of fine hairs (pili) projecting from its surface (Fig. 2.11).* Similarly, uropathogenic *E. coli* (UPEC) adhere to uroepithelial cells by means of a well-characterised pilus. The bladder is not easily infected in the male; the urethra is 20 cm long, and generally bacteria need to be introduced via an instrument such as a catheter to reach the bladder. The female urethra is much shorter, only about 5 cm long, and more readily traversed by microorganisms; it also suffers from a dangerous proximity to the anus, the source of intestinal bacteria. Urinary infections are about 14 times as common in women, and most women have urinary tract infections at some time. Bacteruria,† however, often occurs without frequency, dysuria, or other symptoms. Even the urethral deformations taking place during sexual intercourse may introduce infection into the female bladder.‡ Spread of infection to the kidney is promoted by the refluxing of urine from bladder to ureter that occurs in some young females.

Urine, as long as it is not too acid, provides a fine growth medium for many bacteria and the entire urinary tract is more prone to infections when there is interference with the free flow and flushing action of urine, or when a 'sump' of urine remains in the bladder after urination. Urinary infections are thus associated with structural abnormalities of the bladder, ureter, etc., with stones, or with an enlarged prostate that prevents complete emptying of the bladder. Incomplete emptying also leads to urinary infection in pregnant women, and this

---

* The gonococcus is soon killed in urines that are acid (<pH 5.5), and this helps explain why the bladder and kidneys are not invaded. The prostate is at times affected and the gonococcus accordingly grows in the presence of spermine and zinc, materials that are present in prostatic secretions and that would inhibit many other bacteria.

† By the time it has been voided and tested in the laboratory, urine always contains bacteria. For routine purposes it is not regarded as significant unless there are more than 10^5 bacteria (ml urine)$^{-1}$. But many women have frequency and dysuria with smaller numbers of bacteria in urine and in some cases, perhaps, the infection has spread no further than the urethra.

‡ The importance of sexual activity is often assessed by comparing nuns or prostitutes with 'ordinary' women. Bacteruria is 14 times commoner in ordinary women than in nuns, and in one study, sexual intercourse was the commonest precipitating factor for dysuria and frequency in young women. On the other hand, an innocent bubble bath may facilitate spread of faecal organisms into the urethra.

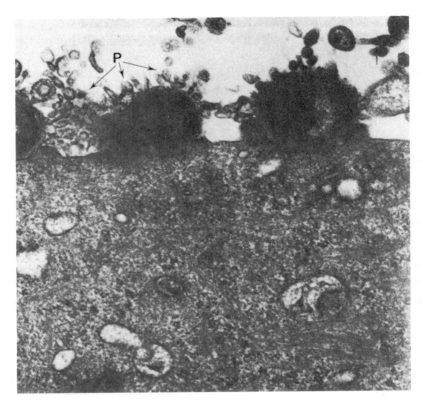

Fig. 2.11 Electron micrograph showing gonococci closely attached to the surface of a human urethral epithelial cell: 40–50 pili (P) project from the gonococcal surface. Adherence of *Neisseria gonorrhoeae* to urethral mucosal cells: an electron microscope study of human gonorrhoea. (Reproduced from Ward, M. E. and Watt, P. J. (1972). *J. Infect. Dis.* **126**, 601–604.)

is partly due to the sluggish action of muscles in the bladder wall. But the bladder is more than an inert receptacle for infected urine, and responds with inflammation and secretory antibody production. The normal bladder wall, moreover, appears to have some intrinsic but poorly understood antibacterial activity. Uropathogenic strains of *E. coli* bind to epithelial cells lining the bladder, which respond by exfoliating. As a host defence this is not enough because the bacteria then invade deeper tissues.

The vagina has no particular cleansing mechanism and would appear to present an ideal site for colonisation by commensal microorganisms. During reproductive life, however, from puberty until the menopause, the vaginal epithelium contains glycogen because of the action of circulating oestrogens. Doderlein's bacillus (a lactobacillus) colonises the vagina, metabolising the glycogen to produce lactic acid. The lactic acid gives a vaginal pH of about 5.0, and together with other products of metabolism inhibits colonisation by all except Doderlein's bacillus and a select number of bacteria, including various nonpyogenic streptococci and diphtheroids. Normal vaginal secretions contain up to 10^8 bacteria ml^{-1}. Other microorganisms are unable to establish infections, except the specialised ones that are therefore responsible for venereal diseases. Oestrogens thus generate an antimicrobial defence mechanism just at the period of life when contaminated objects are being introduced into the vagina. Before puberty and after the menopause, the vaginal epithelium lacks glycogen, the secretion is alkaline, and bacteria from the vulva, including staphylococci and streptococci, can become established.

The ascent of microorganisms from vagina to uterus is blocked at the cervix because of the downward flow of mucus and the action of cilia, together with local production of lysozyme. Once the cervical barrier has been interfered with, after abortion, miscarriage, childbirth or the presence of an intrauterine contraceptive device, invasion of the uterus, fallopian tubes, etc., becomes easier. Gram-negative intestinal bacteria, group B streptococci or anaerobes are likely culprits. The cervix is less of a barrier to those expert invaders *N. gonorrhoeae* and *C. trachomatis*.

Conjunctiva

The conjunctiva is kept moist and healthy by the continuous flow of secretions from lachrymal and other glands. Every few seconds the lids pass over the conjunctival surface with a gentle but firm windscreen wiper action. Although the secretions (tears) contain lysozyme (see Glossary) and other antimicrobial substances such as defensins (see Glossary), their principal protective action is the mechanical washing away of foreign particles. Microorganisms alighting on the conjunctiva are treated like inanimate particles of dirt or dust and swept away via the tear ducts into the nasal cavity. Clearly there is little or no opportunity for initiation of infection in the normal conjunctiva unless microorganisms have some special ability to attach to the conjunctival surface. The conjunctiva, however, suffers minor injuries whenever we get 'something in the eye', and these give opportunities for infection, as would defects in the cleansing mechanisms due to lachrymal gland or lid disease. The *Chlamydia* responsible for inclusion conjunctivitis and for that greatest eye infection in history, trachoma,* are masters in the art of conjunctival infection. They attach to heparan sulphate-type receptors on cell surfaces, doubtless also taking advantage of breaches in the defence mechanisms. The conjunctiva is also infected from the

---

* World-wide 500 million people are infected, and 5 million blinded by it.

'inside' during the course of measles, when the virus spreads via the circulation and is somehow seeded out to conjunctival blood vessels (see p. 141).

The conjunctiva is infected by mechanically deposited rather than by airborne microorganisms. Flies, fingers and towels play an important role in diseases such as trachoma, and it is significant that the types of *Chlamydia trachomatis* that cause urethritis (see p. 61) also often infect the eye, presumably being borne from one to the other by contaminated fingers. Certain enteroviruses (enterovirus 70, coxsackie virus A24) cause conjunctivitis, and conjunctivitis due to adenovirus 8 is one of the many diseases that can be caused by the physician (iatrogenic diseases). It is transmitted from one patient to the next by the instruments used in extracting foreign bodies from the eye. Microorganisms present in swimming baths have a good opportunity to infect the conjunctiva, water flowing over the conjunctiva depositing microorganisms and at the same time causing slight mechanical and chemical damage. Both the *Chlamydia* and adenovirus 8 have been transmitted in this way. During the birth of an infant, gonococci or *Chlamydia* from an infected cervix can be deposited in the eye to cause severe neonatal conjunctivitis. Certain free-living protozoa (*Acanthamoeba*) present in soil and sometimes in water supplies, can infect the cornea to cause keratitis. This occurs in India, perhaps because of foreign bodies or other infections in the eye, and also in those wearing contact lenses.

In the two preceding sections, several references have been made to *Chlamydia* as an important agent of occulogenital disease. *Chlamydia* are obligate intracellular parasites, sometimes referred to as 'energy parasites'. Not much is known about the details of the infection process *in vivo* but a great deal has been learnt about the biology of infection using cultured epithelial cells *in vitro* as model systems. Both of the two major species (*Chlamydia trachomatis* and *C. psittaci*) attach to host cells, enter by endocytosis, avoid lysosomes, and initiate their complex replication cycle, leading to development of characteristic inclusion bodies within infected cells. While there are intrinsic strain differences in ability to infect cells productively, it has now been beautifully demonstrated that the route of entry into cells has a profound effect on the ability of organisms to replicate. The elegant work of Pearce and colleagues in Birmingham has shown that there are two routes of entry into cells – microfilament-dependent (phagocytic) and pinocytic – for this pathogen involving, one must assume, two different receptor systems. The important point is that while the pinocytic route is a more efficient means of entry, the phagocytic route results in a tenfold greater level of productive infection. Elementary bodies (EBs) enter and differentiate into reticulate bodies (RBs), the replicative form of the organism, which then differentiate into more infectious EBs. The replication of RBs is controlled in a highly complex manner by the availability of nutrients – energy components (ATP) and in particular amino acids, a point to which we shall return in Ch. 10. Owing to the lack of a genetic transfer system for *Chlamydia*, the rate of progress in identifying the molecular components which mediate and regulate these complex processes has been vastly slower compared to the study of other bacterial systems. However, a protein has now been identified which, in a manner yet to be fully understood, plays an important role in the early stages of infection. It is a lipoprotein which has sufficient partial sequence homology and several important properties in common with the macrophage infectivity potentiator (Mip) protein of *Legionella pneumophila* (a peptidyl-prolyl *cis/trans* isomerase, described in Ch. 4) to warrant its designation, chlamydial-like Mip.

The Normal Microbial Flora

The commensal microorganisms that live in association with the body surfaces of man have repeatedly been referred to in this chapter. It has been calculated that the normal individual houses about 10^{12} bacteria on the skin, 10^{10} in the mouth and nearly 10^{14} in the alimentary canal. For comparison there are about 10^{13} cells in the body. Most of these are highly specialised bacteria,* utilising available foods, often with mechanisms for attachment to body surfaces, and looking very much as if they have an evolutionary adaptation to a specific host.

Is the normal microbial flora of any value?

There is no doubt that intestinal microorganisms play a vital role in the nutrition of many herbivorous animals. The caecum of the rabbit and the rumen of the cow were referred to on p. 30. The most important beneficial effect in man is probably the tendency of the normal microbial flora to exclude other microorganisms. Intestinal bacteria such as *E. coli*, for instance, fail to establish themselves in the normal mouth and throat, and disturbances in the normal flora induced by long

---

* The specialised secretion of the genital mucosa of both sexes (smegma), has its own resident bacterium, *Mycobacterium smegma*, which often contaminates urine. Skin residents include certain yeasts, *Pityrosporum ovale* and *Pityrosporum orbiculare*. *Pityrosporon ovale* appears to be responsible for that widespread but humble human condition, dandruff. It is a good parasite, present on most male scalps, feeding on dead skin scales with minimal inconvenience to the host. Fascinating mites (*Demodex folliculorum* and *brevis*) reside unobtrusively in hair follicles or sebaceous glands, feeding on epithelial cells and on sebum. These mites are present in all human beings, and their spectacular success as parasites is reflected by a healthy person's astonishment when shown an adult mite attached to the base of his plucked eyelash. Other mites of the same genus parasitise horses, cattle, dogs, squirrels, etc.

courses of broad-spectrum antibiotics may permit the overgrowth of *Candida albicans* in the mouth or staphylococci in the intestine. In one unusual experiment none of 14 volunteers given 1000 *Salmonella typhi* by mouth developed disease (see also p. 25), but one of four did so when the antibiotic streptomycin was given at the same time. Streptomycin probably promoted infection by its bacteriostatic action on commensal intestinal microorganisms. It is known that other *Salmonella* infections of the intestine persist for longer when antibiotics are given.

The composition of the intestinal flora in man is complex, with several hundred different species recovered from the colon, but there are only a small number of predominant types of bacteria and these are mostly anaerobic. The picture is greatly influenced by diet; for instance, *Sarcina ventriculi*, an intestinal bacterium, is virtually confined to vegetarians, in whom it is present in large numbers. Because of their numbers, the intestinal bacteria have considerable metabolic potential (said to be equal to that of the liver) and products of metabolism can be absorbed. For instance, intestinal bacteria are important in the degradation of bile acids, and glycosides such as cascara or senna taken orally are converted by bacteria into active forms (aglycones) with pharmacological activity. Metabolic products occasionally cause trouble. Substances like ammonia are normally absorbed into the portal circulation and dealt with by the liver, but when this organ is badly damaged (severe hepatitis) they are able to enter the general circulation and contribute to hepatic coma. Adult Bantus, Australian aborigines, Chinese, etc. differ from Anglo-Saxons in that the small intestinal mucosa fails to produce the enzyme lactase. This is presumably related to the fact that these people do not normally drink milk as adults. If lactose is ingested, it is metabolised by the bacteria of the caecum and colon, with the production of fatty acids, carbon dioxide, hydrogen, etc., giving rise to flatulence and diarrhoea.

The resident bacteria are highly adapted to the commensal life, and under normal circumstances cause minimal damage. They are present throughout life, and avoid inducing the inflammatory or immune responses that might expel them. In the normal individual, the only other microorganisms that can establish themselves are by definition 'infectious'. These sometimes cause disease and are eventually eliminated. In other words, if it is inevitable that the body surfaces are colonised by microorganisms, it can be regarded as an advantage that colonisation should be by specialised nonpathogenic commensals. Human infants like other infants are born germ-free, and the microbial colonisation of skin, throat, intestine, etc. during and after birth forms a fascinating story.

The traditional way to obtain evidence about the function of something is to see what happens when it is removed. There have been many studies on germ-free animals, including mice, rats, cats, dogs and monkeys. The mother is anaesthetised shortly before delivery, and infants are delivered by caesarean section into a germ-free environment or 'isolator' and supplied with sterile air, food and water. Germ-free individuals, not unexpectedly, have a less well-developed immune system, because of the absence of microorganisms. Antigens are present in food, but the intestinal wall is thinner, and immunoglobulin synthesis occurs at about 1/50th of the rate seen in ordinary individuals. Germ-free animals that are coprophagous (rabbits, mice) also show a great enlargement of the caecum, which may constitute a quarter of the total body weight. It can cause death when it undergoes torsion. The caecum rapidly diminishes to normal size when bacteria are fed to the germ-free individual. Otherwise, the germ-free individual seems better off and generally has a longer life span. Even caries is not seen, because this requires bacteria (see pp. 39–42).

At one time it was a fashionable belief that the normal intestinal microorganisms produced 'toxins' that were harmful, and large segments of colon were removed from patients with diseases attributed to the action of these toxins. Toxins, especially endotoxin, are indeed absorbed from the intestine, but under normal circumstances this is not now thought to have harmful effects. On the other hand, there have been suggestions that carcinogenic substances, formed from the cholic acid in bile by intestinal bacteria, are important in cancer of the intestine, especially when the bowel contents move slowly (e.g. on low-fibre diets) and carcinogens have longer encounters with epithelial cells. Also, bacterial overgrowth in the stomach results in increased production of nitrites which can combine with amines to form carcinogenic nitrosamines.

It must be remembered that pathogenic as well as commensal microorganisms are absent from the germ-free animal, and in experimental animals it is possible to eliminate only the specific microbial pathogens, leaving the normal flora intact. This can be done by obtaining animals (mice, pigs, etc.) by caesarean section and rearing them without contact with others of the same species, but not in a germ-free environment. Alternatively germ-free animals can be selectively contaminated with commensal microorganisms. These specific pathogen-free (SPF) animals have increased body weight, longer life span and more successful reproductive performance, with more litters, larger litters and reduced infant mortality. Furthermore, it has long been known that chickens, pigs, etc. grow larger when they receive broad-spectrum antibiotics in their food, presumably because certain unidentified microorganisms are eliminated. But even if we were to conclude that the normal microbial flora, as opposed to the pathogens, on the whole does more harm than good, this conclusion, although of great interest, would have little practical significance. Colonisation by commensal microorganisms is the unavoidable fate of all normal individuals, and the germ-free life will remain an impossibly artificial condition; expensive, technically demanding and psychologically crippling for an intelligent animal.*

The elimination of specific pathogenic microorganisms, however, is a less theoretical matter. Specific pathogen-free mice are routinely maintained in laboratories and are much superior to non-SPF animals, as mentioned above. The population of the developed countries of the world (USA, Canada, northern Europe) can be likened to SPF mice, most of the serious microbial pathogens having been eliminated by vaccines, quarantine and other public health measures, or kept in check by good medical care and antibiotics. The peoples of the developing countries of the world, on the other hand, are comparable to the conventionally reared, non-SPF mice, who are exposed to all the usual murine pathogens. The comparison is complicated by the often inadequate diet of those in the developing world. A World Health Organisation (WHO) survey of 23 countries showed that in developing countries the common pathogenic infections such as diphtheria, whooping cough, measles and typhoid have respectively 100, 300, 55 and 160 times the case mortality seen in developed countries. Compared with those in the developed countries those in the developing countries often tend to be smaller, with a shorter life span, and poorer reproductive performance (abortions, neonatal and infantile mortality). They are the non-SPF people.

Opportunistic infection

There is one important consequence of the existence of the normal microbial flora. These microorganisms are present as harmless commensals, and are normally well behaved. If, in a given individual, this balance is upset by a decrease in the normal level of resistance, then the commensal bacteria are generally the first to take advantage of it. Thus damage to the respiratory tract upsets the balance and enables normally harmless resident bacteria to grow and cause sinusitis or pneumonia. Minor wounds in the skin enable skin staphylococci to establish small septic foci, and skin sepsis is particularly common in poorly controlled diabetes. This is probably due to defective chemotaxis and phagocytosis in polymorphs, which show impaired energy metabolism. High concentrations of blood sugar and the presence of ketone bodies may play a part, but a more direct effect of diabetes is suggested by the observation that adding insulin to diabetic

---

* A boy who developed aplastic anaemia when 9 years old was maintained in a 2.5 m × 2.7 m germ-free type isolator, shielded from contact with the microbial hazards of the outside world. Life was not easy, although he felt less abnormal when he was able to wear his protective astronaut-type suit at a science fiction convention. He was spared from infection and died at the age of 17 years, from complications of repeated blood transfusions.

polymorphs *in vitro* rapidly restores their bactericidal properties. Commensal faecal bacteria infect the urinary tract when introduced by catheters, and commensal streptococci entering the blood from the mouth can cause endocarditis if there are abnormalities in the heart valves or endocardium. The tendency of commensal bacteria to take opportunities when they arise and invade the host is universal. These infections are therefore called opportunistic infections.

Opportunistic infections are common nowadays. This is partly because many specific microbial pathogens have been eliminated, leaving the opportunistic infections relatively more numerous than they were. Also, modern medical care keeps alive many people who have impaired resistance to microbial infections. This includes those with congenital immunological or other deficiencies, those with lymphoreticular neoplasms, and a great many patients in intensive care units or in the terminal stages of various illnesses. Modern medical treatment also often requires that host immune defences are suppressed, as after organ transplants or in the treatment of neoplastic and other conditions with immunosuppressive drugs. Also, certain virus infections (e.g. cat leukaemia, AIDS in man) can cause a catastrophic depression of immune responses (see Ch. 7). In each case opportunistic microorganisms tend to give trouble.

There are other opportunistic pathogens in addition to the regular commensal bacteria. *Candida albicans*, a common commensal, readily causes troublesome oropharyngeal or genital ulceration. *Pseudomonas aeruginosa* is essentially a free-living species of bacteria, sometimes present in the intestinal tract. In hospitals it is now a major source of opportunistic infection. This is because it is resistant to many of the standard antibiotics and disinfectants, because its growth requirements are very simple, and because it is so widely present in the hospital environment. It multiplies in eyedrops, weak disinfectants, corks, in the small reservoirs of water round taps and sinks, and even in vases of flowers. *Pseudomonas aeruginosa* causes infection especially of burns, wounds, ulcers, and the urinary tract after instrumentation.* It is a common cause of respiratory illness in patients with cystic fibrosis.† When resistance is very low, it can spread systematically through the body, and nowadays this is a frequent harbinger of

---

* *Pseudomonas* demonstrated its versatility by causing a profuse rash in users of a hotel jacuzzi (whirlpool). The bacteria multiplied in the hot, recirculated, inadequately treated water, and probably entered the skin via the orifices of dilated hair follicles.

† Cystic fibrosis, the most common fatal hereditary disease in Caucasians (about 1 in 20 carry the gene), involves defects in mucus-producing cells. The lung with its viscid mucus becomes infected with *Staphylococcus aureus* and *Haemophilus influenzae*, but the presence of *Pseudomonas aeruginosa* is especially ominous. *Pseudomonas* strains from cystic fibrosis patients often produce a jelly-like alginate rather than the regular 'slimy' type of polysaccharide (see Table 4.1), and this may physically interfere with the action of phagocytes. Lung damage is largely due to the action of bacterial and phagocytic proteases.

immunological collapse. Viruses also act as opportunistic pathogens. Most people are persistently infected with cytomegalovirus, herpes simplex virus, varicella zoster virus, etc. (see Ch. 10), and these commonly cause disease in immunologically depressed individuals. Cytomegalovirus, for instance, is activated within the first 6 months after most renal transplant operations, as detected by a rise in antibody titre, and may cause hepatitis and pneumonia. The fungal parasite *Pneumocystis carinii* is an extremely common human resident, normally of almost zero pathogenicity, but can contribute to pneumonia in immunosuppressed individuals. *Clostridium difficile* is another example of an opportunistic pathogen which causes a spectrum of disease (ranging from antibiotic-associated diarrhoea to fatal pseudomembranous colitis) sometimes after a course of antibiotics. Resident spores do not normally germinate in the presence of a normal microflora; antibiotic-induced imbalance in the latter creates the conditions for rapid vegetative growth of *C. difficile* and release of toxins (see Ch. 8).

Exit of Microorganisms from the Body

After an account of the entry of microorganisms into the body, it seems appropriate to mention their exit. General principles were discussed in the first chapter. Nearly all microorganisms are shed from the body surfaces (Fig. 2.1). The transmissibility of a microorganism from one host to another depends to some extent on the degree of shedding, on its stability, and also on its infectiousness, or the dose required to initiate infection (see Table 11.1). For instance, when ten bacteria are enough to cause oral infection (*Shigella dysenteriae*), the disease will tend to spread from person to person more readily than when 10^6 bacteria are required (salmonellosis). The properties that give increased transmissibility are not the same as those causing pathogenicity. There are strains of influenza virus that are virulent for mice, but which are transmitted rather ineffectively to other mice, transmissibility behaving as a separate genetic attribute of the virus. For other microorganisms also, such as staphylococci and streptococci, transmissibility may vary independently of pathogenicity. Types of transmission are illustrated in Fig. 2.12.

Respiratory tract

In infections transmitted by the respiratory route, shedding depends on the production of airborne particles (aerosols) containing microorganisms. These are produced to some extent in the larynx, mouth and

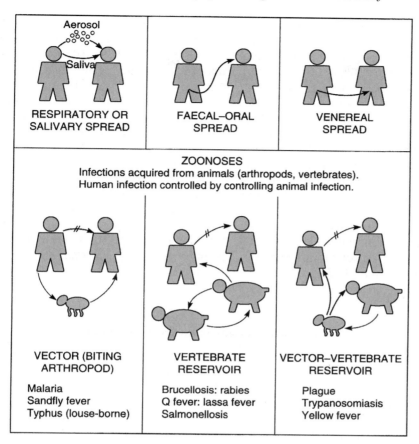

Fig. 2.12 Types of transmission of infectious agents. Respiratory or salivary spread – not readily controllable. Faecal–oral spread – controllable by public health measures. Venereal spread – control is difficult because it concerns social factors. Zoonoses – human infection controlled by controlling vectors,* or by controlling animal infection.†

* Unexpected results can come from enthusiastic vector control. A 1962 outbreak of Bolivian haemorrhagic fever in the town of San Joachin, Bolivia, appears to have been an indirect result of mosquito control. DDT present in the insect population entered the small lizards that ate them and then accumulated in the livers of the local cats that ate the lizards. The cats died with lethal DDT concentrations in liver, and this allowed bush mice that were asymptomatically infected with Bolivian haemorrhagic fever (Machupo) virus to invade human dwellings. People were infected via mouse urine and suffered 15% mortality. The disease outbreak was terminated by setting hundreds of mouse traps.

† Although man to man transmission does not usually occur in the zoonoses, direct contact with blood or secretions from infected individuals occasionally leads to infection of nurses, doctors, etc. (e.g. Lassa fever).

throat during speech and normal breathing. Harmless commensal bacteria are thus shed, and more pathogenic streptococci, meningococci and other microorganisms are also spread in this way, especially when people are crowded together inside buildings or vehicles. There is particularly good aerosol formation during singing and it is always dangerous to sing in a choir with patients suffering from pulmonary tuberculosis. Microorganisms in the mouth, throat, larynx and lungs are expelled to the exterior with much greater efficiency during coughing; shedding to the exterior is assured when there are increased mucus secretions and the cough reflex is induced. Tubercle bacilli in the lungs that are carried up to the back of the throat are mostly swallowed and can be detected in stomach washings, but a cough will project bacteria into the air.*

Efficient shedding from the nasal cavity depends on an increase in nasal secretions and on the induction of sneezing. In a sneeze (Fig. 2.13) up to 20 000 droplets are produced† and during a common cold, for instance, many of them will contain virus particles. The largest droplets (1 mm diameter) fall to the ground after travelling 4 m or so, and the smaller ones evaporate rapidly, depending on their velocity, water content and on the relative humidity. Many have disappeared within a few feet and the rest, including those containing microorganisms, then settle according to size. The smallest (1–4 mm diameter), although they fall theoretically at 0.3–1.0 m h^{-1}, in fact stay suspended indefinitely because air is never quite still. Particles of this size are likely to pass the turbinate baffles (see above) and reach the lower respiratory tract. If the microorganisms are hardy, as in the case of the tubercle bacillus, people coming into the room later on can be infected. Many other microorganisms are soon inactivated by drying of the suspended droplet or by light, and for transmission of measles, influenza, the common cold or the meningococcus, fairly close physical proximity is needed. Conversely, foot and mouth disease virus spreads by air and wind over surprisingly long distances.‡

---

* Mycobacterium leprae multiplies in nasal mucosa and 10^8 bacilli a day can be shed from the nose of patients with lepromatous leprosy. The bacteria are shed as plentifully as from patients with open pulmonary tuberculosis, and also survive in the dried state.

† Most of the droplets in fact originate from the mouth, but larger masses of material ('streamers') as well as droplets are expelled from the nose when there is excess nasal secretion. A cough, in contrast, produces no more than a few hundred particles. Talking is also a source of airborne particles, especially when the consonants, f, p, t and s are used. It is perhaps no accident that the most powerfully abusive words in the English language begin with these letters, so that a spray of droplets (possibly infectious) is delivered with the abuse.

‡ Pigs infected with foot and mouth disease virus excrete in their breath 100 million infectious units each day. With relative humidity of more than 65% the airborne virus survives quite well, and can be carried in the wind across the sea from France to the Channel Islands or England where cattle, who inhale 150 m^3 air a day, become infected. Outbreaks of this disease are often explained by studying air trajectories and other meteorological factors. In humans, legionellosis (see Glossary) can spread by air over shorter distances. An outbreak in Glasgow affected 33 people and had its source in a contaminated industrial cooling tower, cases occurring downwind up to a distance of 1700 m.

Fig. 2.13 Droplet dispersal following a violent sneeze. Most of the 20 000 particles seen here are coming from the mouth. The authors used oblique illumination, to give a dark-field effect, and high speed (1/30 000 s flash) photography. Particles as small as 5–10 mm could be seen, images are larger than actual particle size, and objects out of focus are magnified. (Reproduced from Jennison, M. W. (1947). 'Aerobiology', p. 102, AAAS No. 17, Washington, DC.)

Shedding from the nasal cavity is much more effective when fluid is produced and, among the viruses that are shed from this site, evolution has favoured those that induce a good nasal discharge.* In the crowded conditions of modern life, with unprecedented numbers of susceptible individuals in close physical proximity and with only temporary nasal immunity (see Ch. 6), there is rapid selection for the virus strains that spread most effectively. There are more than 100 antigenically different common cold viruses, and there are signs that these infectious agents are entering their golden age, with little hope of control by vaccination or chemoprophylaxis.

---

* Nasal secretions are inevitably deposited (directly or via handkerchiefs) onto hands, which can then be a source of infection. Contamination of other people's fingers, and thus of their nose and conjunctiva, might be as important as aerosols in the transmission of these infections.

Saliva

Microorganisms reach the saliva during upper or lower respiratory tract infections, and may be shed during talking and other mouth movements as discussed above. Certain viruses, such as mumps, Epstein–Barr virus (EBV), herpes simplex and cytomegalovirus in man infect the salivary glands. Virus is present in the saliva, and shedding to the exterior takes place in infants and young children by the contamination of fingers and other objects with saliva. Adolescents and adults who have escaped infection earlier in life exchange a good deal of saliva in the process of kissing, particularly 'deep' kissing. In developing countries, EBV infects mainly infants and children, and at this age causes little or no illness. In developed countries, however, infection is often avoided during childhood, and primary infection with EBV occurs at a time of life when sexual activity is beginning. At this age it gives rise to the more serious clinical conditions included under the heading of glandular fever. In animals also, saliva is often an important vehicle of transmission, depending on social and sexual activities such as licking, nibbling, grooming, fighting. Rabies, foot and mouth disease virus, and the various types of cytomegalovirus and other herpes viruses may be present in large amounts in saliva.*

Spitting is an activity practised only by man and a few animals including camels, chameleons and certain snakes. Chimpanzees soon learn to do it. Microorganisms resistant to drying, such as the tubercle bacillus, can be transmitted in this way. The expectorated material contains saliva together with secretions from the lower respiratory tract. In the days when pulmonary tuberculosis was commoner, spitting in public places came to be frowned upon and there were laws against it. It is perhaps better for the chronic bronchitic to discharge his voluminous secretions discreetly into a receptacle rather than swallow them, but the expectoration of mere saliva in public places, now becoming commoner again, is a regrettable reversion to the unaesthetic days of the spittoon.

Skin

Shedding of commensal skin bacteria takes place very effectively. Skin bacteria are mostly shed attached to desquamated skin scales, and an average of about 5×10^8 scales, 10^7 of them carrying bacteria, are shed per person per day, the rate depending very much on phys-

---

* Rabies virus, for instance in the wolf, enhances its own transmission by invading the limbic system of the brain. This alters the behaviour of the infected animal, making it more aggressive, more likely to roam, and thus more likely to bite another individual.

ical activity. The fine white dust that collects on surfaces in hospital wards consists to a large extent of skin scales. The potentially pathogenic *Staphylococcus aureus* colonises especially the nose (nose-picking area), fingers and perineum. Shedding takes place from the nose and notably from the perineal area. Males tend to be more effective perineal shedders than females, and this is partly hormonal and partly because of friction in this area; shedding can be prevented by wearing occlusive underpants. A good staphylococcal shedder can raise the staphylococcal count in the air from less than 36 m^{-3} to 360 m^{-3}. Although people with eczema or psoriasis shed more bacteria from the skin, it is not known why some normal individuals are profuse shedders; the phenomenon is important for cross infection in hospitals.

For microorganisms that cause skin lesions (see Table 5.2), however, shedding to the environment is not necessarily very important. Shedding takes place only if the skin lesion breaks down, as when a vesicle ruptures or if the microorganism penetrates through to the outer layers of the epidermis (wart virus). Even then, spread of infection is often by direct bodily contact, as with herpes simplex, syphilis or yaws, rather than by shedding into the environment.

Intestinal tract

All microorganisms that infect the intestinal tract are shed in faeces. Those shed into the bile, such as hepatitis A (a gut picornavirus) and typhoid bacilli in the typhoid carrier, also appear in the faeces. Microorganisms swallowed after growth in the mouth, throat or respiratory tract can also appear in the faeces, but most of them are not resistant to acid, bile and other intestinal substances and are inactivated. Faeces are the body's largest solid contribution to the environment,* and although the microorganisms in faeces are nearly all harmless commensals, it is an important source of more harmful microorganisms. During an intestinal infection, intestinal contents are often hurried along and the faeces become fluid. There is no exact equivalent to the sneeze, but diarrhoea certainly leads to increased

---

* Herbivorous animals make a bigger and less well-controlled contribution than do human beings. The output of a pig is about three times and a cow ten times that of a man. We are less fussy about the disposal of animal sewage and this can be important for instance in transfer of salmonellosis. The amount from an individual animal seems less important than its quality and site of deposition when we consider the appalling canine contribution to public parks and paths in dog-ridden cities.

Gas is another intestinal product, and a few hundred millilitres depart from the anus and mouth of the normal person each day. About half is nitrogen from swallowed air, the rest being mostly methane (CH_4). Microbial fermentation in the gut forms H_2 and CO_2, which methanogenic bacteria convert to CH_4. This is particularly prominent after ingestion of beans, which have a polysaccharide not handled by digestive enzymes of humans.

faecal contamination of the environment and spread to other individuals. In animal communities and in primitive human communities, there is a large-scale recycling of faecal material back into the mouth. Contamination of food, water and living areas ensure that this is so, and the efficiency of this faecal–oral movement is attested to by the great variety of microbes and parasites that spread from one individual to another by this route. If microorganisms shed into the faeces are resistant to drying and other environmental conditions, they remain infectious for long periods. Protozoa such as *Entamoeba histolytica* produce an especially resistant cyst which is the effective vehicle of transmission, and *Clostridia* spp. and *B. anthracis* form resistant spores that contaminate the environment and remain infectious for many years. The soils of Europe are heavily seeded with tetanus spores from the faeces of domestic animals, and these spores can infect the battlefield wound or the gardening abrasion to give tetanus. Viruses have no special resistant form for the hazardous journey to the next host, but they show variable resistance to thermal inactivation and drying. Poliovirus, for instance, is soon inactivated on drying.

Many microorganisms are effectively transmitted from faeces to mouth after contamination of water used for drinking. The great water-borne epidemics of cholera are classical examples,* and any faecal pathogen can be so transmitted if it survives for at least a few days in water. In densely inhabited regions faecal contamination of water is inevitable unless there is adequate sewage disposal and a supply of purified water. Two hundred years ago in England there were no water closets and no sewage disposal, human excrement was deposited in the streets. There was nowhere else to put it, although one enterprising Londoner in 1359 was fined twelve pence for running his sewage by a pipe into a neighbour's cellar. Water supplies came from rivers and from wells, of which there were more than 1000 in London. Efficient sewage disposal and piped water supplies are a comparatively recent (nineteenth-century) development. Nowadays the map of the London sewage system resembles that of the London Underground (subway) system. Water for domestic use is collected into vast reservoirs before being shared out to tens of thousands of individuals. This would give great opportunities for spread once pathogens entered the water supply, but water purification and chlorination ensures that this spread remains at almost zero level. Life in present-day urban society depends on the large-scale supply of pure water and the large-scale disposal of sewage. Both are complex and vital public services of which

---

* Dr John Snow, a London physician, charted the cases of cholera on a street map during an outbreak in 1854. After observing that all cases had used water from the same pump in Broad Street, Soho, he removed the handle of this pump. The outbreak terminated dramatically, and the mode of transmission was thus demonstrated nearly 40 years before Koch identified the causative organism.

the average citizen or physician is profoundly ignorant. Largely as a result of these developments the steady flow of faecal materials into the mouth that has characterised much of human history has been interrupted.

Urinogenital tract

Urine can contaminate food, drink and living space, and the same things can be said as have been said about faeces. Urine in the bladder is normally sterile, and is only contaminated with skin bacteria as it is discharged to the exterior. The pathogens present in urine include a specialised group that are able to spread through the body and infect the kidney or bladder. The leptospiral infections of rats and other animals are spread in this way, sometimes to man. *Leptospira** survive in water, can penetrate the skin, and people are infected following contact with contaminated canals, rivers, sewage, farmyard puddles and other damp objects. Polyomavirus spreads naturally in colonies of mice after infecting tubular epithelial cells in the kidney and being discharged to the exterior in urine. Mice carrying lymphocytic choriomeningitis virus shed the virus in urine and can thus infect people in mouse-infested dwellings. Humans infected with their own polyomavirus, or with cytomegalovirus, excrete the virus in urine. Urinary carriers of typhoid have a persistent infection in the bladder, especially when the bladder is scarred by *Schistosoma* parasites, and typhoid bacilli are shed in the urine.

Microorganisms shed from the urethra and genital tract generally depend for transmission on mucosal contacts with susceptible individuals. Herpes simplex type 2 can infect the infant as it passes along an infected birth canal during delivery, and gonococci or *Chlamydia* infect the infant's eye in the same way. Venery, however, gives far greater opportunities for spread, as was discussed in Ch. 1. If there is a discharge, organisms are carried over the epithelial surface and transmission is more likely to take place.

The transmission of microorganisms by mucosal contact is determined by social and sexual activity (see also p. 3). In animals, licking, nuzzling, grooming and biting can be responsible for the transmission of microorganisms such as rabies and herpes viruses. In recent years there have been major changes in man's social and sexual customs, and this has had an interesting influence on certain infectious diseases. Generally speaking there has been less mucosal contact in the course

---

* There are more than 20 different serotypes carried by mice, rats, swine, dogs, cattle and leptospirosis is the most widespread zoonosis in the world. In the UK nowadays, cases of rat-borne leptospirosis occur in the bathers, canoeists, etc. who use canals and rivers, rather than sewer workers or miners, and leptospirosis from cattle continues to cause a mild disease in farmers and cowmen.

of regular social life. In modern societies saliva is exchanged less freely between children (as noted on p. 56) or within a family, and children are more likely to escape infections that are spread via saliva such as those due to EBV. Some of the so-called 'genital' warts (e.g. HPV 16) seem to be transmitted also between schoolchildren via saliva, possibly indicating a more ancient method of spread.

Things are different when we consider sexual activity. For adolescents and adults, mucosal contacts are possibly increasing in frequency, but more importantly they are being made with a greater number of different partners. Sexual activity is now considered less sinful, and the fact that it is safer (pregnancy avoidable and disease treatable) means that multiple partners are commoner than they used to be. Furthermore, infectious agents are transmitted with much greater efficiency now that many couples use oral rather than mechanical contraceptives. All these things have led to a great flowering of sexually transmitted diseases, which with respiratory infections are now the commonest communicable diseases in the world. Their incidence is rising. The four most frequent sexually transmitted diseases in England today are nonspecific urethritis (largely due to *Chlamydia*), gonorrhoea, candidiasis, and genital warts.* AIDS has had an impact on sexual promiscuity. HIV originated in central (sub-Saharan) Africa where it is spread by (vaginal) heterosexual intercourse. In developed countries it is still mostly a disease of male homosexuals, drug addicts, and haemophiliacs. In these countries promiscuity has already been curtailed, as indicated by falling gonorrhoea infection rates. AIDS is estimated to affect over 30 million people by the year 2000, and most of it is in Africa, Asia and Latin America, where it continues to spread heterosexually. It is hoped that the threat of an infection, with no vaccine and no treatment, in which virtually all those infected develop AIDS and die, will act as a restraining influence on heterosexual promiscuity and encourage the use of barrier contraceptives.†

A list of sexually transmitted diseases is given in Table 2.3. Even the more serious diseases such as syphilis and gonorrhoea have been difficult to control. A small number of sexually active individuals, if they evade the public health network, can be relied upon to infect many others.

*This is not to say that promiscuity is a new thing. The well-charted sexual adventures of Casanova (1725–1798) brought him four attacks of gonorrhoea, five of chancroid, and one of syphilis, while Boswell (1740–1795) experienced 19 episodes of (mainly gonococcal) urethritis. Of course, these activities were not restricted to those who became famous or wrote books. But the extraordinary increase in man's mobility has transformed social life and, together with the factors mentioned above, has had a major impact on the sexual transmission of infectious diseases.

† Condoms have been shown to reliably retain herpes simplex virus, HIV, *Chlamydia* and gonococci in simulated coital tests of the syringe and plunger type.

Table 2.3. Principal sexually transmitted diseases in man [a]

	Microorganism	Disease	Comments
Viruses	Herpes simplex type 2	Genital herpes	Very common – reactivates
	Human papillomavirus	Genital warts	Very common – involvement in cervical and penile cancer makes them more than ornamental appendages
	HIV-1[b]	AIDS	Most cases are in the Third World and are spread by heterosexual (vaginal) intercourse. In the First World most common in male homosexuals and transmitted by anal intercourse
	Hepatitis B	Hepatitis	Spread mainly in male homosexuals
Chlamydia	*C. trachomatis* (types D–K)	Nonspecific urethritis	Responsible for more than half of cases; causes eye infection in newborn
	C. trachomatis (types L1–L3)	Lymphogranuloma inguinale	Ulcerating papule plus lymph node suppuration. Commoner in tropics and subtropics
Mycoplasmas	*Ureoplasma* spp.	Nonspecific urethritis	Importance not clear. Require 10% urea for growth, which would direct them to urogenital tract
Bacteria	*Neisseria gonorrhoeae*	Gonorrhoea	Acute and more severe urethritis in male; chronic pelvic infection in female; eye infection in newborn
	Treponema pallidum	Syphilis	Syphilis was name of infected shepherd in Frascator's poem (1530) describing disease
	Haemophilus ducreyi	Chancroid	Genital sore, lymph-node suppuration, commoner in subtropics
	Calymmato-bacterium granulomatis	Granuloma inguinale	Commoner in subtropics. Ulcerative lesions
Fungi	*Candida albicans*	Vulvovaginitis (balanoposthitis in male)	Asymptomatic vaginal carriage common
Protozoa	*Trichomonas vaginalis*	Vulvovaginitis (urethritis in male)	Disease worse in female (compare gonorrhoea)

[a] Also common are pediculosis pubis (caused by the crab louse *Phthirus pubis*) and genital scabies (caused by the scabies mite *Sarcoptes scabiei*). More than half of all infections occur in people under the age of 24 years. In addition, there are special 'at risk' groups, such as tourists, long-distance lorry drivers, seamen, homosexuals.
[b] Human immunodeficiency virus.

Because almost all mucosal surfaces in the body can be involved in sexual activity, microorganisms encounter a number of interesting opportunities to infect new bodily sites. Thus, *Neisseria meningitidis*, a resident of the nasopharynx, is occasionally recovered from the cervix, the male urethra and the anal canal. *Neisseria gonorrhoeae* infects the throat and the anal region. *Chlamydia* can at times be recovered from the rectum and pharynx as well as the urethra. Genito-oro-anal contacts in sexually promiscuous communities give chances for intestinal microorganisms to spread between individuals in spite of good sanitation and sewage disposal.* For example, there have been examples of sexual transmission of *Salmonella, Giardia lamblia*, hepatitis A, pathogenic amoebae and *Shigella*, constituting what has been referred to as the 'gay bowel syndrome'.†

Blood

Most of the microorganisms that are transmitted by blood-sucking arthropods such as mosquitoes, fleas, ticks, sandflies or mites, have to be present in blood. This is true for arthropod-borne viruses, rickettsiae, malaria, trypanosomes and many other infectious agents. In these diseases transmission is biological (see p. 20). The microorganism is ingested with the blood meal, multiplies in the arthropod and then is discharged from the salivary gland or intestinal tract of the arthropod to infect a fresh host. To infect the arthropod vector, the blood of the vertebrate host must contain adequate amounts of the infectious agent. Microorganisms can be said to have been shed into the blood. A few microbes that are shed into blood (hepatitis B, hepatitis C) are transmitted not by biting arthropods but by modern devices such as needles, syringes, and blood transfusions. Presumably other routes such as saliva and mucosal contact are also significant. Could these viruses have arisen from ancestors that were spread by biting arthropods?

Miscellaneous

Microorganisms rarely occur in semen, which is not designed by nature for shedding to the environment. Perhaps it is because of the superb opportunities for direct mucosal spread during venery that

---

* In Western societies intestinal pathogens can also spread by more innocent pathways, as when amoebiasis was transmitted to 15 patients who received colonic irrigation in a clinic in Colorado.

† It should be noted that these conditions, like other sexually transmitted infections, are confined to homosexuals of the male variety. Female homosexuals, in contrast, enjoy more discreet, less promiscuous, relationships which are infection-free because that necessary instrument for transmission, the penis, is absent.

only an occasional microorganism, such as cytomegalovirus in man, has made use of semen as a vehicle for transmission. Milk, in contrast, is a fairly common vehicle for transmission. Mumps virus and cytomegalovirus are shed in human milk, although perhaps not very often transmitted in this way, but the mammary tumour viruses of mice are certainly partly transmitted via milk. Cows' milk containing *Brucella abortus*, tubercle bacilli or Q fever rickettsia is a source of human infection.

No shedding

In a very few instances transmission takes place without any specific shedding of microorganisms to the exterior. Anthrax, for instance, infects and kills susceptible animals, and the corpse as a whole then contaminates the environment. Spores are formed aerobically, where blood leaks from body orifices, and they remain infectious in the soil for very long periods. It seems that spores are only formed during the terminal stages of the illness or after death, so that death of the host can be said to be necessary for the transmission of this unusual microorganism. Again, kuru (see p. 35) is only transmitted after death when the infectious agent in the brain is introduced into the body via mouth, intestine or fingers during cannibalistic consumption of the carcass.

Finally, certain microorganisms such as leukaemia and mammary tumour viruses spread from parent to offspring directly by infecting the egg or the developing embryo. If sections from mice congenitally infected with lymphocytic choriomeningitis (LCM) virus are examined after fluorescent antibody staining, infected ova can be seen in the ovary (Fig. 5.1). Also ovum transplant experiments show that similar infection occurs with murine leukaemia virus, and the embryos of most strains of mice have leukaemia virus antigens present in their cells. All progeny from the originally infected individuals are infected and there is no need for shedding to the exterior. Some other mode of spread would be necessary if there were to be infection of a fresh lineage of susceptible hosts.

References

Amin, I. I., Douce, G. R., Osborne, M.P. and Stephen, J. (1994). Quantitative studies of invasion of rabbit ileal mucosa by *Salmonella typhimurium* strains which differ in virulence in a model of gastroenteritis. *Infect. Immun.* **62**, 569–578.

Blaser, M. J. (1993). *Helicobacter pylori*: microbiology of a 'slow' bacterial infection. *Trends Microbiol.* **1**, 255–260.

Bolton, A. J., Martin, G. D., Osborne, M. P., Wallis, T. S. and Stephen, J. (1999). Invasiveness of *Salmonella* serotypes Typhimurium, Choleraesuis and Dublin for rabbit terminal ileum in vitro. *J. Med. Microbiol.* **48**, 800–810.

Bolton, A. J., Osborne, M. P. and Stephen, J. (2000). Comparative study of invasiveness of Salmonella serotypes Typhimurium, Choleraesuis and Dublin for Caco-2 cells, HEp-2 cells and rabbit ileal epithelia. *J. Med. Microbiol.* **49**, 503–511.

Bolton, A. J., Osborne, M. P., Wallis, T. S. and Stephen, J. (1999). Interaction of *Salmonella choleraesuis, Salmonella dublin* and *Salmonella typhimurium* with porcine and bovine terminal ileum *in vivo. Microbiology* **145**, 2431–2441.

Buckley, R. M. *et al.* (1978). Urine bacterial counts after sexual intercourse. *N. Engl. J. Med.* **298**, 321–323.

Curtiss, R. III, MacLeod, D. L., Lockman, H. A., Galan, J. E., Kelly S. M. and Mahairas, G. G. (1993). Colonization and invasion of the intestinal tract by *Salmonella. In* 'The Biology of *Salmonella*' (F. C. Cabello, C. E. Hormaeche, P. Mastroeni and L. Bonina, eds), pp. 191–198. Plenum Press, New York.

Cutler, C. W., Kalmer, J. R. and Genco, C. A. (1995). Pathogenic strategies of the oral anaerobe, *Porphyromonas gingivalis. Trends Microbiol.* **3**, 45–51.

Donaldson, A. I. (1983). Quantitative data on airborne foot and mouth disease virus: its production, carriage and deposition. *Phil. Trans. R. Soc. London, B* **302**, 529–534.

Duguid, J. P. (1946). The size and duration of air carriage of respiratory droplets and droplet nuclei. *J. Hyg., Camb.* **44**, 471.

Frankel, G., Phillips, A. D., Rosenshine, I., Dougan, G., Kaper, J. B. and Knutton, S. (1998). Enteropathogenic and enterohaemorrhagic *Escherichia coli*: more subversive elements. *Mol. Microbiol.* **30**, 911–921.

Gaastra, W. and Svennerholm, A. M. (1996). Colonization factors of human enterotoxigenic Escherichia coli (ETEC). *Trends Microbiol.* **4**, 444–452.

Gordon, H. A. and Pesti, L. (1972). The gnotobiotic animal as a tool in the study of host–microbial relationships. *Bact. Rev.* **35**, 390–429.

Hartland, E. L., Batchelor, M., Delahay, R. M., Hale, C., Matthews, S., Dougan, G., Knutton, S., Connerton, I. and Frankel, G. (1999). Binding of intimin from enteropathogenic *Escherichia coli* to Tir and to host cells. *Mol. Microbiol.* **32**, 151–158.

Haywood, A. M. (1994). Virus receptors: adhesion strengthening, and changes in viral structure. *J. Virol.* **68**, 1–5.

Hinds, C. J. (1985). Medical hazards from dogs. *Brit. Med. J.* **291**, 760.

Hoepelman, A. I. M. and Tuomanen, E. I. (1992). Minireview. Consequences of microbial attachment: directing host cell functions with adhesins. *Infect. Immun.* **60**, 1729–1733.

Kaper, J. B. and Hacker, J., eds (1999). 'Pathogenicity Islands and Other Mobile Virulence Elements'. ASM Press, Washington, DC. An excellent book, which includes articles on *E. coli, Yersinia, Salmonella, Shigella, V. cholerae, H. pylori, Dichelobacter nodosus*, Listeriae, Bacilli, Clostridia, Staphylococci and Streptococci.

Kerr, J. R. (1999). Cell adhesion molecules in the pathogenesis of and host defence against microbial infection. *J. Clin. Pathol. Mol. Pathol.* **52**, 220–230.

Ketley, J. M. (1997). Pathogenesis of enteric infection by *Campylobacter. Microbiology* **143**, 5–21.

Lentz, T. L. (1990). The recognition event between virus and host cell receptor: a target for antiviral agents. *J. Gen. Virol.* **71**, 751–766.

Lodge, J. M., Bolton, A. J., Martin, G. D., Osborne, M. P., Ketley, J. M. and Stephen, J. (1999). A histotoxin produced by Salmonella. *J. Med. Microbiol.* **48**, 811–818.

Mackowiak, P. A. (1982). The normal microbial flora. *N. Engl. J. Med.* **307**, 83.

Mims, C. A. (1981). Vertical transmission of viruses. *Microbiol. Rev.* **45**, 267–286.

Mims, C. A. (1995). The transmission of infection. *Rev. Med. Microbiol.* **6**, 2217–2227.

Nataro, J. P. and Kaper, J. B. (1998). Diarrheagenic *Escherichia coli. Clin. Microbiol. Rev.* **11**, 142–201.

Newhouse, M. *et al.* (1976). Lung defense mechanisms. *N. Engl. J. Med.* **295**, 990, 1045.

Noble, W. C. (1981). 'Microbiology of the Human Skin', 2nd edn. Lloyd-Luke, London.

Nomoto, A., ed. (1992). Viral receptors and cell entry. *Semin. Virol.* **3**, no. 2.

Reynolds, D. J. and Pearce, J. H. (1991). Endocytic mechanisms utilized by Chlamydiae and their influence on induction of productive infection. *Infect. Immun.* **59**, 3033–3039.

Sansonetti, P. J. and Phalipon, A. (1999). M cells as ports of entry for enteroinvasive pathogens: mechanisms of interaction, consequences for the disease process. *Semin. Immunol.* **11**, 193–203.

Scully, C. (1981). Dental caries: progress in microbiology and immunology. *J. Infection* **3**, 101–133.

Shuster, S. (1984). The aetiology of dandruff and mode of action of therapeutic agents. *Brit. J. Dermatol.* **111**, 235–242.

Svanborg, C. (1993). Resistance to urinary tract infection. *N. Engl. J. Med.* **329**, 802.

Vasselon, T., Mounier, J., Prevost, M. C., Hellio, R. and Sansonetti, P. J. (1991). Stress fiber-based movement of *Shigella flexneri* within cells. *Infect. Immun.* **59**, 1723–1732.

Virji, M. (1997). Mechanisms of microbial adhesion; the paradigm of *Neisseriae. In* 'Molecular Aspects of Host–Pathogen Interactions' (M. A. McCrae, J. R. Saunders, C. J. Smyth and N. D. Stow, eds). Society for General Microbiology, Symposium 55, pp. 95–110. Cambridge University Press, Cambridge.

Wallis, T. S. and Galyov, E. E. (2000). Molecular basis of *Salmonella*-induced enteritis. *Molec. Microbiol.* **36**, 997–1005.

Wallis, T. S., Starkey, W. G., Stephen, J., Haddon, S. J., Osborne, M. P. and Candy, D. C. A. (1986). The nature and role of mucosal damage in relation to *Salmonella typhimurium*-induced fluid secretion in the rabbit ileum. *J. Med. Microbiol.* **22**, 39–49.

Wilcox, R. R. (1981). The rectum as viewed by the venereologist. *Br. J. Ven. Dis.* **57**, 1–6.

Wooldridge, K. G. and Ketley, J. M. (1997). *Campylobacter* host cell interactions. *Trends Microbiol.* **5**, 96–102.

Zhang, J. P. and Stephens, R. S. (1992). Mechanisms of *C. trachomatis* attachment to eukaryotic cells. *Cell* **69**, 861–869.

3

Events Occurring Immediately After the Entry of the Microorganism

Growth in Epithelial Cells

Some of the most successful microorganisms multiply in the epithelial surface at the site of entry into the body, produce a spreading infection in the epithelium, and are shed directly to the exterior (Table 3.1). This is the simplest, most straightforward type of microbial parasitism. If the infection progresses rapidly and microbial progeny are shed to the exterior within a few days, the whole process may have been completed before the immune response has had a chance to influence the course of events. It takes at least a few days for antibodies or immune cells to be formed in appreciable amounts and delivered to the site of infection. However, we may underestimate the time of appearance of antibodies, as the first antibodies formed are immediately complexed with the microorganism and no free antibody appears until antibody is present in excess. With a variety of respiratory virus infections, especially those caused by rhinoviruses, coronaviruses, parainfluenza viruses and influenza viruses, epithelial cells are destroyed, and inflammatory responses induced, but there is little or no virus invasion of underlying tissues. The infection is terminated partly by nonimmunological resistance factors, and partly because most locally available cells have been infected. Interferons are important resistance factors. They are low molecular weight proteins, coded for by the cell, and formed in response to infection with nearly all viruses (see Ch. 9). The interferon formed by the infected cell is released and can act on neighbouring or distant

Table 3.1. Microbial infections that are generally confined to epithelial surfaces of the body

Microbe	Respiratory tract and conjunctiva	Urinogenital tract	Skin	Intestinal tract
Viruses	Influenza Parainfluenza 1–4 Rhinoviruses Coronaviruses	Certain papilloma viruses	Papilloma viruses (warts) Molluscum contagiosum	Rotaviruses of man, mouse, etc.
Chlamydias	Trachoma Inclusion conjunctivitis	Nonspecific urethritis	—	—
Mycoplasma	*Mycoplasma pneumoniae* (atypical pneumonia)	T strains (nonspecific urethritis)	—	—
Bacteria	*Bordetella pertussis Corynebacterium diphtheriae Streptococci*	Gonococcus	Staphylococci *Corynebacterium minutissimum*[a]	Most Salmonellae; Shigellae *Campylobacter* sp.
Rickettsias	—	—	—	—
Fungi	*Candida albicans* (thrush)	*Candida albicans*	*Trichophyton* spp. (athlete's foot, ringworm, etc.)	—
Protozoa	—	*Trichomonas vaginalis*	—	*Entamoeba coli Giardia lamblia*

[a] This bacterium commonly infects the stratum corneum and causes erythrasma, a scaly condition of the axilla, groin and between toes.

cells, protecting them from infection. Freshly formed virus particles from the first infected epithelial cell enter the fluids bathing the epithelial surface and are borne away to initiate fresh foci of infection at more distant sites. Interferons too can reach these sites, and as more and more interferon is formed on the epithelial sheet, more and more cells are protected, so that the infectious process is slowed and finally halted. Other unknown antiviral factors probably play a part, and the immune response itself comes into action in the final stages. Interferons are produced a few hours after infection of the first epithelial cell at a site where they are needed and without the delay characteristic of the immune response. The immune response provides resistance to subsequent re-infection, but it does not appear to be of primary importance in recovery from respiratory infections of this type.

The spread of infection is very rapid on epithelial surfaces that are covered with a layer of liquid because of the ease with which the microorganism in the fluid film encounters cells and is disseminated over the surface. This is true for the respiratory infections mentioned above, and also for infections of intestinal epithelium, such as those caused by the human diarrhoea viruses. The argument does not apply, however, to local infections of the skin. In this case, where the microorganism is not carried across the epithelial surface in a liquid film to establish fresh foci of infection, the whole process takes a much longer time. Papilloma (wart) viruses, for instance, cause infection in a discrete focus of epidermal cells; indeed a wart consists of a clone of cells produced by the division of a single initially infected cell. The inevitably slow evolution of single virus-rich lesions means that immune responses have the opportunity to play a more important part in the infection. Only small amounts of interferon are produced and these viruses are in any case often relatively resistant to its action. But wart virus escapes the attention of the immune system as follows. In the basal layer of the epidermis, adjacent to the antibodies and immune cells that arrive from dermal blood vessels, the virus infection is incomplete; in this layer of the epidermis, virus antigens are not formed on the cell surface and no virus particles are produced. The infected basal cell is therefore not recognised and not a target for the immune response. As the cells move further away from these immune forces, approaching the epidermal surface and becoming keratinised, more and more virus is produced for liberation to the exterior. Neither antibodies nor immune cells are present on this dry surface to influence virus multiplication and shedding.

The respiratory viruses described above have a hit and run type of infection of epithelial cells, and are very successful parasites. We may ask if there is anything that prevents them invading subepithelial tissues and spreading systemically in the host. A number of other viruses, including measles and chickenpox, infect inconspicuously via the respiratory tract, then spread systemically through the body and only emerge again to cause widespread respiratory infection and shedding to the exterior after a prolonged incubation period. The limitation of rhinoviruses and human coronaviruses to the surface of the upper respiratory tract is at least partly determined by their optimum growth temperature. Many of them replicate successfully at 33°C, the temperature of nasal epithelium, but not very well at the general body temperature, 37°C. Thus, they do not spread systemically nor to the lung. Viruses of the influenza and parainfluenza groups can infect the lung as well as the nasal mucosa, but they are generally limited to the epithelial surfaces. The limitation is not absolute. Occasionally in adults and more often in infants, influenza and parainfluenza viruses spread to infect the heart, striated muscle or the central nervous system.

The spread of infection from epithelial surfaces is also controlled by the site of virus maturation from cells. Influenza and parainfluenza viruses are liberated (by budding) only from the free (external) surface of epithelial cells, as is appropriate for infection limited to surface epithelium. A similar restriction in the topography of budding is seen with rabies virus in the infected salivary gland (p. 133). However, vesicular stomatitis virus (p. 423) is released only from the basal surface of the epithelial cell, from whence it can spread to subepithelial tissues and then through the body; topographical restriction in the site of virus release from epithelial cells reflects the polarisation of function in these cells, which in turn depends on the maintenance of tight junctions between them.

Many bacterial infections are more or less confined to epithelial surfaces (Table 3.1). This is a feature, for instance, in diphtheria and streptococcal infections of the throat, gonococcal infections of the conjunctiva or urethra and most *Salmonella* infections of the intestine. To a large extent this is because host antibacterial forces, to be described at a later stage, do not permit further invasion of tissues. Under most circumstances these bacteria are not able to overcome the host defences, but gonococci and streptococci, at least, often spread locally through tissues and occasionally systemically through the body. Gonococci cause a patchy infection of the columnar epithelium of the male urethra, reaching subepithelial tissues 3–4 days after infection; the yellow discharge consists of desquamated epithelial cells, inflammatory exudate, leucocytes and gonococci. Some recent studies on how gonococci adapt to this *in vivo* situation are described in Ch. 7. Subepithelial spread probably takes the infection to other parts of the urethra and to local glands.

Most Gram-negative bacteria have only a very limited ability to invade a given host. In man, *E. coli, Proteus* spp. and *Pseudomonas aeruginosa* are only capable of invasion when defences are impaired or when bacteria are inadvertently introduced into a suitable site in the body (see Ch. 2). They cause systemic infection in debilitated, malnourished, or immunosuppressed patients; they produce sepsis in the uterus after abortion, and when they are introduced into the body by intravascular devices or catheters. Certain Gram-negative bacteria penetrate the intestinal epithelium but get no further, as in *Shigella* dysentery and salmonellosis (see p. 299). One or two highly specialised Gram-negative bacteria penetrate intestinal epithelium, enter lymphatics and spread systemically through the body to cause enteric or typhoid fever (*Salmonella typhi* and *paratyphi*).

A few bacteria show a temperature restriction similar to that described above for rhinoviruses, which prevents anything more than local spread. For instance, the lesions in leprosy (*Mycobacterium leprae*) are confined to cooler parts of the body (skin, superficial nerves, nasal mucosa, testicles, etc.). Other mycobacteria (*Mycobacterium ulcerans* and *M. marinum*) occur in water and enter human skin through superficial abrasions, especially in warm countries, and cause chronic skin ulcers. These bacteria, which also infect fish, have an optimum growth temperature of 30–33°C and remain restricted to the skin.

Fungi of the dermatophyte group (ringworm, athlete's foot*) infect skin, nails and hair, but are restricted to the dead keratinised layers of epithelium. Fungal antigens are absorbed from the site of infection and immune (including allergic) responses are generated, which at least partly account for the failure to invade deeper tissues.

Intracellular Microorganisms and Spread Through the Body

Some of the important microorganisms that regularly establish systemic infections after traversing epithelial surfaces are listed in Table 3.2.

There is one important distinction between intracellular and extra-cellular microorganisms. If an obligate intracellular microbe is to spread systemically from the body surface, it must first enter the blood or lymph. This means gaining access to the lumen of a subepithelial lymphatic or blood vessel, either as a free microorganism, or alternatively after entering a mobile cell (leucocyte) that will carry it to other parts of the body. The microorganism cannot replicate until it reaches a susceptible cell, and the absence or shortage of such cells except at the body surface would prevent or seriously hinder its spread through the body. Thus rotaviruses and rhinoviruses replicate at the epithelial surface but cannot infect leucocytes, and in any case would be unlikely

Table 3.2.   Examples of infections in which microorganisms enter across epithelial surfaces and subsequently spread through the body

Microbe	Respiratory tract and conjunctiva	Urinogenital tract	Skin	Intestinal tract
Viruses	Measles Rubella Varicella	Herpes simplex 2	Arboviruses	Enteroviruses Certain adenoviruses
Bacteria	Psittacosis	Lympho- granuloma venereum	—	—
	Myobacterium tuberculosis Yersimia pestis	*Treponema pallidum*	*Bacillus anthracis*	*Salmonella typhi*
	Q fever	—	Typhus	Q fever?
Fungi	Cryptococcosis Histoplasmosis	—	Maduromycosis	Blastomycosis
Protozoa	Toxoplasmosis	—	Malaria Trypanosomiasis	*Entamoeba histolytica*

* Fungi causing this condition flourish in a moist environment, and athlete's foot is restricted to those who encase their feet in shoes. However, those who do not wear shoes (e.g. tropical Africa) are vulnerable to other fungi that enter skin at sites of injury and cause deeper lesions called mycetomas.

to find susceptible cells elsewhere in the body if they entered blood or lymphatic vessels. Certain viruses (yellow fever, poliovirus) spread through the body to reach susceptible target organs (liver, central nervous system) after free virus particles have entered vessels below the skin or intestinal epithelium. Measles virus and tubercle bacilli infect leucocytes, which carry them through the body to organs such as the liver, spleen, skin and lung.

If, on the other hand, the microbe is able to replicate outside cells and does not have to find a susceptible cell, it can in principle multiply locally, in the blood and lymph, and in whatever part of the body it gets to. Extracellular replication, however, itself conveys a serious disadvantage, because the microorganism is forever naked and exposed to all the antimicrobial forces that the body can summon up. Indeed bacteria and other microorganisms that are capable of extracellular replication generally advertise their presence by releasing a variety of products into surrounding fluids, many of which cause inflammation and thus bring antibacterial agents such as immunoglobulins, complement and leucocytes to the site of the infection. Lymphatics are also dilated, and carry the infecting organisms to lymph nodes for further exposure to antibacterial and immune forces. Intracellular microorganisms in contrast, although exposed to the infected cell's own defence mechanisms, are directly exposed to the general bodily defences only during transit from one infected cell to another. However, if the infected cell is recognised as such by the immune defences, it can be destroyed (see Chs 6 and 9). A number of bacteria and protozoa, such as *Mycobacterium tuberculosis*, *Legionella pneumophila*, *Brucella abortus*, or *Leishmania donovani* carry out much of their multiplication in macrophages that have ingested them. Although they are not obligate intracellular parasites, this shifts the host–microbe battlefield into the cell. The battle is then waged in the infected macrophage, whose antimicrobial powers (Ch. 4) and participation in immune defences (see Chs 6 and 9) become of critical importance.

The infected host has a variety of defences that operate without delay, before the immune response comes into action (see Table 3.3). These 'early' defences are referred to in this and in subsequent chapters, and they are the type of defences that mattered before the immune system had evolved. Many microbes have strategies for interfering with these defences. For example, cells infected with a virus can commit suicide before the virus has completed its growth cycle in the cell. This is called apoptosis (see Glossary) and occurs after reovirus, HIV, and other infections. The fact that viruses (e.g. adenoviruses) have developed mechanisms for inhibiting apoptosis indicates that it plays an important part in defence. Apoptosis occurs also in bacterial infections. When uropathogenic strains of *E. coli* infect bladder epithelium, the host responds by apoptosis of the infected cells. The actual value of this response is not clear.

Table 3.3 Early defences*

Acute phase proteins	See pp. 77–78
Lysozyme	See Glossary
Lactoferrin	See p. 81
Interferons and other cytokines	See Ch. 9
Complement activation (by the alternative pathway)	See Ch. 6
Phagocytosis	See Ch. 4
Natural killer (NK) cells	See Ch. 6
Apoptosis	See Glossary and Ch. 8
Collectins	See end of Ch. 6

*These early or 'innate' defences operate immediately after a microbe has penetrated the body, during that critical period before immune responses have had time to come into action.

Subepithelial Invasion

After traversing the epithelial cell layer, a microorganism encounters the basement membrane. The basement membrane acts as a filter and can to some extent hold up the infection, but its functional integrity is soon broken by inflammation or epithelial cell damage. The invading microorganism has now reached the subepithelial tissues (Fig. 3.1), and here it is exposed to three important host defence systems. These

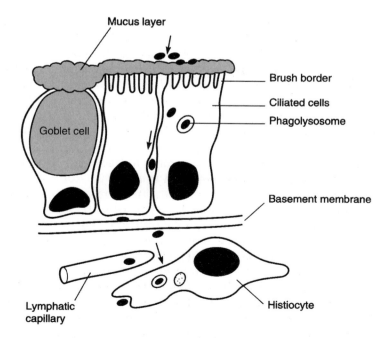

Fig. 3.1 Microbial invasion across an epithelial surface.

are (1) the tissue fluids, (2) the lymphatic system leading to the lymph nodes, and (3) phagocytic cells.

These three host defence mechanisms are of supreme importance and come into play whatever part of the body is infected, whether the nasal mucosa, meninges, urethra, cardiac muscle or liver lobule. Each depends for its action on the inflammatory response, because this response brings the phagocytes and serum factors to the site of infection and promotes drainage from the site by the lymphatic system. Therefore a short account of the inflammatory response will be given, and after this, each of the three antimicrobial factors will be considered separately. Table 3.2 shows some of the important microorganisms that regularly spread through the body in spite of these antimicrobial factors.

The inflammatory response

The capillary blood vessels supplying a tissue bring oxygen and low molecular weight materials to the cells, taking away carbon dioxide and metabolic or secretory products. There is also a constant passage of plasma proteins and leucocytes from capillaries into normal tissues, and these are returned to the blood via the lymphatic system after entering lymphatic capillaries (see below). Indeed, their presence in tissues is inferred from their presence in lymphatics draining these tissues. The cells are nearly all T lymphocytes, which leave blood capillaries by actively passing through endothelial cells. After moving about and performing any necessary tasks in the tissues, the lymphocytes penetrate lymphatic capillaries and thus enter the lymph. The lymph, with its content of proteins and cells, then passes through the local lymph nodes and generally at least one more lymph node before entering the thoracic lymph duct and being discharged into the great veins in the thorax or abdomen. Blood lymphocytes also enter lymph nodes directly and in larger numbers through post-capillary venules. The constant movement of lymphocytes from blood to tissues or lymph nodes, and back via lymphatics to the blood again, is called *lymphocyte recirculation*. Circulating lymphocytes are mostly T-cells, and in the course of their continued entries into tissues and lymph nodes they have regular opportunities to encounter any microbial antigens that may be present. There is in fact a regular monitoring of tissues by T-lymphocytes, and this is referred to as *immune surveillance*.

The various plasma proteins occur in the tissues in much the same proportion as in plasma, the actual concentrations depending on the structure of the capillary bed. As determined by concentrations in local lymphatics, the leaky sinusoids of the liver let through 80–90% of the plasma proteins into liver tissue, the less leaky fenestrated capillaries of the intestine admit 40–60% into intestinal tissues, and capillaries of skeletal muscle with their continuous lining only 10–30% (see Fig. 3.2). Thus, immunoglobulins, complement components, etc. occur regularly

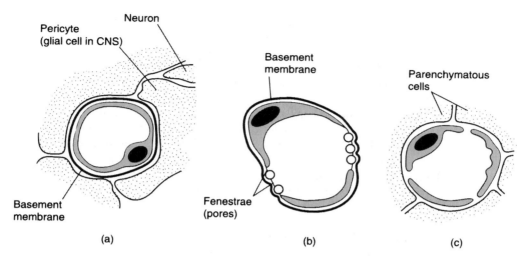

Fig. 3.2 Diagram to show types of blood–tissue junction in capillary, venule, or sinusoid. (a) Continuous endothelium (transport of tissue nutrients and metabolites): central nervous system, connective tissue, skeletal and cardiac muscle, skin, lung. (b) Fenestrated endothelium (transport of secreted, excreted or digested materials): renal glomerulus, intestinal villi, choroid plexus, pancreas, endocrine glands. (c) Sinusoid (reticuloendothelial system): liver, spleen, bone marrow, adrenal, parathyroid.

in normal tissues, but in lesser concentrations than in blood. There is some discrimination against very large molecules because the largest immunoglobulins (IgM) do not leave the blood vessels and are not detectable in afferent lymph.

There is a prompt and vigorous change in the microcirculation when tissues are damaged or infected. Capillaries and post-capillary vessels are dilated, gaps appear between endothelial cells, and the permeability of these vessels increases, allowing leakage from the blood of a protein-rich fluid. Increased amounts of immunoglobulins, complement components and other proteins are then present in tissues and fibrinogen, for instance, may be converted into fibrin so that a diffuse network of fibrils is laid down. Circulating leucocytes (especially polymorphs and monocytes) adhere to endothelial cells, and this is followed by active passage (diapedesis) of leucocytes between endothelial cells and out into tissues. The affected part now shows the four cardinal signs of inflammation, being RED and WARM (vasodilation), SWOLLEN (vasodilation, cell and fluid exudate) and often PAINFUL (distension of tissues, presence of pain mediators).

Lymphatic capillaries also become dilated, taking up the inflammatory fluids and carrying them to local lymph nodes. There is a greatly increased turnover of plasma components in the inflamed tissue. Initially, the predominant cell is the polymorph, a reflection of the situation in the blood, but polymorphs only live for a day or two in tissues, and as the acute inflammatory state subsides mononuclear cells become more prominent, especially macrophages, which phagocytose dead polymorphs and tissue debris.

The initial stages of the inflammatory response tend to be much the same, whatever the nature of the tissue insult, and this is partly because the changes are caused by the same mediators of acute inflammation. These include histamine (released from mast cells lying close to blood vessels), kinins (polypeptides derived from precursors in plasma; see Glossary) and products of complement activation by the alternative pathway (C3a and C5a; see p. 177). Some of the kinins are highly active and kallidin, for instance, a decapeptide formed from kallidinogen (an α_2 globulin) is about 15 times more active (on a molar basis) than histamine in causing inflammation. Most bacteria form inflammatory materials during their growth in tissues, but these are not very potent compared with the activation of C3 and other molecules by carbohydrates (e.g. polysaccharides) present on bacterial surfaces (see Fig 6.6). Macrophages, when they are stimulated, release a variety of inflammatory mediators and, in addition, immune-mediated inflammation results from interaction of microbial antigen with antibody (via C3a and C5a) or reaction of antigen with IgE antibody on mast cells. The final mediators include molecules such as TNF (tumour necrosis factor), ICAM-1 (intercellular adhesion molecule-1) and ELAM-1 (endothelial cell leucocyte adhesion molecule-1). Inflammatory responses, like other powerful tissue responses, must be controlled and terminated, and the mediators of inflammation not only have a variety of inhibitors but are also inactivated locally (e.g. kinins inactivated by kininases). At a later stage prostaglandins (a family of 20-carbon fatty acid molecules) and leukotrienes (a group of biologically active lipids) come into play.* They are produced from leucocytes, endothelial cells and platelets, and they both mediate and control the response.

If inflammation is due to infection with one of the pyogenic bacteria and the infection continues, then the continued supply of inflammatory and chemotactic products from the multiplying bacteria (see below) maintains vasodilation and the flow of polymorphs to the affected area. It is a polymorph exudate.† There is an increase in the number of circulating polymorphs, because of an increase in the rate of release from the bone marrow. The bone marrow holds a vast reserve supply with 20 times as many polymorphs as are present in the blood. If the tissue demand continues, the rate of production in the bone marrow is increased, and circulating polymorphs may remain elevated in persis-

---

* The terminolgy becomes complicated. Eicosanoids are produced by metabolism of arachidonic acid and include leukotrienes, prostaglandins, thromboxanes, and lipoxins.

† Human polymorphs can be labelled with the gamma-emitter indium-111 and reinfused into a patient. Within 30 s they localize in hidden foci and inflammation, which can then be located with a body scanner.

tent bacterial infections such as subacute bacterial endocarditis. Polymorph production in the bone marrow is regulated by certain colony-stimulating factors, and it is a serious matter if something goes wrong and the marrow supplies are exhausted. A fall in circulating polymorphs (neutropenia) during a bacterial infection is of ominous significance.

Viruses produce inflammatory products in tissues in the form of necrotic host cell materials or antigen–antibody complexes, but these are less potent than bacterial products, and the acute inflammatory response is of shorter duration, polymorphs being replaced by mononuclear cells. Mononuclear infiltrates are also favoured in virus infections because the infected tissues themselves are often one of the sites for the immune response, with mononuclear infiltration and cell division.

After extravasation from blood vessels, leucocytes would not automatically move to the exact site of infection. Polymorphs show random movement in tissues, and also a directional movement (chemotaxis) in response to chemical gradients produced by chemotactic substances. Monocytes show little or no random movement, but they too respond to similar chemotactic substances. Chemotactic substances such as leukotrienes, C3a and C5a (see above) are formed during the inflammatory response itself. Also, many bacteria, such as *Staphylococcus aureus* or *Salmonella typhi*, form chemotactic substances, and thus automatically betray their presence and attract phagocytic cells. It would obviously be an advantage to an infectious agent if no inflammatory or chemotactic products were formed, but for most large microorganisms (bacteria, fungi, protozoa) these products seem an almost inevitable result of microbial growth and metabolism.

The early stages of the inflammatory response in particular are known to have an important protective effect against microorganisms. In experimental staphylococcal skin infections, for instance, if the early inflammatory response is inhibited by adrenalin, and the early delivery of plasma factors and leucocytes to the site of infection thus reduced, bacteria multiply more rapidly and produce a more severe lesion. Perhaps it is not surprising that *Staphylococcus aureus* can suppress the early inflammatory response, probably by means of its α-haemolysin which causes local vasoconstriction.* Other staphylococcal products inhibit the movement of polymorphs, and lesions are thus enhanced.

If inflammation becomes more severe or widespread, it is generally modulated by increased output of corticosteroid hormones (see pp. 380–384), but at the same time it is backed up by a general metabolic response in the body. This is called the *acute phase response*. The liver releases about 30 different proteins, including C-reactive protein

---

* By shutting down local blood vessels, the α-haemolysin can increase the severity of staphylococcal mastitis in cows and sheep, leading to gangrenous mastitis (black udder).

(see pp. 77–78) and serum amyloid protein, which undergo 1000-fold increases in concentration, as well as mannose-binding protein, hapto-globulins (α_2-glycoproteins), protease-inhibitors, and fibrinogen. The exact function of these *acute phase proteins* is not clear, but they are protective; they fix complement, opsonize, and inhibit bacterial pro-teases. Their presence is associated with an increased erythrocyte sedi-mentation rate (ESR; see p. 322). The patient may develop headache, muscle pains, fever, anaemia, with decreased iron and zinc and increased copper and ceruloplasmin in the serum. Proteins in muscle are broken down, partly to provide energy required during fever and fasting, and partly to provide amino acids needed by proliferating cells and for the synthesis of immunoglobulins and acute phase proteins.

Many of the features of the acute phase response appear to be due to the action of interleukin-1 (see Glossary), and also IL-6 and TNF released from macrophages and lymphocytes. It is a complex response, which on the whole would be expected to serve useful purposes, although some less obviously beneficial 'side effects' may be unavoidable.

Tissue fluids

Tissue fluids normally contain variable amounts of plasma proteins, including IgG antibodies as discussed above. In the absence of specific antibodies and complement, tissue fluids make a good culture medium for most bacteria, but bacterial multiplication almost inevitably causes some inflammation. Powerful inflammatory events are set in motion when molecules on the bacterial surface (e.g. endotoxin) activate the alternative complement pathway. Larger amounts of IgG as well as activated complement components, will then be present in tissue fluids. At a later stage, secretory products from phagocytes (lysosomal enzymes, oxygen radicals, lactoferrin, etc.) will also be present, and finally tissue breakdown products and additional antimicrobial sub-stances liberated from dead platelets, polymorphs and macrophages.

Lymphatics and lymph nodes

A complex network of lymphatics lies below the epithelium at body surfaces. After reaching subepithelial tissues, foreign particles of all kinds, including microorganisms, rapidly enter lymphatic capillaries after uptake by or passage between lymphatic endothelial cells. There is a particularly rich superficial plexus of lymphatics in the skin and in the intestinal wall. Microorganisms scratched or injected into the skin inevitably enter lymphatics almost immediately. The intestinal lymphatics not only take up microorganisms that have breached the epithelial surface, but also have an important role in the uptake of fat in the form of chylomicrons.

Microorganisms in peripheral lymphatics are borne rapidly (within minutes) to the local lymph nodes strategically placed to deal with the flow of lymph before it returns to the blood. The rate of flow of lymph is greatly increased during inflammation, when there is increased exudation of fluid from local blood vessels and the lymphatics are dilated. Microorganisms carried to the node in the lymph are exposed to the macrophages lining the marginal sinus (Fig. 3.3), and these cells take up particles of all types from the lymph and thus filter it. The efficiency of filtration depends on the nature of the particles, on the physiological state of the macrophages, and also on the particle concentration and flow rate, the efficiency falling off at high particle concentrations or high flow rates (see Ch. 5).

All infecting microorganisms are handled in the same way and delivered via lymphatics to the local lymph node. When there has already been microbial multiplication at the site of initial infection, very large numbers may be delivered to the node. The efficiency of the node as a defence post depends on its ability to contain and destroy microorganisms rather than allow them to replicate further in the node and spread to the rest of the body. The antimicrobial forces are the macrophages of the node, the polymorphs and serum factors accumulating during inflammation, and the immune response which is initiated in the node. Under normal circumstances, as the first trickle of microorganisms reaches the node, the most important event is the

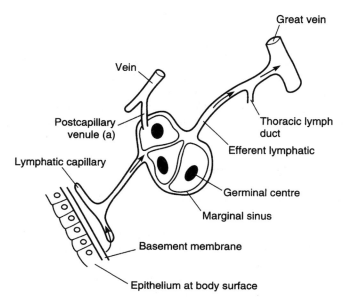

Fig. 3.3 Diagram of the lymphatic system, showing pathways from body surface to venous system; (a) indicates a site of lymphocyte circulation from blood to lymph node and back to blood (see Fig. 6.5).

encounter with macrophages in the marginal sinus. Microorganisms escaping phagocytosis by these cells enter the intermediate sinuses where they run the gauntlet of a further set of macrophages before leaving the node. If there is an inflammatory reaction in the node, a substantial migration of polymorphs into the sinuses greatly increases the phagocytic forces and thus the filtering efficiency. There is usually a further node to be traversed before the lymph is discharged into the venous system.

As well as functioning as filters, the lymph nodes, of course, are sites where the immune response comes into play. Soon after infection, as inflammatory products of microbial growth arrive in the node, there is some swelling and inflammation. The microbial antigens, some of which are already associated with antigen-presenting cells encountered at the body surface (see p. 151), generate an immune response, and there is further swelling of the node as cells divide and additional lymphoid cells are recruited into the node from the blood. The ability of viruses and other intracellular microorganisms to bypass the defences of the node and spread to the bloodstream is discussed in Ch. 5.

Phagocytic cells

Specialised phagocytic cells are divided into two main types: the macrophages, scattered through all the major compartments of the body (see Ch. 4) and the circulating neutrophil polymorphonuclear granulocytes (polymorphs, or microphages). The phagocytic cells to which microbes are exposed in the subepithelial tissues are the local macrophages (histiocytes) and also the cells arriving from the small blood vessels during inflammation. These comprise the blood monocytes which become macrophages after extravasation, and the polymorphs.

From the time of the Russian zoologist, Elie Metchnikoff, who described phagocytosis in 1883, the importance of the phagocyte in defence against disease organisms has been accepted, and children have learnt of the white blood cells that act both as scavengers and policemen, removing debris, foreign particles and microorganisms. Because of the central importance of the phagocytic defence mechanisms, the subject will receive a chapter to itself.

Nutritional Requirements of Invading Microbes

In addition to being able to resist host defence mechanisms, a pathogenic organism – be it an obligate intracellular, facultative intracellular or extracellular pathogen – must also overcome the problem of obtaining essential nutrients if it is to be successful. Two examples illustrate this point. Iron is essential for bacterial growth but the concentration of free iron in body fluids is too low (*ca.* 10^{-18} M in serum) to support growth. In the host, iron is bound by both intracellular (ferritin, haemosiderin and haem) and extracellular (transferrin in serum and milk, and lactoferrin in milk) Fe-binding proteins with high association constants for iron. *Listeria monocytogenes* may have direct access to host iron, because it produces a soluble reductant which will remove iron from transferrin. Otherwise, two main systems have been identified whereby organisms could overcome this hurdle; it is not always easy to identify which particular system is operative *in vivo*. One involves the synthesis of low molecular weight compounds called siderophores, which have an extraordinary affinity for iron. At the same time the bacteria express new outer membrane proteins that act as receptors for the Fe-siderophore complexes so that Fe is taken up into the cell. *E. coli* has been extensively studied, and three classes of siderophore have been recognised; ferrichrome, the hydroxymates and aerobactin. *Salmonella* and *Shigella* spp. and *Pseudomonas aeruginosa* produce more than one siderophore. *Salmonella* is also known to synthesise receptors for siderophores other than its own, which could be advantageous when present with other organisms, particularly in the competitive environment of the gut. *Corynebacterium diphtheriae, Staphylococcus aureus, Streptococcus pyogenes* and *Mycobacteria* spp. all probably produce siderophores. The mycobacterial siderophores are different. Mycobactins are lipid soluble and membrane associated, and exochelins are water soluble, extracellular and are the more important of the two.

The other system for dealing with the Fe shortage involves the synthesis of new outer membrane proteins which interact directly with the host's own Fe-binding proteins. This is the method used by *Neisseria meningitidis* and *N. gonorrhoea* which are able to scavange iron directly from the host.

Perhaps the most dramatic example of Fe uptake and storage is exhibited by *Yersinia. Yersinia pestis* is the agent of bubonic plague and has been responsible for devastating epidemics throughout human history. This pathogen persists in wild rodent populations in many parts of the world except Australia, and is transmitted to humans by the bites of fleas. The blockage of the proventriculae of fleas by *Y. pestis* forces infected fleas to bite and subsequently regurgitate the infected blood meal into the bite-wound of a new host. The ensuing bacteraemia in rodents completes the rodent–flea–rodent cycle essential for *Y. pestis* spread. However the ecology, pathogenicity and host range of *Yersinia pseudotuberculosis* and *Yersinia enterocolitica* are quite different from *Y. pestis*. These are orally transmitted from contaminated food or water. As described in Ch. 2, they invade Peyer's patches and disseminate to mesenteric lymph nodes (where they multipy extracellularly) and occasionally beyond causing septicemic plague-like infections; normally these infections are self-limiting. Despite their disease-causing differences, they do have some common pathogenic strategies, in particular the mechanism(s) for acquiring Fe. *Yersinia* have two important sets of pathogenicity genes: the 70 kb virulence plasmid encodes the antiphagocytic genes to which we will return in Ch. 4, and chromosomal virulence genes. The latter encode 'invasin' (involved in interaction with Peyer's patches) and the *pgm* (pigmentation) locus. Virulent *Y. pestis* strains accumulate huge quantities of exogenous haemin to form pigmented colonies on haemin agar (hence the alternative acronym, *hms*). Contiguous with the *hms* is the *ybt* gene cluster responsible for the synthesis of the siderophore yersiniabactin. In *Y. tuberculosis hms* and *ybt* are not contiguous, and in *Y. enterocolitica*, *hms* is absent. Inactivation of *hms* renders *Y. pestis* avirulent and unable to develop blockages in fleas, whereas inactivation of *ybt* in *Y. enterocolitica* hugely reduces virulence for laboratory animals. In fact, the *ybt* gene cluster has been designated the 'high pathogenicity island' (HPI) because biotype IB strains of *Y. pestis*, *Y. tuberculosis* and *Y. enterocolitica* ('New World strains) all possess *ybt* biosynthetic genes and kill mice with very low infectious doses. In contrast, biotypes 2–5 (Old World strains) are much less virulent for mice and do not possess *ybt* genes. It is of interest that HPI has now been found in some pathotypes of *E. coli, Klebsiella, Enterobacter*, and *Citrobacter.*

The other nutrients in short supply in the mammalian host are aromatic amino acids like tryptophan. Interestingly, the expression of the *trp* operon (which comprises the genes responsible for the synthesis of tryptophan) is controlled by Fe as well as by tryptophan levels. A functional aromatic biosynthetic pathway is absolutely vital for growth *in vivo* since aromatic amino acids are generated via chorismic acid. The latter is a branch point at which the biosynthesis of *p*-amino benzoic acid (PABA) also begins. PABA is required as a precursor of folic acid; it is as a competitor for PABA that the sulphonamide drugs work. By introducing lesions in one, preferably two, key genes in this system (*aroA* and *aroD*), it has been possible to attenuate strains of *S. typhimurium* such that their initial invasive properties are unaltered but their ability to grow *in vivo* is severely restricted. Such crippled constructs can be manipulated to carry genes encoding heterologous antigens and have been used experimentally to deliver protective antigenic stimuli to a whole range of heterologous antigens.

References

Baumannn, H. and Gaulle, J. (1994). The acute phase response. *Immunol. Today* **15**, 74–78.

Chatfield, S., Li, J. L., Sydenham, M., Douce, G. and Dougan, G. (1992). *Salmonella* genetics and vaccine development. *In* 'Molecular Biology of Bacterial Infection. Current Status and Future Perspectives'. (C.

E. Hormaeche, C. W. Penn and C. J. Smyth, eds), Society for General Microbiology Symposium 49, pp. 299–312. Cambridge University Press, Cambridge.

Everett, H. and McFadden, G. (1999). Apoptosis; an innate immune response to virus infection. *Trends Microbiol.* **7**, 160–165.

Rakin, A., Schubert, S., Pelludat, C., Brem, D. and Heesemann, J. (1999). The high-pathogenicity island of *Yersiniae*. *In* 'Pathogenicity Islands and Other Mobile Virulence Elements' (J. B. Kaper and J. Hacker, eds), pp 79–90. ASM Press, Washington, DC.

Rodriguez-Boulan, E. and Sabatini, D. D. (1978). Asymmetric budding of viruses in epithelial monolayers. Model systems for epithelial polarity. *Proc. Natl. Acad. Sci. U.S.A.* **75**, 5071–5075.

Roitt, I. M., Brostoff, J. and Male, D. (1998). 'Immunology', 5th edn. Mosby, London.

Ryan, G. B. and Majno, G. (1977). Acute inflammation. *Am. J. Pathol.* **86**, 183–276.

Stadnyk, A. W. and Gauldie, J. (1991). The acute phase response during parasitic infection. *Immunol. Today* **7**, A7–A12.

Taussig, M. J. (1994). 'Processes in Pathology and Microbiology', 3rd edn., Blackwell Scientific Publications, Oxford.

Tucker, S. P. and Compans, R. W. (1993). Virus infection of polarized epithelial cells. *Adv. Virus Res.* **42**, 187–247.

Yoffey, J. M. and Courtice, F. C. (1970). 'Lymphatics, Lymph and the Lymphomyeloid Complex'. Academic Press, London and New York.

4

The Encounter with the Phagocytic Cell and the Microbe's Answers

The phagocyte is the most powerful and most important part of the host defences that can operate without delay against the invading microorganism after the epithelial surface has been breached. There are two types of specialised phagocytic cells, the macrophage and the polymorphonuclear leucocyte. In the subepithelial tissues there are local resident macrophages (histiocytes), and as soon as an inflammatory response is induced, polymorphonuclear leucocytes arrive in large numbers after passing through the walls of small blood vessels. The inflammatory cells include also monocytes and lymphocytes.

Polymorphonuclear leucocytes arise in the bone marrow and are continuously discharged in vast numbers into the blood. The 3×10^{10} polymorphs that are present in normal human blood carry out their functions after leaving the circulation and entering sites of inflammation in tissues. These cells are nondividing, live only for a few days and each day about 10^{11} disappear from the blood, even in the absence of significant inflammation. Indeed, at any given time about half of them are adherent to or moving slowly along the walls of capillaries and post-capillary venules. This daily loss is balanced by entry into the blood from the bone marrow and, to make some provision for sudden demands, the bone marrow contains an enormous reserve of about 3×10^{12} polymorphs.

Monocytes are circulating precursors of macrophages. They arise from stem cells in the bone marrow, and as soon as they leave the circulation and begin to carry out their phagocytic duties in tissues they become macrophages. Macrophages are widely distributed throughout the body, but they are not as numerous as polymorphs, and there are no great reserves of macrophages in tissues. Fixed macrophages line the blood sinusoids of the liver (Kupffer cells), spleen, bone marrow and adrenals, and monitor the blood for effete cells, microorganisms or other foreign particles. Macrophages lining lymph sinuses in lymph nodes monitor the lymph, and the alveolar macrophages in the lung monitor the alveolar contents. The peritoneal and pleural cavities also contain large numbers of macrophages. Macrophages, in fact, are strategically placed throughout the body to encounter invading microorganisms. Alveolar macrophages deal with those entering the lung when they are deposited in the alveolus, beyond the mucociliary defences. Those in subepithelial sites in the skin, intestine, etc. meet invading microorganisms once epithelial surfaces are breached, and those lining lymphatics and blood sinusoids come into play if there is spread of the infection via lymph or blood. Macrophages in the peritoneal and pleural cavities form a first line of defence in these vulnerable cavities.

Lymphocytes have an immunological function to be described more fully later. When they encounter microbial antigens to which they are by nature or by previous experience sensitised, they undergo profound changes. B- (bursa-derived) cells are stimulated to differentiate into antibody-producing (plasma) cells and T- (thymus-derived) cells differentiate into lymphoblasts, divide, and release lymphokines in the course of carrying out the cell-mediated immune response. Lymphokines induce further inflammatory or immunological changes and have profound effects on the function of macrophages, activating them, promoting their accumulation in the tissue and indeed 'focusing' them onto the site of infection (see Ch. 6).

Phagocytosis is a basic type of cell function, and is not restricted to macrophages and polymorphs. For instance, epidermal cells in the skin take up injected carbon particles, and intestinal epithelial cells and vascular endothelial cells also ingest certain marker particles, but this is on a very restricted scale compared with the professional phagocytes. Mere phagocytosis is not enough. If the ingestion of a microorganism is to be of service to the infected host, it must be followed by killing and preferably intracellular digestion of the microorganism. The specialised phagocytic cells are therefore equipped with a powerful array of antimicrobial weapons and lysosomal enzymes.

Cell Biology of Phagocytosis

Pinocytosis (the uptake of fluid and solutes) and receptor-mediated endocytosis share a clathrin-based mechanism which usually is independent of actin polymerisation. In contrast, phagocytosis is involved in the uptake of larger particles, is usually clathrin-independent and occurs by an actin-dependent mechanism. Phagocytosis is a highly complex phenomenon the molecular details of which are now beginning to be unravelled. A brief summary of some of the main features of this rapidly developing area is given as it will help our understanding of the material here and in Chapter 8.

Receptors/ligands

Before phagocytosis can take place there must be preliminary (non-specific) attachment of the object to the phagocytic cell surface, followed by firmer specific attachment. The firm attachment and certainly the ingestion of particles is facilitated by serum substances called opsonins which then act as ligands. Opsonins include acquired or naturally occurring antibodies, and complement. Opsonin comes from a Greek word meaning sauce or seasoning; it makes the microbe more palatable and more easily ingested by the phagocyte. Initiation of internalisation involves interaction between opsonins on the particle and surface receptors on the phagocyte. Receptor/ligand complexes comprise the initial link in the complex train of signal transduction events in the cell which in the case of the phagocyte ultimately involves the cytoskeleton. Two of the best characterised receptors in macrophages are the complement receptor 3 (CR3) which binds C3bi on complement-opsonised targets and Fc gamma receptors (FcγRs) which bind to antibody-coated targets (see below). Type I phagocytosis is FcγR-mediated, constitutively active, results in pseudopod extension and membrane ruffling, and is also accompanied by the respiratory burst and the production of pro-inflammatory signals (of which more below). In contrast, Type II phagocytosis is CR3 mediated, requires additional signals, results in the organism sinking into the cell and does not activate the respiratory burst. A third (as yet undesignated) type exemplified by the internalisation of *Legionella pneumophila* involves the coiling of the plasma membrane around the bacterium before the formation of a single vacuole. There are other receptors such as the mannose receptor which recognises a broad range of pathogens expressing branched mannose and fucose structures on their surfaces. In addition there are many receptors on macrophages (for example, the scavenger receptors) for recognising apoptotic cells.* In man, all IgG immunoglobulins except the IgG2 subclass attach to (are cytotrophic for) polymorphs. Phagocytic cells thus have a special affinity for

---

* Apoptosis is programmed cell death and is crucial in the development and homeostasis of all muticellular organisms. An enormous number of cells undergo apoptosis and are removed by macrophages but these dying cells are very rarely seen *in vivo* since their removal does not involve activation of pro-inflammatory responses.

microorganisms coated with antibody. If the C3b complement fragment is present on the microbe (following antibody-mediated or alternative pathway activation of complement), it can react with CR3 receptors on the cell and thus provide further assistance for attachment and ingestion. Ordinary tissue cells differ from professional phagocytes and do not have these specific receptors; they consequently fail to adsorb and ingest opsonised microorganisms.

Downstream effectors

The detailed mechanisms by which initial membrane events are transduced into engulfment of an organism are not fully known but several biochemical systems are involved in the ultimate polymerisation of actin to generate the phagosome. Only one is discussed here, the rho family of GTPases,† the role of which in signal transduction is schematised in Fig. 4.1. Cdc42 and Rac GTPases are used for Type I, and Rho GTPase for Type II phagocytosis. As we shall see it is precisely these systems that some pathogens interfere with in their interaction with phagocytic and other cells.

Phagocytosis in Polymorphonuclear Leucocytes

Polymorphs generally carry out their functions after leaving the bloodstream, but they can under certain circumstances adhere to the endothelium of small blood vessels, especially in the lungs, and act as 'fixed' phagocytes. This happens, for instance, when Gram-negative bacteria or endotoxin enter the bloodstream, and probably depends on the action of complement.

There are three types of polymorphonuclear leucocytes: the neutrophils, the basophils and the eosinophils, each serving separate functions and distinguished by the staining reactions of their prominent cytoplasmic granules. The granules are lysosomes, consisting of membrane-lined sacs containing enzymes and other materials.

The neutrophils are the most numerous, comprising 70% of the total leucocytes in blood, and are the cells generally referred to as 'polymorphs'. They have no mitochondria but plenty of glycogen as an

---

† The rho family of GTPases belong to a large family of proteins which bind GTP and slowly hydrolyse GTP to GDP. They are involved, together with accessory proteins, in controlling many cell functions by initiating (in their active GTP-bound form), the first step in complex biochemical cascades, the outcome of which reflects the differentiated nature of the cell. It is beyond the scope of this book to review this huge and rapidly developing topic. The historical acronyms used for these small GTPases are hopelessly confusing, unlike current acronym usage which is designed to give some indication of function.

energy source, which can be used under anaerobic conditions. They contain two or three types of granules, whose enzymes include peroxidase, alkaline phosphatase, acid phosphatase, ribonuclease, deoxyribonuclease, nucleotidases, glucuronidase, lysozyme and cathepsins. In addition neutrophils (and macrophages) contain cationic peptides which contain around 30–33 amino acids, are rich in cysteine and arginine and have a specific antibiotic-like activity by virtue of their pore-forming activity. They are called defensins, comprise *ca.* 30–50% of granular protein or 5% of total cellular protein, and are active against a range of pathogens as diverse as *Staphylococcus aureus, Pseudomonas aeruginosa, E. coli, Cryptococcus neoformans* and *Herpes simplex* virus. Defensin or defensin-like substances have been found in the skin of frogs (magainins) and the small intestine of the mouse (cryptdin). Another important constituent is the Fe-binding protein lactoferrin which binds Fe over a wide pH range.

The eosinophils (1% of the leucocytes) are less effective than polymorphs in the phagocytosis and killing of microbes but they are especially active in the phagocytosis of immune complexes. For every circulating eosinophil, there are 300–500 in the extravascular tissues, and they are especially numerous in the submucosal tissues of the intestinal and respiratory tracts. Their granules contain, in addition to

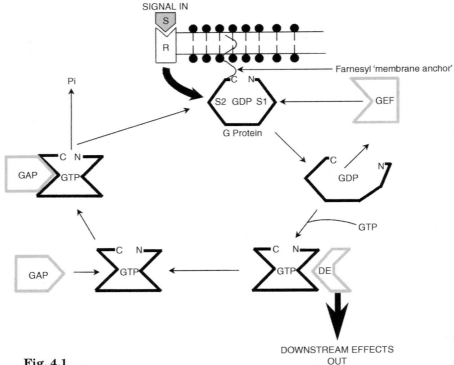

Fig. 4.1

various enzymes, blockers of the inflammatory mediators (histamine, kinins and serotonin), five distinct cationic proteins and a major basic protein generated by immune complexes (see Ch. 6). A rise in the number of circulating eosinophils is a feature of certain parasitic and allergic diseases, and they are attracted into tissues by eosinophil chemotactic factor released from mast cells. They bear C3b and Fc receptors and attach to and kill certain parasites (e.g. schistosomula) that are coated with specific antibody, probably by discharge of the major basic protein.*

Basophils make up 0.5% of the total blood leucocytes, and their granules are especially rich in histamine and heparin. They closely resemble the mast cells seen in submucosal tissues and round blood

---

* The polymorph also may attach in large numbers to the surface of the schistosomula, which is too big to be phagocytosed. Killing takes place if enough damage is inflicted in this combined assault. When polymorphs kill opsonised *Trichinella spiralis*, dozens of them may be seen attached to the surface of the parasite.

Fig. 4.1 Transduction of cell signalling involving monomeric G proteins. A signalling ligand (S) binds to its receptor (R) which initiates the activation of membrane-bound G protein. G protein is so-called because it binds GDP or GTP nucleotides; it is also a GTPase which will slowly hydrolyse bound GTP to GDP. G protein has two domains, Switch 1 (S1) and Switch 2 (S2). In its non-activated GDP-bound state, S1 and S2 are off. The next step involves interaction of S1 with GEF (guanine exchange factor) which facilitates removal of GDP for which G protein has a high affinity. This allows binding of GTP which is present in the cytoplasm at a concentration ten times higher than GDP. Binding of GTP induces conformational changes in G protein which allows interaction with the first component (DE; downstream effector) involved in initiating a cascade of downstream events. This stimulation of downstream effects continues until GTP is hydrolysed to GDP by the endogenous GTPase activity of G protein. The slow rate of endogenous hydrolysis is increased 1000-fold by interaction of GAP (GTPase-activating protein) with S2. G protein is thus returned to its inactive state. The efficiency of this process will depend on a number of factors including the number of receptor molecules (R) in the target cell. Specificity will be determined by the particular member of the G protein family involved (Rho, Ras, etc.) and in broad terms by the differentiated nature of the cell which in turn will determine the nature of DE and the ensuing biochemical cascade. Attempts to elucidate the determinants of the specificity of the downstream effects is one of the most actively researched areas in cell biology, discussion of which is inappropriate for this book. However, two examples are given relevant to our subject. First, is the induction of phagocytosis (see text). Second, some bacterial toxins (see Ch. 8) interfere with this cyclical activation/deactivation of G proteins. Toxins A and B of *Clostridium difficile* glucosylate a serine residue in S1, thereby permanently inactivating G protein by preventing the action of GEF. *Escherichia coli* CNF1 toxin and *Bordetella bronchiseptica* DNT toxin both deamidate a glutamine residue in the active form of G protein such that its GTPase function is lost, hence it is permanently switched on. See text for possible effects on tight junctions of epithelial cells and recruitment of phagocytes.

vessels, and bear Fc receptors for IgE antibody. When antigen binds to IgE antibody on their surface the granules are discharged, and this leads to various 'allergic' inflammatory changes.

After extravasation from blood vessels, polymorphs would not automatically congregate at the exact site of microbial infection without any guidance. When seen in time-lapse cinephase movies, they display very active cell movement, travelling at up to 40 mm min^{-1}. One type of movement is random, in all planes, and to some extent this would bring the cells to the scene of infection. They also show chemotaxis, which is a directional cell movement in response to chemical gradients formed by the release of certain chemotactic materials in tissues. Many soluble bacterial products attract polymorphs in this way, as do the mediators generated by C3 and C5 components after antigen–antibody interactions (see Ch. 6), and various substances derived from host tissue.

Although phagocytosis is strikingly enhanced by these opsonins, and sometimes depends entirely on them, phagocytosis occurs during an infection, before antibodies have been formed, and this is a vital part of the 'early' defence system. A variety of objects including starch grains, yeasts, bacteria and polystyrene particles, are adsorbed to the polymorph surface and are phagocytosed without apparent need for antibodies. For example, mannose-binding lectin in serum reacts with carbohydrates on many bacteria, viruses, and fungi, and can opsonize them after attaching to a specific receptor on the phagocyte surface or activating the alternative pathway of complement. Familial deficiency of mannose-binding protein means susceptibility to meningococcal disease.

Phagocytosis is a familiar event in physical terms (Fig. 4.2). The infolding of the plasma membrane to which particles are attached is due

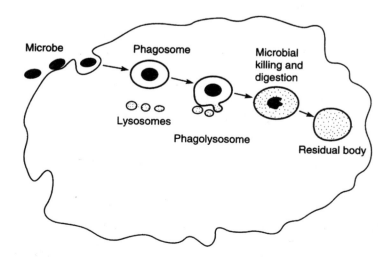

Fig. 4.2 Diagram to show phagocytosis and intracellular digestion.

to the contraction of actin and myosin filaments ('muscles') anchored to a skeleton of microtubules in the cytoplasm. As outlined above, the process is triggered off by the attachment of particles to the receptors on the plasma membrane. Phagocytosis is associated with energy consumption involving oxidation of glucose via the hexosemonophosphate pathway – the respiratory burst. There is a 10–20-fold increase in the respiratory rate of the cell. There is also an increased turnover of membrane phospholipids. This is hardly surprising, because the multiple infoldings of the cell surface during active phagocytosis, in which up to 35% of the plasma membrane may be internalised, obviously requires synthesis of extra quantities of cell membrane.

As a result of phagocytosis, microorganisms are enclosed in membrane-lined vacuoles in the cytoplasm of the phagocytic cell, and subsequent events depend on the activity of the lysosomal granules (Fig. 4.2). These move towards the phagocytic vacuole (phagosome), fuse with its membrane to form a phagolysosome, and discharge their contents into the vacuole, thus initiating the intracellular killing and digestion of the microorganism. The loss of lysosomal granules is referred to as degranulation. The process of ingestion, killing and digestion of a nonpathogenic bacterium by polymorphs can be followed biochemically by radioactive labelling of various bacterial components, and structurally by electron microscopy. When *E. coli* are added to rabbit polymorphs *in vitro*, phagocytosis begins within a few minutes. Nearly all polymorphs participate, each one ingesting 10–20 bacteria. Polymorph granules then move towards the phagocytic vacuoles and fuse with them, delivering their contents into the vacuoles. The pH of the vacuoles becomes acid (pH 3.5–4.0), and this alone has some antimicrobial effect. Bacteria are killed (in the sense that they can no longer multiply when freed from the phagocytic cell) a minute or two later, before there is detectable biochemical breakdown of bacteria. Digestion then proceeds, first the bacterial cell wall components (detectable by the release from bacteria of radioactively labelled amino acids) and subsequently the contents of the bacterial cell. By electron microscopy the bacterial cell wall appears 'fuzzy' rather later, after about 15 min. The early killing is presumably associated with impaired functional integrity of the bacterial cell wall, the gross digestion of the corpse being detectable biochemically at a later stage, and changes in ultrastructural appearances later still.

The biochemical basis for the killing of bacteria and other microorganisms by polymorphs is complex, comprising various components. Although some of these components kill bacteria when added to them *in vitro*, their significance in the phagocyte is often not known.

(1) Generation of reactive oxygen intermediates (ROI), outlined in Fig. 4.3. The brief burst of respiratory activity that accompanies phagocytosis is needed for killing rather than for phagocytosis itself, and membrane-associated NADPH oxidase is activated after phagocytosis has occurred. The following events taking place within the vacuole are

OXYGEN-DEPENDENT

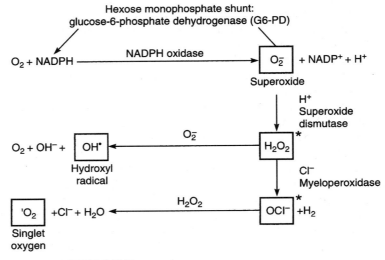

OXYGEN INDEPENDENT

* Acid pH
* Lysozyme–dissolves the cell wall of certain Gram-positive bacteria
  Cationic proteins*–bactericidal activity
  Lactoferrin
  Vitamin B_{12}-binding protein ⎤ Bacteriostatic activity?
  Acid hydrolases–post-mortem digestion of microorganisms?

Fig. 4.3 Antimicrobial mechanisms in the neutrophil polymorph (* = probably important in killing).

important. The oxygen produced gives rise to superoxide by the addition of one electron, and two superoxide molecules may interact (dismutate) and form hydrogen peroxide, either spontaneously or with the help of superoxide dismutase. The hydrogen peroxide in turn can be reduced to give the hydroxyl radical (OH·). It can also undergo myeloperoxidase-mediated halogenation to generate hypochlorite (OCl⁻) which not only disrupts bacterial cell walls by halogenation, but also reacts with H_2O_2 to form singlet oxygen, which is possibly antimicrobial. Thus, free hydroxyl (OH·) and superoxide (O_2^-) radicals, H_2O_2, OCl⁻ and singlet oxygen ($'O_2$) are all produced in polymorphs in the membrane of the phagosome, mostly by means of an electron transport chain, and involving cytochrome b. But it is not clear whether some or all of these products are responsible for killing or whether it also depends on other activities of the electron transport chain.

(2) Oxygen-independent killing mechanisms. Oxygen-dependent killing is not the whole story. Polymorphs often need to operate at low oxygen tension, for instance where relatively anaerobic bacteria are multiplying, and such microorganisms are killed quite effectively in

the absence of oxygen. There are a number of possible mechanisms. First, within minutes of phagocytosis the pH within the vacuole falls to about 3.5 and this would itself have an antimicrobial effect. Also the granules delivered to the phagocytic vacuole contain certain antimicrobial substances. There are 'specific' granules and 'azurophil' granules, as well as the regular lysosomes. These contain, as mentioned above, not only myeloperoxidase, but also lactoferrin, lysozyme, a vitamin B_{12}-binding protein, a variety of cationic proteins and acid hydrolases. Lactoferrin, which binds iron very effectively, even at a low pH (see p. 387) would not kill but would deprive the phagocytosed microorganism of iron. The cationic proteins bind to bacteria and, under alkaline conditions, have a pronounced antibacterial action; they must act early, before the pH becomes acid. The most potent of them is bactericidal/permeability-increasing protein (BPI), which is active at picomolar concentrations. It binds to lipopolysaccharide (LPS) on Gram-negative bacteria, damages their surface and inhibits their growth. Animals given BPI are protected against a wide range of Gram-negative bacteria. Exposure to BPI induces expression of a range of proteins in *Salmonella* and enteropathogenic *E. coli* (EPEC) including BipA. The latter is a remarkable protein belonging to the class of small GTPases involved in signal transduction (see above) and is a new type of virulence regulator. It is involved in resisting the cytotoxic effect of BPI, modelling of the EPEC-induced pedestal and flagella-mediated motility.

The acid hydrolases probably function by digesting the organisms after killing. The enzyme lysozyme hydrolyses the cross-links of the giant peptidoglycan molecules that form most of the cell wall of Gram-positive cocci (Fig. 4.4). The cell wall is rapidly dissolved and the bacteria killed. Gram-negative bacteria have an additional lipopolysaccharide component incorporated into the outer surface of the cell wall, and this gives these bacteria relative resistance to the action of lysozyme.*

Fusion of lysosomal granules with phagosomes is the prelude to intracellular digestion in phagocytes, and is closely comparable with the process by which a free-living protozoan such as *Amoeba* digests its prey. In both cases, the phagocytic vacuole becomes the cellular stomach. Under certain circumstances, polymorph granules fuse with the cell surface rather than with the phagocytic vacuole, and the contents of the granule are then discharged to the exterior, producing local concentrations of lysosomal enzymes in tissues and often giving rise to severe histological lesions. Antigen–antibody complexes induce this type of response in polymorphs, and the resultant tissue damage is exemplified in the blood vessel wall lesions in a classical Arthus

---

* Granule proteins generally have to bind to the bacterial surface if killing is to occur, and a longer polysaccharide chain makes binding less effective.

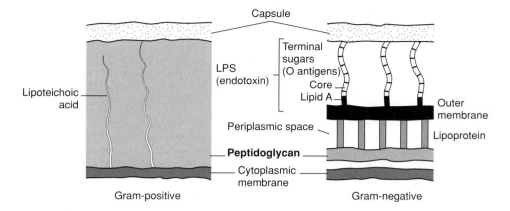

Fig. 4.4 Comparison of Gram-positive and Gram-negative bacterial cell walls. Pili and flagella (the latter bearing H antigens in Gram-negative bacilli) are not shown. Peptidoglycan has lipoteichioic acid molecules extending through it, ± teichoic acid linked to peptidoglycan. The capsule may be protein or poly-saccharide and is the site of the K antigen of Gram-negative bacilli.

response. On other occasions lysosomes fuse with the phagocytic vacuole before phagocytosis is completed and the vacuole internalised. Lysosomal enzymes then pass to the exterior of the cell to give what is referred to as 'regurgitation after feeding'. This occurs after exposure to certain inert particles or to antigen–antibody complexes. Since poly-morphs live for no more than a day or two, their death and autolysis inevitably leads to the liberation of lysosomal enzymes into tissues. When this occurs on a small scale, macrophages ingest the cells and little damage is done, but on a larger scale the accumulation of necrotic polymorphs and other host cells, together with dead and living bacteria, and autolytic and inflammatory products, forms a localised fluid product called pus. This product, resulting from the age-old battle between microorganism and phagocyte, can be thin and watery (streptococci), thick (staphylococci), cheesy (*Mycobacterium tuberculosis*), green (*Pseudomonas aeruginosa* pigments), or foul-smelling (anaerobic bacteria). Before the advent of modern antimicro-bial agents, a staphylococcal abscess could contain more than half a litre of pus.

Phagocytosis in Macrophages

The processes of adsorption, ingestion and digestion of microorganisms in macrophages are in general similar to those in polymorphs, but there are important differences.

Macrophages exhibit great changes in surface shape and outline, but do not have the polymorph's striking ability to move through tissues. They show chemotaxis, but the chemotactic mediators are different from those attracting polymorphs. This contributes to the observed local differences in macrophage and polymorph distribution in tissues. Macrophages also have a different content of lysosomal enzymes, which varies with the species of origin, the site of origin in the body and the state of activation (see Ch. 6). They do not contain the cationic proteins found in polymorph granules, but they do contain defensin peptides and an equivalent, but not the same, oxygen-dependent antimicrobial system. This gives rise to differences in their ability to handle ingested microorganisms. Thus, although the fungus *Cryptococcus neoformans* is phagocytosed by human polymorphs and then killed by chymotrypsin-like cationic proteins and the oxygen-dependent system, the same fungus survives and grows readily after phagocytosis by human macrophages. The antimicrobial armoury of human polymorphs also gives them a major role in the killing of the fungus *Candida albicans*, whereas macrophages are much less effective. Indeed, for many bacteria polymorphs show a bactericidal activity that is superior to that of monocytes and macrophages. This is because opsonised phagocytosis is often more rapid in polymorphs, and there is a greater generation of the antibacterial species of oxygen mentioned on pp. 91–92. On the other hand, macrophages live for long periods (months, in man) compared with polymorphs (days, in man). Polymorphs are very much 'end cells', delivered to tissues with a brief life span and limited adaptability, whereas macrophages are capable of profound changes in behaviour and biochemical make up in response to stimuli (see Ch. 6). When polymorphs have discharged their lysosomal granules into phagosomes the cells, rather than the granules, are renewed. Macrophages, on the other hand, retain considerable synthetic ability, so that they can be stimulated to form large amounts of lysosomal and other enzymes. Also, because of their longer life in tissues, it is common to see macrophages loaded with phagocytic vacuoles whose contents are in all stages of digestion and degradation. Certain materials, particularly the cell walls of some bacteria, are only degraded very slowly or incompletely by macrophages.

Macrophages, like poylmorphs, express receptors for the Fc portion of IgG and IgM immunoglobulins, and complement, so that immune complexes or particles coated with immunoglobulins and complement are readily adsorbed. Macrophages also have the ability to recognise and adsorb to their surface various altered and denatured particles, such as effete or aldehyde-treated erythrocytes. However, mere adsorption of microorganisms to the cell surface does not necessarily lead to phagocytosis. Certain mycoplasma, for instance, attach to macrophages and grow to form a 'lawn' covering most of the cell surface, but are not phagocytosed unless antibody is present. Macrophages are also secretory cells, and liberate about 60 different products ranging from lysozyme to collagenase. These may be important in antimicrobial defence as well as in immunopathology (see Ch. 6).

Another important difference is the ability of many macrophages, especially when activated, to generate reactive nitrogen intermediates (RNI); the nitric oxide (NO) pathway (Fig. 4.5). Among its many activities (on the vascular system, on neurons, on platelets, etc.), NO is microbicidal, being effective against a range of organisms including mycobacteria and *Leishmania* spp. Bacteria produce an enzyme (NO dioxygenase) that detoxifies NO, and if this capacity is removed they become exquisitely sensitive to NO. Paradoxically, it is doubtful if tetrahydrobiopterin is made by human macrophages and the role of the NO pathway in the antimicrobial function in human macrophages *in vivo* is not clear. However, other nonimmunological cells (fibroblasts, endothelial cells, hepatocytes and cerebellar neurons) are known to generate RNI, although less markedly than macrophages, and RNI may represent an important basic mechanism of local resistance against intracellular pathogens.

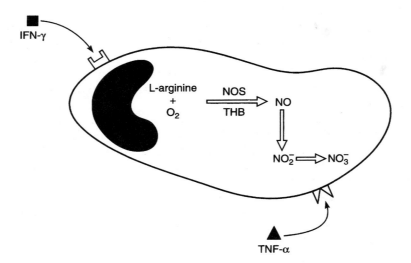

Fig. 4.5 Schematic representation of the nitric oxide pathway in murine macrophages. Nitric oxide synthetase (NOS) mediates the addition of O_2 to the guanidino N of L-arginine to form NO. This is rapidly converted to NO_2^- and NO_3^-. Precisely which RNI is involved and by what mechanism killing takes place is not clear. Tetrahydrobiopterin (THB) is an essential cofactor for NOS but this is not present in human macrophages. The pathway is blocked by the arginine analogue N^G-monomethyl- L-arginine. The process is subject to modulation by several cytokines but two seem to be very important. The synthesis of NOS is activated by interferon-γ (IFN-γ) and the subsequent steps optimised by tumour necrosis factor-α (TNF-α). The latter may arise from the macrophage stimulated by IFN-γ in the first place – an autocrine effect.

After the microorganism has been killed, the subsequent disposal of the corpse is only of concern to the host. Most microorganisms are readily digested and degraded by lysosomal enzymes. But the microbial properties that give resistance to killing sometimes also give resistance to digestion and degradation, because the cell walls or capsules of certain pathogenic bacteria are digested with difficulty. Group A streptococci, for instance, are rapidly killed once they have been phagocytosed, but the peptidoglycan–polysaccharide complex in the cell wall resists digestion, and streptococcal cell walls are sometimes still visible in phagocytes a month or so after the infection has terminated.* The waxes on the outer surface of certain mycobacteria are not readily digested by lysosomal enzymes and it is possible that this is why such bacteria (e.g. *Mycobacterium lepraemurium*) are difficult to kill. Although saprophytic mycobacteria have a similar type of covering, it may have particular properties in *Mycobacterium lepraemurium*.

Microbial Strategy in Relation to Phagocytes

As has been discussed earlier, microorganisms invading host tissues are first and foremost exposed to phagocytes, and the encounter between microbe and phagocyte has played a vital role in the evolution of multicellular animals, all of which, from the time of their origin in the distant past, have been exposed to invasive microorganisms. The central importance of this ancient and perpetual warfare between the microbe and the phagocytic cell was clearly recognised by Metchnikoff over a 100 years ago.

Microorganisms that readily attract phagocytes, and are then ingested and killed by them, are by definition unsuccessful. They fail to cause a successful infection. Phagocytes, when functioning in this way, have an overwhelming advantage over such microorganisms. Most successful microorganisms, in contrast, have to some extent at least succeeded in interfering with the antimicrobial activities of phagocytes, or in some other way avoiding their attention. The contest between the two has been proceeding for so many hundreds of millions of years that it can be assumed that, if there is a possible way to interfere with or otherwise prevent the activities of phagocytes, then some microorganisms will almost certainly have discovered how to do this. Therefore the types of interaction between microorganisms and phagocytes will be considered from this point of view.

Microbial factors that damage the host or actively promote the spread of infection are often called agressins. Such factors have a 'toxic'

---

* Because the capsules or cell walls of streptococci, pneumococci, mycobacteria, *Listeria* and other bacteria pose problems for lysosomal enzymes and are not readily digested in phagocytes, bacterial fragments are sometimes retained in the host for long periods. This can lead to interesting pathological or immunological results (see Ch. 8).

activity that is demonstrable in a suitable test system. In many instances, however, microbial factors inhibit the operation of host defence mechanisms without actually doing any damage. There is no 'toxic' activity, and they have been called impedins. Until relatively recently it was exceedingly difficult to ascribe, with confidence, definitive roles to many factors produced by pathogenic bacteria, particularly in cases like staphylococci which produce a large number of putative virulence determinants. Often this was (maybe still is) due to the lack of really suitable animal models and the fact that injection of a bolus of purified toxin often proved lethal. For too long this tended to direct attention away from the potentially more relevant effects of such toxins in sublethal amounts on host defence mechanisms, in particular on phagocytes. However, an increasing number of studies have been carried out on isogenic mutants resulting in a picture which at least approximates to the *in vivo* situation.

Microbes that are noninfectious for man are dealt with and destroyed by the phagocytic defence system just as in the case of the nonpathogenic bacteria in polymorphs as described above. Nearly all microorganisms, indeed, are noninfectious and it is only a very small number that can infect the vertebrate host, and an even smaller number that are significant causes of infection in man. The ways in which microorganisms meet the challenge of the phagocyte will be classified, for simplicity (Fig. 4.6, Table 4.1).

Table 4.1.   Showing types of interference with phagocytic activities

Microorganism[a]	Type of interference[b]	Mechanism or responsible factor
Streptococcus pyogenes	Kill phagocyte Inhibit polymorph chemotaxis Resist phagocytosis Resist digestion	Streptolysin[c] induces lysosomal discharge into cell cytoplasm Streptolysin M substance on fimbriae; hyaluronic acid capsule
Staphylococci	Kill phagocyte Inhibit opsonised phagocytosis Resist killing	Leucocidin induces lysosomal discharge into cell cytoplasm Protein A blocks Fc portion of Ab; polysaccharide capsule in some strains Cell wall peptidoglycan; production of catalase?
Bacillus anthracis	Kill phagocyte Resist killing	Lethal factor (LF) of tripartite toxin Capsular polyglutamic acid
Haemophilus influenzae *Streptococcus pneumoniae* *Klebsiella pneumoniae*	Resist phagocytosis (unless Ab present) Resist digestion	Polysaccharide capsule

Microorganism[a]	Type of interference[b]	Mechanism or responsible factor
Pseudomonas aeruginosa	Kill phagocyte	Exotoxin A kills macrophages; also cell-bound leucocidin
	Resist phagocytosis (unless Ab present) Resist digestion	'Surface slime' (polysaccharide)
Escherichia coli	Resist phagocytosis (unless Ab present) Resist killing Kill macrophages	O antigen (smooth strains) K antigen (acid polysaccharide) K antigen
Salmonella spp.	Resist phagocytosis (unless Ab present) Resist killing; survival in macrophages Kill phagocyte	Vi antigen Secreted products of SPI[d]-2 Secreted products of SPI-1
Clostridium perfringens	Inhibit chemotaxis Resist phagocytosis	θ-toxin Capsule
Cryptococcus neoformans	Resist phagocytosis	Capsular polyuronic acid
Treponema pallidum	Resist phagocytosis	Capsular polysaccharide
Yersinia pestis	Kill phagocyte	Yop virulon proteins
Mycobacteria	Resist killing and digestion Inhibit lysosomal fusion	Cell wall component Unknown
Brucella abortus	Resist killing	Cell wall substance
Toxoplasma gondii	Inhibit attachment to polymorph Inhibit lysosomal fusion	Unknown Unknown
Plasmodium berghei	Resist phagocytosis	Capsular material

[a] Often it is only the virulent strains that show the type of interference listed.
[b] Sometimes the type of interference listed has been described only in a particular type of phagocyte (polymorph or macrophage) from a particular host, but it generally bears a relationship to pathogenicity in that host.
[c] Streptolysin (SLO) is a haemolysin which will lyse red cells, platelets and kill phagocytes *in vitro*. However its role *in vivo* is far from clear due to a lack of a good animal model for group A streptococcal infections. The situation with respect to another streptococcal haemolysin (SLS) is even less clear. See Ch. 8 for discussion of streptococcal toxins.
[d] SPI, *salmonella* pathogenicity island.

Inhibition of chemotaxis or the mobilisation of phagocytic cells

Various substances released from bacteria attract phagocytes, but their activity is generally weak. Other bacterial substances react with complement to generate powerful chemotactic factors such as C5a. Microorganisms can avoid the attentions of phagocytic cells by inhibiting chemotaxis, and as a result of this the host is less able to focus polymorphs and macrophages into the exact site of infection.

Some bacterial toxins (see above) inhibit the locomotion of polymorphs and macrophages. The streptococcal streptolysins which kill phagocytes can suppress polymorph chemotaxis in even lower concentrations, apparently without adverse effects on the polymorph. Random motility is not affected. *Clostridium perfringens* θ toxin has a similar

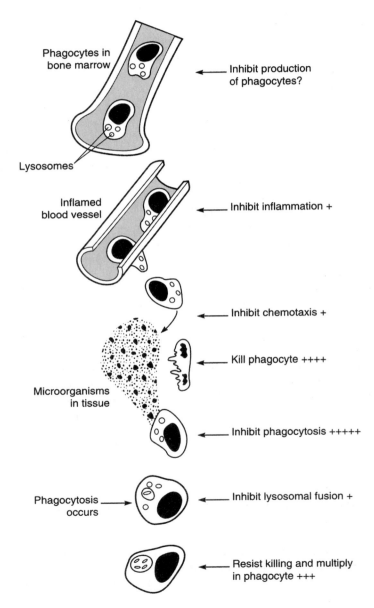

Fig. 4.6 Antiphagocytic strategies available to microorganisms. The extent to which strategies are actually used by microorganisms are indicated by pluses.

action on polymorphs. There are good methods available for the quantification of chemotaxis, and it is possible that other pathogenic bacteria will be shown to produce inhibitors.

Both polymorphs and macrophages arise from stem cells in the bone marrow, and their rate of formation is greatly increased during infection, so that blood leucocyte counts reach 2–4 times normal levels. This is associated with increased blood levels of certain factors that stimulate colony formation by leucocyte precursors. Four of these colony-stimulating factors, which are glycoproteins active at very low concentrations (10^{11}–10^{13} M), have been identified. They are produced in many tissues, and their concerted action is needed for the production and final differentiation of polymorphs (also eosinophils) and macrophages. They also help control the activity of differentiated cells. Little is known of the factors regulating their production, or of the influence of microbial products. Clearly, if it were possible for microorganisms to release substances that inhibited the formation or action of colony-stimulating factors, and thus seriously impair the phagocytic response to infection, some of them might be expected to do so. There is a decrease rather than an increase in blood polymorphs during certain infections such as typhoid and brucellosis, but there is so far no evidence that this is due to effects on colony-stimulating factors. *Listeria monocytogenes* is so-called because it causes an increase in circulating monocytes by means of a cell-wall component of 22 kDa, but the significance of this is not known.

However, there are two bacterial toxins which could affect the physical arrival of polymorphonuclear cells (PMNs) to an infection site. It has been suggested that the activity of CNF1 (an *E. coli* toxin) by virtue of its activating effect on Rho (a G protein essential in formation of epithelial tight junctions, see Fig. 4.1) could maintain the closure of tight junctions in gut epithelia, thereby decreasing the numbers of PMNs arriving at the focus of infection by blocking their traverse through the paracellular route. By contrast the effect of *C. difficile* toxins A and B would be the complete opposite: by inhibition of Rho (see Fig. 4.1), tight junctions would not be maintained thereby increasing the ease with which inflammatory cells arrive at the site of *C. difficile* infection, a highly characteristic feature of such infections.

Inhibition of adsorption of microorganism to surface of phagocytic cell

Many microorganisms tend to avoid phagocytosis without being obviously toxic for phagocytes. As a rule, it is not possible to distinguish between a failure to adsorb and a failure to ingest the microorganism. Since our understanding of adsorption is so slight, our understanding of failure to adsorb is equally inadequate. Yet the distinction can sometimes be made, as when pilated (virulent) gonococci attach to polymorphs but are not ingested or killed. *Mycoplasma hominis* remains extracellular when added to human polymorphs *in vitro*, and it appears that there is no firm adsorption of the mycoplasmas to the polymorph surface, although in the presence of antibody to the mycoplasmas there is adsorption, ingestion and digestion. The reason for the failure in adsorption is not clear, but it may be because the mycoplasmas damage the polymorph, which shows increased oxidation of glucose and defective killing of phagocytosed *E. coli*. If polymorphs are added to the protozoan parasite *Toxoplasma gondii* (see Glossary) *in vitro*, the mobile polymorphs are seen to turn aside from the toxoplasmas, indicating perhaps a failure of attachment. Antibody-coated or dead toxoplasmas, on the other hand, are successfully phagocytosed and digested by polymorphs. Macrophages, it may be noted, ingest the live parasite and support its growth (see below).

Many viruses will not attach to, and therefore cannot infect, a cell unless a specific receptor is present on the cell surface (see Ch. 2). When the virus cannot grow in the phagocyte it would be an advantage to avoid being taken up and destroyed, but so far it has not been possible to associate avoidance of phagocytosis with virus pathogenicity. When, however, a virus infects and grows in the phagocytic cell this may be an important part of the infectious process (see Ch. 5), especially if the phagocyte is so little affected that it carries the infecting virus from one part of the body to another.

Inhibition of phagocytosis – opsonins

Microbial products that kill phagocytes (see below) may at lower concentrations interfere with their locomotion or their phagocytic activity, for instance by inhibiting protein synthesis. A more direct challenge to the phagocyte is provided by the various microorganisms whose surface properties prevent their phagocytosis. As mentioned above, it is not usually possible to distinguish between inhibition of adsorption to the phagocytic cell and inhibition of phagocytosis which follows adsorption.

Many important pathogenic bacteria bear on their surface substances that inhibit phagocytosis (see Table 4.1). Clearly it is the bacterial surface that matters. The phagocyte physically encounters the surface of the microorganism, just as the person knocked down by a car encounters the hard metal exterior of the vehicle, and the phagocyte has no more immediate interest in the internal features or antigens of the microorganism than the person knocked down has of the upholstery or luggage inside the car. Resistance to phagocytosis is sometimes due to a component of the bacterial cell wall, and sometimes it is due to a capsule enclosing the bacterial wall, secreted by the bacterium. Classical examples of antiphagocytic substances on the bacterial surface include the M proteins (fibrous structures) of streptococci and the polysaccharide capsules of pneumococci. These bacteria owe their success to their ability to survive and grow extracellularly, avoiding uptake by phagocytic cells. Certain M proteins on the surface and the pili of streptococci are undoubtedly associated with resistance to phagocytosis and with virulence, but it is not clear how these act (see also p. 204). Streptococci appear to slither off the surface of polymorphs that are attempting to engulf them, suggesting the absence of specific adsorption factors. When the M protein is covered with antibody (opsonised), the Fc portion of the antibody molecule attaches to receptors on the polymorph surface and phagocytosis takes place. It must be remembered that, although attachment to the phagocyte is to be avoided, it is advantageous to attach to cells at the body surface (see pp. 12–19).

The polysaccharide capsule of the pneumococcus is likewise associated with resistance to phagocytosis and with virulence. It takes less than ten encapsulated virulent bacteria to kill a mouse after injection into the peritoneal cavity, but 10 000 bacteria are needed if the capsule is removed by hyaluronidase. As with pathogenic streptococci, phagocytosis takes place more readily via Fc and C3b receptors, when the bacterial surface has been coated with specific antibody and C3b deposited. Unencapsulated strains of bacteria are coated with C3b without the need for antibody, after activation of the alternative pathway (see p. 177), but this is inhibited by neuraminic acid components of the capsule. If a mouse is rendered incapable of forming antibodies to the capsule, infection with a single bacterium is then enough to cause death. It is not clear why the capsule confers resistance to phagocytosis; perhaps its slimy polysaccharide nature makes the phagocytic act difficult for purely mechanical reasons. Although antibody is needed for phagocytosis in a fluid medium, it is known that phagocytosis takes place without antibody on the solid surface lining of an alveolus or lymphatic vessel (or on a piece of filter paper!), where the physical act of phagocytosis is favoured, and the phagocyte can 'corner' and get round the bacterium. Pathogenic bacteria with similar polysaccharide capsules include *Haemophilus influenzae* and *Klebsiella pneumoniae*. Patients with agammaglobulinaemia have repeated infections with streptococci and these encapsulated bacteria. Their polymorphs fail to take up and destroy bacteria because opsonising antibodies cannot be produced. Polysaccharide capsules are not necessarily associated with virulence, since they occur in free-living nonparasitic bacteria. Presumably they have functions other than the antiphagocytic one, perhaps giving protection against phages and colicins (see Glossary).

Anthrax and plague bacteria also have capsules that are associated with virulence. Bacteria of the *Bacteroides* group are normally commensal, but can form abscesses, often together with other microorganisms, and they have polysaccharide capsules. Pathogenic strains of *E. coli* and *Salmonella typhi* have thin capsules consisting of acidic polysaccharide (K antigen), which in some way make phagocytosis difficult. Perhaps this is because (in the absence of antibody) the encapsulated strains do not activate complement via the alternative pathway, and are therefore poorly opsonised. Gram-negative bacteria also have cell walls containing a lipopolysaccharide complex (endotoxin), and the somatic (O) antigens occur in the polysaccharide side chains (Figs 4.4; 8.15). Bacteria with certain types of O antigen have a colonial form designated as smooth, and they show an associated virulence, with resistance to phagocytosis except in the presence of antibody. Rough colonial forms lack these particular antigens, which are determined by immunodominant sugars in the polysaccharide side chains, and are not virulent, showing no resistance to phagocytosis.

The parasitic trypanosomes causing African sleeping sickness circulate in the blood, from which they are transmitted to fresh hosts by biting tsetse flies. The bloodstream forms have a pronounced surface coat with an outer carbohydrate layer, which perhaps inhibits phagocytosis of the parasites by reticuloendothelial cells (see Ch. 5) and enables the parasitaemia to continue.

There is one other everyday example of a possible mechanism for inhibition of phagocytosis. Virulent strains of staphylococci produce a coagulase that forms fibrin strands when added to plasma in the presence of certain accessory plasma factors. The fibrin network may help form a wall round the staphylococcal infection site, but the advantage to the bacteria is not clear. Infiltration by polymorphs still occurs on a large scale and, at least for streptococci, virulence is associated with breakdown rather than formation of tissue barriers. Could it be that the staphylococcal coagulase precipitates fibrin in the immediate vicinity of the bacteria, thus impeding the final access of phagocytic cells, as well as depositing host fibrin on the bacterial cell wall, which therefore presents a less foreign surface to phagocytic cells? However, coagulase-negative strains, created by site-specific mutation, are just as virulent as their parents in several mouse models; coagulase might be important in other models, but this has not been tested.

Some microorganisms pose purely mechanical problems for the phagocytic cell without specifically preventing phagocytosis. There are difficulties with motile microorganisms, whether motility is due to flagella (Gram-negative bacteria, *Trichomonas vaginalis*) or to amoeboid movement (*Entamoeba histolytica*). Immobilising antibodies may be necessary. The sheer size of a microorganism can be a problem. A single macrophage will be unable to phagocytose a large microorganism, and macrophages attempting to phagocytose the advancing tip of fungal hyphae are just carried along by the hyphal growth. In such situations several macrophages must cooperate and if necessary form syncytial giant cells, as in the response to fibres and other large foreign objects.

As mentioned above, both polymorphs and macrophages have specific surface receptors for the Fc fragment of IgG and IgM antibodies and also for the C3b product of complement activation (see Ch. 6). This ensures that microorganisms coated with antibody or complement are opsonised. Cells other than polymorphs and macrophages lack these receptors, and here attachment and phagocytosis of particles coated with antibody is not promoted but even inhibited. Opsonised microbes are not only taken up but also killed more rapidly in the phagocyte. For instance, in the early stages of typhus, the rickettsiae multiply in macrophages after phagocytosis, but later, when antibodies have formed, the antibody-coated rickettsiae are rapidly phagocytosed and killed, and eventually digested.

Opsonisation without specific antibody takes place following deposition of C3b on the bacterial surface after activation of the alternative complement pathway (see p. 177) and attachment to the C3b receptor on the phagocyte. It is an important host defence early in infection, before antibodies are formed, and the following can be considered as microbial 'strategies' to prevent this type of opsonisation. Encapsulated strains of *Staphylococcus aureus* appear to activate and bind complement without the need for antibody, but are not opsonised and phagocytosed. It is thought that C3b is somehow hidden by the bacterial capsule and cannot attach to C3b receptors on phagocytes. In the case of Group A streptococci the outer covering of M protein prevents complement activation by the alternative pathway. Strains of *E. coli* with K1 capsular polysaccharide are pathogenic for newborn infants and show an associated resistance to opsonisation by the alternative complement pathway. Finally, gonococci become resistant to killing by normal serum (presumably involving alternative pathway activation) if they have a receptor for an IgG antibody present in normal serum that blocks the killing action.

There are a number of ingenious ways in which bacteria and other microorganisms avoid inactivation by host antibodies, or even avoid eliciting antibodies (see Ch. 7). One example will be given here, since it involves phagocytosis (Fig. 4.6). A substance called protein A is present in the cell wall of *Staphylococcus aureus*. Each molecule of protein A binds strongly to two molecules of IgG via the Fc portion, and there are about 80 000 binding sites on each bacterium. It is tempting to suppose that such a molecule is not there by accident, and that it interferes with the opsonisation and phagocytosis of staphylococci. Similar IgG-binding molecules are present on many streptococci. Experiments in which mice were infected subcutaneously or intraperitoneally with *S. aureus*, showed that protein A-deficient constructs were less virulent; constructs which were believed to express β toxin and overexpress protein A were more virulent than their parents in a mouse mastitis model. Viruses of the herpes group code for Fc receptors, induced on the surface of infected cells, and this is a further indication that antibody-binding molecules are useful to infecting microorganisms. The antibody molecules are not only bound in a useless 'upside-down' position to the microbe or the infected cell, but also, by their presence at this site, they interfere with the access of specific antimicrobial antibodies or cells.

Inhibition of fusion of lysosome with phagocytic vacuole

Clearly if the phagocytosed microorganism is not exposed to intracellular killing and digestive processes, it has the opportunity to survive and multiply. In the case of mycobacteria there is evidence that they enter macrophages via C3b receptors without inducing the respiratory burst (see above). *M. leprae* also has phenolic glycolipid-1 on its surface, which scavenges ROI, thereby protecting the pathogen. Those parts of the oxygen-dependent killing mechanisms that do not require myeloperoxidase (a lysosomal enzyme, see p. 92) will operate without fusion of lysosomes with the phagocytic vacuole, but if there were a way in which fusion could be prevented (and hence less expression of the full killing complement of the phagocyte), some microorganisms might be expected to have accomplished this. It occurs when virulent *Mycobacterium tuberculosis* is ingested by mouse macrophages. There is a failure of lysosomal fusion. Many phagocytic vacuoles remain free from lysosomal enzymes; inhibition of fusion is prevented by the secretion of ammonium chloride. Virulent *S. typhimurium* also inhibits fusion and divides within unfused vacuoles. This is in contrast to the events after uptake of nonvirulent *M. tuberculosis*, when lysosomal fusion is general, phagocytic vacuoles receive lysosomal contents, and bacilli are killed. The forces that move lysosomes towards vacuoles and then cause fusion are not known, so that little can be said about mechanisms except that a soluble inhibitor is presumably released from vacuoles by the virulent bacteria. This is not a general inhibition of fusion in the phagocyte, but a failure to fuse with the particular vacuoles containing the microorganism.

The intracellular protozoan parasite *Toxoplasma gondii* (see Glossary) is phagocytosed by macrophages, inducing its own engulfment by actively inserting a specialised 35 nm diameter cylinder into the macrophage.* But in a large proportion of the vacuoles there is no lysosomal fusion, and the toxoplasmas multiply, eventually killing the cell. Mitochondria and lengths of endoplasmic reticulum surround these vacuoles, presumably in response to chemical stimuli arising from the toxoplasmas, and perhaps playing a part in nourishment of the parasite. The adenylate cyclase toxin of *B. pertussis* which increases intracellular cAMP, inhibits phagosome–lysosome fusion and leads to an increase in growth in macrophages; toxin-defective bacteria show a hundred-fold fall in growth, compared to the parent strain. Other instances of nonfusion of lysosomes are known, such as when the fungus *Aspergillus flavus* enters the alveolar macrophages in susceptible (cortisone-treated) mice, when *Chlamydia psittaci* enters macro-

---

* *Toxoplasma gondii* can also invade a large variety of nonphagocytic cells. Little is known about attachment mechanisms or receptors, but the parasite secretes substances that help penetration, and the process is an active one, help being given by the host cell!

phages in culture, or when *Staphylococcus aureus* is phagocytosed by Kupffer cells in the perfused liver *in vitro*. Inhibition of fusion is an active process, and does not generally occur when microorganisms are killed or coated with antibody beforehand.

Escape from the phagosome

After capture in a phagosome, a microorganism can still evade anti-microbial forces by escaping at an early stage from the phagosome and entering the cytoplasm. There are now good examples of this phenom-enon. We have already met with *Shigella* which can escape from vacuoles and spread to adjacent cells. When *Listeria monocytogenes* infects mouse macrophages only a proportion of incoming bacteria escape into the cytoplasm. The bacteria are taken into phagosomes which are demonstrably acidified, conditions necessary for the activity of listeriolysin, a vital virulence determinant mediating escape from the vacuole. Within the first hour following phagocytosis most bacteria are killed in the phagosomal compartment due to the transfer of lyso-somal enzymes to about two-thirds of these vacuoles. By electron microscopy only 14% of the total number of organisms are found in the cytoplasm which includes those that had just escaped and those that had already started to multiply. Multiplication is rapid thereafter with a doubling time of 40 min, with clear evidence of actin-mediated spread to adjacent cells. Two phospholipases C (PLC-A and PLC-B) are also involved in this process. PLC-A negative mutants are able to invade but not replicate in mouse peritoneal macrophages. Phospholipase A seems to render the phagolysosomal membrane susceptible to the damaging effects of listeriolysin O, thereby allowing the organisms to escape. PLC-B is apparently required to accomplish escape from the double membrane in order to infect adjacent cells, in a manner similar to that described for *Shigella* (see Ch. 2). For viruses, escape involves fusion of the virus envelope with the phagosome membrane so that the nucleocapsid core is set free in the cytoplasm, and this is discussed on p. 114. There is evidence that the phenomenon also occurs with *Mycobacterium leprae, Rickettsia mooseri*, and the trypomastigote form of *Trypanosoma cruzi*. These can be seen free, often multiplying, in the cytoplasm of macrophages. Escape is generally prevented when the microorganism is coated with antibody.

Resistance to killing and digestion in the phagolysosome

Many successfully infectious microorganisms resist killing and diges-tion in the phagocytic vacuole. For those whose multiplication is for the most part extracellular, this ability to survive rather than suffer death and dissolution in the phagocyte may possibly add to their success in the infected host. Other microorganisms, however, are specialists in intracellular growth and some of them grow in phagocytes. Certain viruses depend for their success on infecting the phagocyte after avoiding killing and digestion in the phagolysosome; macrophages rather than polymorphs are important (see below). In the case of reoviruses, exposure to lysosomal enzymes actually initiates the 'uncoating' of the virus particle in the cell and thus helps virus multiplication. The cells susceptible to reoviruses, however, are not necessarily specialised phagocytes. Many other viruses have specialised mechanisms for entering susceptible nonphagocytic cells (see receptors, below); their fate in phagocytic cells is not necessarily important. Polio- and rhinoviruses, for instance, are taken up, killed and digested in phagocytic cells, but they nevertheless successfully infect target cells in the upper respiratory tract and alimentary canal and are shed profusely from these sites.

Bacteria, as a result of phagocytosis, enter phagocytic cells more commonly than any other type of host cell, and intracellular bacteria cannot establish a successful infection unless they resist killing and then grow in the phagocyte. Thus, macrophages are important sites of bacterial growth in infections with *Mycobacteria, Brucella, Listeria, Trypanosoma, Nocardia* and *Yersinia pestis*.* In some instances the microorganism escapes from the phagosome or inhibits lysosomal fusion (see above), but *Mycobacterium lepraemurium, Listeria monocytogenes, Y. pestis* and virulent strains of *Salmonella typhimurium* can grow in the phagosome in spite of lysosomal fusion. Polymorphs are less important sites of microbial growth, partly because of their short life span, but their powerful lysosomal enzymes take a heavy toll of ingested bacteria that show no particular resistance to killing and digestion.

Once a microorganism has been phagocytosed the most important thing is whether or not it is killed in the phagocyte. It should be remembered that, by definition, microorganisms are dead when they are incapable of multiplying. When nonvirulent *E. coli* is phagocytosed by polymorphs it is soon killed, but bacterial macromolecular machinery proceeds for a while after death of the bacterium. Most microorganisms are killed after phagocytosis, but the bacteria or protozoa that infect phagocytes must allow themselves to be taken up by these cells, and their success hinges in the first place on their resistance to killing (Table 4.1).

Many pathogenic bacteria show a degree of resistance to killing and sometimes also to digestion in the phagolysosome, as indicated in Table

---

* *Yersinia pestis* is the causative agent of the plague (from the Latin *plaga*, a blow), an often lethal infection transmitted to man by fleas from infected rats or other rodents, which can also spread from man to man by the respiratory route. In the fourteenth-century epidemics of the Black Death, it is estimated to have killed a third of the people of Europe. The bacteria are able to grow in the phagolysosome of macrophages when Ca^{2+} concentrations reach low levels (<100 mM), and produce several potent toxins.

4.1. Catalase, by destroying H_2O_2 might protect bacteria from killing, and catalase-rich strains of staphylococci and *Listeria monocytogenes* show better survival inside polymorphs. Superoxide dismutase, on the other hand, generates H_2O_2, but there is no correlation between production of this enzyme by bacteria and their survival in polymorphs.

In the case of *Salmonella*, the basis of resistance to intracellular killing is beginning to be unravelled. A number of transposon mutants were generated, showing reduced virulence in macrophages. Some of them were more susceptible to oxidation (implying that wild-type *Salmonella* have the genetic capacity to resist the initial respiratory burst) and others were less resistant than the wild type to the effects of defensins (already alluded to in this chapter). This latter property is due to the ability of the organism to sense the hostile environment of the phagosome (probably low pH) by the two-component regulator system phoP/phoQ (see Ch. 11). Another gene *pagC* (*phoP/Q* activated gene) has been identified as conferring intraphagocytic resistance by an as yet unknown mechanism. It has also been shown that mutations in the heat shock (stress) protein HtrA reduce the virulence of *S. typhimurium* in mice, almost certainly by reducing their ability to survive the oxidative stress imposed by macrophages. However, a complementary approach is the exploitation of 'proteomics'* which in this case involves comparison of the protein profiles obtained from organisms grown inside and outside macrophages. This shows that the process of successful adaptation to intracellular conditions is far from simple. It is now clear that infection of macrophage-like cell lines with *S. typhimurium* results in the altered expression of large numbers of proteins. The numbers vary between laboratories, probably reflecting the use of different bacterial strains and cell lines and the limits of present technology; some are upregulated and others are downregulated. Among those expressed are members of the heat shock protein family, better described as stress proteins and highly conserved in both prokaryotic and eukaryotic cells, which somehow enable stressed cells to survive. This phenomenon, production of stress proteins, is known to occur with other intracellular pathogens, including *Chlamydia* and mycobacteria. As far as *S. typhimurium* is concerned, a *Salmonella* homologue of EPEC BipA (see above) is also likely to be important.

Growth in the Phagocytic Cell

The ways in which microorganisms avoid being phagocytosed and killed have been discussed above. An equally satisfactory victory over

---

* Proteome: the sum total of all the proteins expressed by an organism under a defined set of conditions.

the phagocyte is achieved when the microorganism uses it as a site of growth. The microorganism now allows itself to be phagocytosed, but resists killing and digestion, and then multiplies, deriving nourishment from the phagocytic cell. As was pointed out in the preceding section, polymorphs have such a brief life span that they are rarely important sites for microbial growth. Virulent bacteria tend to remain viable if they are phagocytosed by polymorphs, but intracellular growth is generally slight compared with the growth of bacteria in extracellular fluids. Macrophages, by comparison, live for long periods. Many microorganisms have, as it were, come to accept eventual phagocytosis by macrophages as inevitable, and are able to multiply inside the cell (Table 4.2). They have learnt how to induce the macrophage to protect and feed them, rather than destroy and digest them. Sometimes mitochondria and ribosomes are recruited to the edge of the phagosome, where they perhaps play a part in bacterial nutrition and growth. This ability to grow in macrophages is often a key property of successful invasive microorganisms (see Ch. 5).

Rickettsias, bacteria, fungi and protozoa usually multiply inside phagocytic vacuoles. Nourishment of the parasite takes place across the membrane of the vacuole, and host materials must be made available to the parasite. Certain coccidias, for instance, induce the host cell to extrude material into the vacuole and then take it up by endocytosis (see Glossary). Macrophages parasitised by *Toxoplasma gondii* appear to be giving biochemical support to the invader in a most hospitable

Table 4.2.   Examples of microorganisms that regularly multiply in macrophages

Viruses	Herpes-type virus
	Hepatitis viruses of mice
	HIV
	Measles, distemper
	Poxviruses
	LCM (see Glossary)
	Lactate dehydrogenase-elevating virus of mice
	Aleutian disease of mink (see Glossary)
Rickettsias	*Rickettsia rickettsi*
	Rickettsia prowazeki
Bacteria	*Mycobacterium tuberculosis*
	Mycobacterium leprae
	Listeria monocytogenes
	Brucella spp.
	Legionella pneumophila
Fungi	*Cryptococcus neoformans*
Protozoa	Leishmanias
	Trypanosomes
	Toxoplasmas

fashion. Microvilli from the host cell extend into the vacuole which is surrounded by strips of endoplasmic reticulum and mitochondria. For most of the microorganisms that parasitise macrophages, including leprosy bacilli, tubercle bacilli, *Leishmania* and *Toxoplasma*, little or nothing is known about microbial nutrition inside the cell.

Certain viruses grow in macrophages and in a few instances, such as the highly successful lactate dehydrogenase-elevating virus of mice (see Glossary), the macrophage is the only cell in the body that is infected. Viruses do not generally infect by phagocytosis as this leads to their destruction by lysosomal enzymes, but by endocytosis (nonenveloped and most enveloped viruses), or by fusion with the plasma membrane (some enveloped viruses) as with other types of cell (see below).

Killing the Phagocyte

The most straightforward antiphagocytic approach is to kill the phagocyte, and many successful infectious bacteria do this. Some, as they multiply in tissues, release soluble materials that are lethal for phagocytes. Part of the success of pathogenic streptococci and staphylococci is attributable to their ability to kill the phagocytes that pour into foci of infection. Pathogenic streptococci release haemolysins (streptolysins) which lyse red blood cells and are much more active weight for weight than haemolysins such as bile salts or saponin, but which also have a more important toxic action on polymorphs and macrophages. Within 1–2 min of its addition to polymorphs, streptolysin O causes the polymorph granules to explode and their contents are discharged into the cell cytoplasm. The lysosomal enzymes, when confined to a phagocytic vacuole, help the cell by performing valuable digestive functions, but when enough are released into the cell cytoplasm in this way, they act on cell components and within a minute or two the cytoplasm liquefies and the cell dies. The streptolysin, by damaging the lysosomes, makes them function as 'suicide bags'. Streptolysin S has an even more potent action on membranes. Various haemolysins (α, β, γ) are released also by pathogenic staphylococci, and these too can kill phagocytes. Mutants deficient in either α or β or both are demonstrably less virulent in the mouse mastitis model. These toxins appear to inhibit macrophage chemotaxis and function, and kill the cells. In cattle double mutants were cleared completely from 50% of infected animals and caused much less severe disease. Staphylococcal leucocidin causes discharge of lysosomal granules just as with streptolysin O. *Listeria monocytogenes* secretes listeriolysin which acts like streptolysin. The lethal factor (LF) of the tripartite anthrax toxin is a zinc metallo-protease which is potently cytotoxic for macrophages (see Ch. 8 for a fuller discussion of some of these toxins).

Y. enterocolitica has long been regarded as a paradigm for studying intracellular pathogens. However, it is now abundantly clear that it is essentially an extracellular pathogen with an ability to survive in lymphoid tissue resisting phagocytosis by killing phagocytes. The high pathogenicity island of *Yersinia* was referred to in Ch. 3. Here we deal with the Yop virulon, a 70 kb plasmid encoding more than 50 genes comprising an amazingly sophisticated system for resisting the immune system of the host. *Yersinia* spp. synthesise a Type III secretion apparatus (which has many similarities to the flagella basal body through which external flagellar proteins can be secreted) spanning both inner and outer membranes and which is normally plugged. In addition these organisms have a pool of cytoplasmic molecules ready for secretion through the specialised apparatus. Upon contact with eukaryotic cell membranes, a sensor interacts with a receptor on the bacterial cell causing removal of the stop valve and addition of further components to the secretion apparatus which allows pore-forming fusion with the target cell membrane, thus creating an 'injectosome'. Through this injectosome, preformed effector molecules are introduced into the cell which inhibit phagocytosis and cytokine release and kill the cell.

In general, polymorphs are more readily killed by toxins than are macrophages, possibly because their lysosomes are more easily discharged. Invaders with good lysosomal weaponry of their own, such as virulent strains of the protozoan parasite *Entamoeba histolytica*, can kill polymorphs by mere contact (see Ch. 8). Others exert their toxic action on the phagocyte after phagocytosis has taken place, releasing cytotoxic substances which pass directly through the vacuole membrane and into the cell. The phagocyte can be said to have died of food poisoning. For instance, virulent *Shigellae* kill mouse macrophages after phagocytosis, whereas avirulent *Shigellae* fail to do so, and are themselves killed and digested. Certain *Chlamydia* multiply in macrophages after phagocytosis and destroy the cell by inducing the discharge of lysosomal contents into the cytoplasm. Virulent intracellular bacteria of the *Mycobacterium*, *Brucella* and *Listeria* groups owe much of their virulence to their ability to multiply in macrophages, although the macrophage is often destroyed in the end by mechanisms which with one or two exceptions (see below) are not known.

We conclude this section by describing the interaction of *L. pneumophila* with phagocytes as it brings together many of the aspects dealt with separately in the preceding sections. *L. pneumophila* is essentially a protozoan parasite with an ability to cause a respiratory infection (Legionnaire's disease) in humans; this serious lung infection can be reproduced in guinea pigs. *L. pneumophila* infects Types I and II alveolar epithelial cells and macrophages. It cannot grow in cell-free lung lavage from normal or infected guinea-pigs, is killed by polymorphs, but successfully infects and grows in macrophages.

In the environment, *Legionella* spp. are ubiquitous and parasites of protozoa. Bacterial transmission to humans occurs through droplets generated from environmental sources such as cooling towers and shower heads, but at present we do not know what constitutes the 'infectious particle'. Initial attachment to both protozoa and mammalian cells is mediated by pili and (for mammalian cells) adsorbed C3b. A surface exposed 24 kDa protein (Mip, macrophage infectivity potentiator; a peptidyl-prolyl *cis/trans* isomerase) was the first factor shown to confer an ability on the organism to infect and grow successfully in macrophages both *in vivo* and *in vitro*, but other genes have now been identified. These include *dot* (defect in organelle trafficking), *icm* (intracellular multiplication), *pmi* (protozoa and macrophage infectivity), all of which are required for survival in macrophages, a rather remarkable fact when one considers how evolutionarily distant these two cellular hosts are. To date, only *mil* (macrophage infectivity loci) has been recognised as specifically necessary for survival in macrophages. Immediately after initial entry, the *L. pneumophila* phagosome is surrounded by host cell membranes (see above) and is not routed into the endosome/lysosomal fusion pathway, whereas *dot* mutant *L. pneumophila* phagosomes are. Intracellular replication occurs for which availability of iron is important. Experiments with *L. pneumophila* infection of mouse peritoneal macrophages show that resident macrophages obtained by peritoneal lavage were nonpermissive for bacterial growth, but those obtained (elicited) following intraperitoneal injection of thioglycolate were permissive. The difference in ability to support the growth of the organisms was due to the upregulation of transferrin receptors in the elicited macrophages, which meant that higher intracellular concentrations of Fe were available for growth. The conclusion of this process seems to be dependent on the initial multiplicity of infection (MOI). Initially this would be expected to be low, which gives rise to apoptosis of infected cells and delayed release of a high number of organisms. Secondary infection can then occur with high MOIs at the foci of infection which induces rapid necrosis of host cells. The cytotoxin has not been identified but at least one candidate has been proposed, the zinc metalloprotease designated as tissue destructive protein, cytolysin or major secretory protein (Msp). Msp reproduces the major pathological features of experimental disease in guinea-pigs when introduced into the lung directly (see Ch. 8). It is also known to be produced *inside* infected macrophages and could well contribute to the complex interplay between pathogen and host cell, allowing the pathogen to gain the ascendancy.

Entry into the Host Cell other than by Phagocytosis

Although the usual way in which a particle enters a cell is by phagocytosis, so that the particle is enclosed in a phagocytic vacuole, there are other methods of entry. Electron microscope studies indicate that some bacteria, for instance, adsorb to the cell surface and enter the cytoplasm directly after inducing a local breakdown in the plasma membrane. The plasma membrane is reformed immediately. Shigellas and pathogenic salmonellas appear to enter intestinal epithelial cells in this way, and other bacteria show the same behaviour in tissue culture cells. It may be a less frequent occurrence in specialised phago-cytic cells. Protozoa have a complex structure and can utilise their own lysosomal enzymes to penetrate host cells. Trypanosomes, *Eimeria*,* *Toxoplasma gondii* and *Entamoeba histolytica* enter susceptible cells by active penetration, and the active end of the parasite has vesicles containing lysosomal enzymes that aid the penetration process. When a malaria parasite penetrates a red blood cell, a specialised projection (conoid end) on the malarial merozoite makes contact with the red cell surface. The parasite then injects a lipid-rich material from special glands (rhopteries), and it seems that this material is inserted into the red cell membrane, whose area is thus increased. As the merozoite actively enters the red cell, the membrane stays intact, but there is now enough of it to form an invagination and accommodate the advancing parasite.

If viruses enter cells by phagocytosis, they are destroyed by hydrolytic enzymes. Most nonenveloped and enveloped viruses enter cells by endocytosis, but a few types of enveloped virus enter instead by fusion at the plasma membrane at the cell surface. It commences with the virus attachment proteins binding to a critical number of cell receptor molecules. This triggers endocytosis – the invagination of the plasma membrane into small virus-sized depressions coated on the cytoplasmic side with a cellular protein (clathrin), giving them the name of 'coated pits'. These then detach from the plasma membrane and become vesicles free in the cytoplasm. At this stage viruses still have to release their genome and deliver it across the vesicle membrane into the cytoplasm. Non-enveloped viruses achieve this after destablisation and permeabilisation of the virus particle, which results from interaction with cell receptors alone, or in conjunction with the reduction of the internal pH of the vesicle (to about 5.5–6) by the importation of protons by a cellular pump. The genomes of endo-cytosed enveloped viruses are released into the cytoplasm after fusion of the virion and vesicle membranes. It involves the destabilisation of the lipids of both membranes so that they can fuse together to form a single continuous membrane. All viral envelope proteins have a buried hydrophobic region which at this time enters and disrupts the struc-ture of the membrane as a prelude to fusion. Exposure of this 'fusion peptide' is triggered by the low pH conditions referred to above. Fusion

* Various species of *Eimeria* (a protozoan parasite) cause contagious enteritis, or coccidiosis, in all domestic animals. Ingested oocysts invade intestinal epithelial cells and the entire life cycle with schizonts, merozoites and gametocytes takes place in these cells.

at the plasma membrane by viruses such as HIV and paramyxoviruses (like measles and Sendai viruses), takes place in exactly the same way except that it occurs at neutral pH.

Most of the discussion has dealt with the ways in which microorganisms can avoid intracellular digestion. There are one or two instances where exposure to lysosomal enzymes is actually necessary for the multiplication of microorganisms. Spores of *Clostridium botulinum* are said to germinate in cells only after the stimulus of exposure to lysosomal contents.

Consequences of Defects in the Phagocytic Cell

The importance of the phagocytic cell in defence against microorganisms is illustrated from observations on diseases where there are shortages or defects of phagocytic cells. A serious shortage of polymorphs, with less than 1000 mm^{-3} in the blood (normal 2000–5000 mm^{-3}), is seen in acute leukaemia or after X-irradiation, and predisposes to infection with Gram-negative and pyogenic Gram-positive bacteria. There are also one or two inherited shortages. Blood polymorph counts are about one-tenth of normal in Yemenite Jews, although surprisingly they seem little the worse for it except for a susceptibility to periodontal disease. But certain naturally occurring defects in the function of phagocytes have more serious consequences, and studies of these defects have thrown much light on normal phagocyte function. Unfortunately the defects are often multiple so that interpretation is not easy.

Children with chronic granulomatous disease, usually an X-linked recessive trait, have polymorphs that look normal and show normal chemotaxis and phagocytosis, but there is defective intracellular killing of bacteria. The gene that is abnormal has been cloned, and it appears to code for an essential component in the phagocyte's NADPH-oxidase system. There is no respiratory burst, and the superoxide radical and H_2O_2 are therefore not generated (see Fig. 4.3), and associated with this there is increased susceptibility, especially to staphylococcal and Gram-negative bacterial infections. In spite of undiminished immune responses to infection, patients suffer recurrent suppurative infections with bacteria of low-grade virulence such as *E. coli*, *Klebsiella* spp., staphylococci and micrococci, and usually die during childhood. Polymorphs are present in foci of infection but cannot kill the microorganisms, and are eventually taken up by macrophages, leading to the formation of a chronic inflammatory lesion called a granuloma (see Ch. 8). Interestingly, the patients have normal resistance to streptococcal infections because streptococci are catalase-negative and can themselves generate the H_2O_2 without destroying it. The H_2O_2, with the help of the cell's myeloperoxidase, then generates hypochlorite and singlet oxygen. As might be expected, patients with severe glucose 6-phosphate dehydrogenase (G6-PD) deficiency also suffer from infection with catalase-positive organisms, because they too fail to generate superoxide and H_2O_2. Patients with myeloperoxidase deficiency show delayed killing of bacteria in polymorphs but normal resistance to bacterial infections, and surprisingly an increased susceptibility to *Candida albicans*.

Another example of a polymorph defect is seen in Chediak–Higashi disease, which occurs in mice, mink, cattle, killer whales and man, and here too there is increased susceptibility to certain infections. Polymorphs contain anomalous giant granules (lysosomes) and the basic defect is probably in microfilaments. Phagocytosis and even lysosomal enzyme content appear normal, but there is defective chemotaxis, defective lysosomal fusion and delayed bacterial killing.

As well as shortages or defects in the quality of phagocytic cells, there may also be defects in the delivery system by which phagocytes are focused and assembled in the infectious foci where they are needed. Polymorphs show defective chemotaxis (as well as phagocytosis) in a variety of rare conditions such as the lazy leucocyte syndrome, due to a disorder of the cell membrane, and the actin dysfunction syndrome where actin is not polymerised as normally to form microfilaments so that the 'muscle' system of the cell is defective. Impaired digestion and disposal of microbial antigens is likely to be associated with immunopathology, but little is known of this aspect of phagocyte function.

Abnormalities in phagocyte function are not uncommon in certain acute infections such as Gram-negative bacteraemia and also in otherwise normal patients with recurrent staphylococcal infections. Presumably the abnormality could cause the infections, but at times it reflects the antimicrobial activities of the infectious agent.

Some of the clinical conditions are complex and of varied origin, often with multiple defects. In chronic mucocutaneous candidiasis, for instance, some patients have impaired cell-mediated immunity and others defective macrophage function. They suffer from persistent and at times severe infection of mucous membranes, nails and skin with the normally harmless yeast-like fungus *Candida albicans*.

Summary

In summary, the encounter between the microorganism and the phagocytic cell is a central feature of infection and pathogenicity. Phagocytes are designed to ingest, kill and digest invaders, and the course of the infection depends on the success with which this is carried out.

Virulent microorganisms have developed a great variety of devices for countering or avoiding the antimicrobial action of phagocytes. Although substances produced by or present on microbes may at first sight appear to have a useful function, not all will prove to be of practical importance in the infected host. Microbial killing and digestion in phagocytes is still only partly understood, but it is important to conceive logically of the ways in which microorganisms can avoid being ingested, killed and digested. Most viruses do not infect phagocytes, and the exceptions do so by endocytosis or fusion, not phagocytosis.

References

Abshire, K. Z. and Neidhardt, F. C. (1993). Analysis of proteins synthesised by *S. typhimurium* during growth within a host macrophage. *J. Bact.* **175**, 3734–3743.

Aderem, A. and Underhill, D. M. (1999). Mechanisms of phagocytosis in macrophages. *Annu. Rev. Immunol.* **17**, 593–623.

Armstrong, J. A. and Hart, P. D. (1971). Response of cultured macrophages to *Mycobacterium tuberculosis*, with observations on fusion of lysosomes with phagosomes. *J. Exp. Med.* **134**, 713–740.

Bannister, L. H., Mitchell, G. H., Butcher, G. A. and Dennis, E. D. (1986). Lamellar membranes associated with rhopteries in erythrocyte merozoites of *Plasmodium knowlesi*: a clue to the mechanism of invasion. *Parasitology* **92**, 291–303.

Beaman, L. and Beaman, B. L. (1984). The role of oxygen and its derivatives in microbial pathogenesis and host defence. *Annu. Rev. Microbiol.* **38**, 27–48.

Bogdan, C. and Rollinghoff, M. (1999). How do protozoan parasites survive inside macrophages? *Parasitol. Today* **15,** 730–732.

Caron, E. and Hall, A. (1998). Identification of two distinct mechanisms of phagocytosis controlled by different Rho GTPases. *Science*, **282**, 1717–1721.

Chen, L. M., Kaniga, K. and Galan, J. E. (1996). Salmonella spp are cytotoxic for cultured macrophages. *Molec. Microbiol.* **21**, 1101–1115.

Cirillo, D. M., Valdivia, R. H., Monack, D. M. and Falkow, S. (1998). Macrophage-dependent induction of the Salmonella pathogenicity island 2 type III secretion system and its role in intracellular survival. *Molec. Microbiol.* **30**, 175–188.

De Chastellier, C. and Berche, P. (1994). Fate of *Listeria monocytogenes* in murine macrophages: evidence for simultaneous killing and survival of intracellular bacteria. *Infect. Immun.* **62**, 543–553.

Eissenberg, L. G. and Wyrick, P. B. (1981). Inhibition of phagolysosome fusion is localized to *Chlamydia psittaci*-laden vacuoles. *Infect. Immun.* **32**, 889–898.

Farris, M., Grant, A., Richardson, T. B. and O'Connor, C. D. (1998). BipA: a tyrosine-phosphorylated GTPase that mediates interactions between enteropathogenic Escherichia coli (EPEC) and epithelial cells. *Molec. Microbiol.* **28**, 265–279.

Kaufmann, S. H. E. and Flesch, I. E. A. (1992). Life within phagocytic cells. *In* 'Molecular Biology of Bacterial Infection. Current Status and Future Perspectives' (C. E. Hormaeche, C. W. Penn and C. J. Smyth, eds), pp. 97–106. Society for General Microbiology Symposium 49, Cambridge University Press, Cambridge.

Kwaik, Y. A. (1998). Fatal attraction of mammalian cells to Legionella pneumophila. *Molec. Microbiol.* **30**, 689–695.

Marsh, M. and Pelchen-Matthews, A. (1993). Entry of animal viruses into cells. *Rev. Med. Virol.* **3**, 173–185.

Massol, P., Montcourrier, P., Guillemot, J. C. and Chavrier, P. (1998). Fc receptor-mediated phagocytosis requires CDC42 and Rac1. *EMBO J.* **21**, 6219–6229.

Rechnitzer, C., Williams, A., Wright, J. B., Dowsett, A. B., Milman, N. and Fitzgeorge, R. B. (1992). Demonstration of the intracellular production of tissue-destructive protease by *Legionella pneumophila* multiplying within guinea-pig and human alveolar macrophages. *J. Gen. Microbiol.* **138**, 1671–1677.

Small, P. L. C. *et al.* (1994). Remodelling schemes of intracellular bacteria. *Science* **263**, 637.

Van Epps, D. E. and Anderson, B. R. (1974). Streptolysin O inhibition of neutrophil chemotaxis and mobility: non-immune phenomenon with species specificity. *Infect. Immun.* **9**, 27–33.

White, J. M. (1990). Viral and cellular membrane fusion reactions. *Annu. Rev. Physiol.* **52**, 675–697.

Wilkinson, P. C. (1980). Leucocyte locomotion and chemotaxis: effects of bacteria and viruses. *Rev. Infect. Dis.* **2**, 293–319.

Wright, S. D. and Silverstein, S. C. (1983). Receptors for C3b and C3bi promote phagocytosis but not the release of toxic oxygen from human phagocytes. *J. Exp. Med.* **158**, 2016–2023.

5

The Spread of Microbes through the Body

When invading microorganisms have traversed one of the epithelial surfaces and arrive in subepithelial tissues, it is almost inevitable that they enter local lymphatics, as discussed in Ch. 3. They are then delivered to the filtration system and immune forces in the local lymph node. Sometimes this serves to disseminate rather than confine the infection, and spread from local to regional lymph nodes, and eventually to the blood, takes place. It is also possible, though not common, for the invading microorganism in subepithelial tissues to enter small blood vessels directly. This occurs when the microorganism damages the vessel wall, when viruses grow through the vessel wall, or when the initial act of infection, whether by injury or insect bite (see Ch. 2), introduces microorganisms directly into blood vessels.

Spread via lymphatics and the blood is rapid, enabling microorganisms to reach distant target organs or tissues within a few days. Thus poliovirus and *Salmonella typhi* infect the intestine and enter the bloodstream within a few days, silently reaching the central nervous system (CNS) (polio) or causing disease by multiplication in reticuloendothelial cells and liberation into the blood (typhoid).

Direct Spread

Microorganisms that fail to spread via the circulating body fluids in this way are confined to the site of initial entry into the body, except in so far as there is local extension of the infection into neighbouring tissues. The extent of the local spread depends on the outcome of the encounter between microbe and host defences mentioned in Chs 3 and 4, but some spread is common. Certain bacteria are capable of dramatic spread through tissues causing rapidly fatal disease as, for example, necrotising fasciitis caused by some strains of group A streptococci.

If microbes infect epithelial surfaces that are covered with fluid, their progeny are liberated onto these surfaces and can establish fresh sites of infection elsewhere on the epithelium (see Ch. 2). This form of local spread takes place with respiratory infection, when ciliary action moves microorganisms from the initial site of infection in the direction of the throat, giving fresh opportunities to establish infection *en route*. If cilia are not functioning properly or if there is too much fluid for them to transport, there is a flow by gravity to other parts of the respiratory tract. During upper respiratory tract infections excess nasal secretions tend to drain or are sniffed backwards and downwards, and during lower respiratory infections gravity inevitably plays some part in the movement of excess secretion, especially during sleep (see p. 389). Coughing and sneezing redistribute the infectious agent over uninfected epithelial areas. Sometimes larger amounts of fluid and mucus are coughed up and reinhaled before reaching the back of the throat. Anaesthesia or alcoholic stupor favour the drainage or aspiration of secretions from the throat to the lower respiratory tract.

The same type of local spread takes place easily in the intestine because of the continuous flow of intestinal contents; salmonellas and shigellas that have established initial foci of infection are thereby carried into more distal parts of the intestine. Local spread in the urethra, conjunctiva or vagina is likewise facilitated by movement of microorganisms in the fluid covering the epithelial surface.

The skin, on the other hand, with its dry surface, is less suitable for this type of local spread. It is not an easily infected surface (see Ch. 2) and most mammals are in any case covered with a generous layer of fur which itself impedes the local spread of microorganisms. There are only a few animals that are more or less naked, and these include the pig, elephant, man and rhinoceros. It is in man that there are best opportunities for spread from one part of the skin to another, because microorganisms are readily transferred by the scratching and rubbing activities of the fingers. The constant attention of fingers to the face promotes the local spread of various bacteria, and can help to distribute over the face the staphylococci or streptococci responsible for impetigo. Warts are spread in the skin by scratching or by the rubbing together of naturally opposing areas of skin ('kissing' warts). The skin has a moister covering of sweat and secretions in backwaters such as the armpits or umbilicus, or in the crevices between toes or body folds. Fungi, for instance, spread locally in these areas.

Direct spread also occurs to organs and tissues below the body surfaces, extending the focus of infection and sometimes accounting for striking complications. When infection in the lung spreads towards the surface of this organ it causes inflammation or actual infection of the pleura (pleurisy). Similarly, infection in the appendix or elsewhere in the intestines causes inflammation or infection of the peritoneum (peritonitis) if it spreads towards the serosal surfaces. The direct spread of infection from the middle ear to neighbouring structures gives rise to meningitis, cerebral abscess, etc.

Microbial Factors Promoting Spread

Direct spread of microorganisms through normal subepithelial tissues is not easy, being limited physically by the gel-like nature of the connective tissue matrix. Certain bacteria produce soluble substances with an effect on the physical properties of this connective tissue matrix, and at first sight these substances would appear to promote the spread of infection. Invasive streptococci liberate hyaluronidase, an enzyme that liquefies the hyaluronic acid component of the connective tissue matrix. Streptococcal skin infections, accordingly, often spread rapidly through the dermis causing the condition erysipelas. Hyaluronidase was originally referred to as the 'spreading factor', but it is certainly not a necessary factor, because bacteria such as *Brucella* are highly invasive in the absence of hyaluronidase. Moreover, *Staphylococcus aureus* produces plenty of hyaluronidase without being particularly invasive. Indeed, there is evidence suggesting that unless a microorganism is highly virulent it gains advantages if after entry into the body it first multiplies locally before spreading. A bacterial inoculum that causes a local lesion after intradermal injection often fails to produce a lesion if hyaluronidase is injected with the bacteria. Early local spread perhaps favours the host by diluting out infecting microorganisms and exposing them more effectively to host defences. As if to warn us against naïve conclusions, staphylococci produce a coagulase that lays down a fibrin network in tissues (see Ch. 4), but at the same time form a fibrinolysin. Streptococci also produce a fibrinolysin (streptokinase) in addition to hyaluronidase. Doubtless these enzymes have functions other than those connected with local spread in the purely mechanical sense. Bacteria in general produce a great variety of enzymes, including proteinases, collagenases, lipases and nucleases, but only very few of them have been clearly shown to be of any pathogenic significance. Many of these enzymes presumably have functions related to bacterial nutrition or metabolism rather than a relation to some theoretically important role in the infectious process.

Spread via Lymphatics

Proteins and particles in tissue fluids enter lymphatic capillaries rather than blood capillaries, and are transported to the nearest lymph node. Virus particles or bacteria injected into the skin, for instance, reach local lymph nodes within a few minutes. A rich lymphatic network lies below the epithelium in the nasopharynx, mouth and lung; microorganisms traversing these epithelia enter cervical and pulmonary lymph nodes. The mesenteric nodes receive microorganisms invading from the intestine, and strategically placed nodes occur along the lymphatics draining the urinogenital tract. Lymph nodes receive and monitor the lymph draining most parts of the body. If microorganisms reach the peritoneal cavity, for instance, they are exposed to phagocytosis by large numbers of resident peritoneal macrophages, and also enter subdiaphragmatic lymphatics to reach the retrosternal lymph nodes. A few microorganisms, such as the leprosy bacillus and certain viruses, grow in the endothelium of lymphatic vessels and thus increase their numbers by the time they reach the node in the lymph. The total flow of lymph in the body is considerable, and in the normal man 1–3 litres of lymph enter the blood each day from the thoracic duct. Under certain circumstances the flow rate from an organ is very greatly increased, such as from the pregnant uterus, the lactating mammary gland, or the intestine after a large fatty meal.

Lymph nodes have an important filtering action on account of the phagocytic cells that line the lymph node sinuses, and microorganisms are filtered off rather than allowed to spread to other lymph nodes, the thoracic duct, and the blood (see Fig. 3.3). Inflammatory substances reach the lymph node at an early stage in most bacterial and fungal infections, or when tissues have been damaged in virus infections. As with inflammation elsewhere, blood and lymphatic vessels are dilated and leucocytes are extravasated, so that the node becomes swollen and tender (see Ch. 3). In the normal node the phagocytes are the resident macrophages lining the sinusoids, but in the inflamed node bloodborne polymorphs are also present. The filtering function of lymph nodes is impaired under the following circumstances:

1. When the lymph flow rate is high, as during inflammation of tissues or exercise of muscles. It has long been a practice to manage infected wounds by immobilisation of the affected part of the body, thus reducing lymph flow and increasing filtration efficiency in the draining lymph nodes.
2. When the concentration of particles is high. Early in infection filtration is efficient, but it may be less so at a later stage when increasing numbers of microorganisms arrive at the node following local multiplication at the site of infection. On the other hand, antibodies to the microorganisms reach inflamed infected tissues and lymphatics at a later stage, promoting uptake of microorganisms by phagocytic cells in the node.
3. When phagocytic cells in the node fail to ingest microorganisms. There is discussion in Ch. 4 of the microbial and host factors that impair phagocytosis of microorganisms and thus depress the filtering function of the lymph nodes.

If the lymph node filters out the invading microorganisms but rather than inactivating them supports their multiplication, then the multiplying microorganisms are discharged in the efferent lymph. This occurs with certain bacteria such as *Yersinia pestis, Brucella* and *Rickettsia typhi*. Tubercle bacilli always enter lymphatics at the site of primary infection, and are usually arrested in the local lymph node. Occasionally, especially in children, there is further spread to regional lymph nodes, thoracic lymph duct and blood to give disseminated (miliary) tuberculosis with foci of infection in many organs. Sometimes (e.g. herpes viruses, adenoviruses, measles) the infectious agent multiplies in cells in the node without seriously damaging them, and the infected cells can then migrate through the body and disseminate the infection (see below).

After arrival of the first microorganisms in the lymph node, the local immune response is set in motion. Microorganisms are phagocytosed by macrophages lining the lymph sinuses and antigenic products are subsequently presented to adjacent lymphoid cells. Antibody and cell-mediated immune (CMI) responses are thus built up. Lymphocytes continually circulate through the nodes (see Fig. 6.5) and these are recruited, cell division takes place, and the node enlarges. Within a day or two, immunologically stimulated cells emerge in the efferent lymph to spread the response to distant parts of the body. If the efferent lymph from the stimulated node is cannulated and the emerging cells collected, the immune response can be confined to this node.

Changes in the lymph node reflect the inflammatory and immunological phenomena. At one extreme is the picture seen following infection with an invasive bacterium such as a β-haemolytic streptococcus. Bacteria and inflammatory products of bacterial growth arrive in the local node, which rapidly becomes swollen and tender as blood vessels dilate, and it is distended with inflammatory cells and exudate. The node becomes the site of a battle between host and parasite, and the determinants of microbial virulence apply in the node as in any other tissue. Virulent bacteria tend to kill phagocytes, resist uptake and inactivation by phagocytes, or multiply in phagocytes (see Ch. 4). While the lymph continues to flow, the invasive streptococci (or the typhus rickettsiae, plague bacilli) have the opportunity to exit via the efferent lymphatics to reach the next node and eventually the bloodstream. If the microorganisms are arriving in large numbers from a peripheral site of multiplication, the filtering efficiency of the node falls off and some of these can pass straight through the node. As inflammation and tissue damage in the node become more severe, the flow of lymph ceases, fibrin is formed and the infection is thus localised. Sometimes the swollen node becomes a mere pocket of pus (invasive staphylococci, bubonic plague), a battlefield containing dead and living microorganisms, host cells and inflammatory exudate.

In less severe infections, the changes in the node are less marked, and when entirely avirulent or nonmultiplying microorganisms reach the node, the nodal swelling is barely detectable and is attributable to an uncomplicated immune response. There are many viruses (measles virus, adenoviruses, polioviruses, HIV) that have the ability to repli-cate rather than be destroyed after being filtered out in the lymph node. The infected cells are generally macrophages or lymphocytes, and progeny virus is liberated into the lymph. If the infected cells are not seriously or acutely damaged, they can carry virus to distant parts of the body in the course of their normal migratory movements. This last method of spread is of supreme importance for viruses. As a rule, bacteria or protozoa must be rather virulent and pathogenic if they are to spread easily through the body from a peripheral site of growth, but many relatively avirulent viruses such as varicella-zoster, rubella and mumps accomplish this with consummate ease by way of the lymphatic system. The lymphatic system, faced with a staphylococcus in the dermis, removes it and delivers it to the antibacterial forces assembled in the local lymph node. If, however, the infecting microorganism grows silently in the lymphoid cells or macrophages of the node without setting off the usual alarm system (inflammation), and if immune responses have not yet been initiated the lymph node fails in its func-tion and merely hands on the microorganisms to the efferent lymph, the bloodstream, and thus to other susceptible tissues in the body.

In summary, most bacteria, fungi, protozoa, viruses, etc. are filtered out and inactivated in lymph nodes. Those that are not dealt with in this way are able to spread through the lymphatic system and into the blood. Certain viruses and bacteria grow in cells in lymph nodes, the nodes serving as sources for the dissemination of infected cells through the body.

Spread via the Blood

Blood–tissue junctions

The blood is the most effective vehicle of all for the spread of microbes through the body. After entering the blood they can be transported within a minute or two to a vascular bed in any part of the body. In small vessels such as capillaries and sinusoids where blood flows slowly there is an opportunity for the microorganism to be arrested and to establish infection in neighbouring tissues.

In a systemic virus infection* the epithelial surface of the body is traversed and virus reaches the blood at an early stage, either via the

---

* Cells infected with viruses can be located and identified with great precision by the fluorescent antibody technique. Antibody conjugated to fluorescein is used to stain viral antigen in tissue sections, which are then examined by fluorescence microscopy. Antigen can also be identified by using antibody coupled to peroxidase, the enzyme then being made visible by adding substrate to give a coloured product. Examples of its use in virus infections are shown in Fig. 5.1.

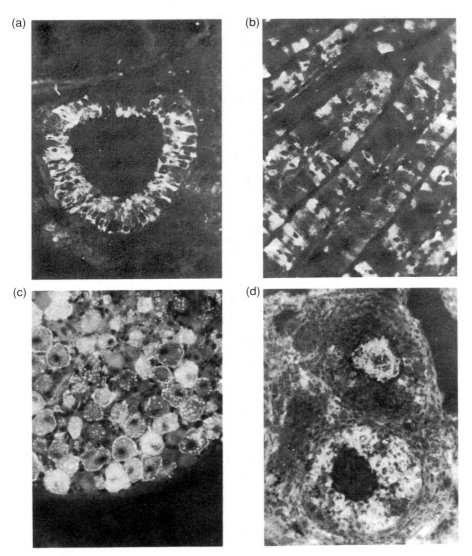

Fig. 5.1 Fluorescent antibody staining of tissue sections from animals infected with various viruses. (a) Small bronchiole from adult mouse lung 24 h after intranasal infection with Sendai (parainfluenza I) virus, showing infection of nearly all epithelial cells. (b) Intestinal epithelium of LCM virus carrier mouse (congenitally infected), showing several villi with many infected epithelial cells. (c) Dorsal root ganglion of a hamster infected with Lagos bat (rabies-like) virus, showing large numbers of infected neurons. (d) Ovary from LCM virus carrier (congenitally infected) mouse (see Glossary). There are two follicles, one with the ovum infected (above), the other with heavy infection of granulosa cells and an uninfected ovum. ((c) is illustrated here by kind permission of Dr F. A. Murphy.)

lymphatics and lymph nodes as discussed above, or after entering a subepithelial blood vessel. It is then spread through the body via the bloodstream, usually without any signs or symptoms. The amount of virus in the blood at any given time may be negligible. This is called a primary viraemia and is a common, silent event often only known to have taken place because of invasion of a distant target organ such as the brain, liver or muscle. After growth of virus in the primary target organs there is sometimes a reseeding of virus into the blood again, to give a secondary viraemia and infection of a fresh set of tissues. The secondary viraemia is of larger magnitude and often easily detectable in blood samples. Thus in measles the infecting virus undergoes minimal growth near the site of infection in the respiratory tract, and then enters the blood via lymphatics and lymph nodes (see Figs 5.2 and 5.3). The viraemia is not detectable, but as a result of it certain organs such as the spleen or liver are infected. After further viral growth in these visceral organs there is a secondary viraemia which seeds virus to the epithelial surfaces of the body, where the growth of virus causes a rash and an enanthem. A similar sequence of events takes place in other infections such as yaws, rubella, mumps, with microbial shedding from skin or in respiratory secretions, saliva, urine, etc. (Fig. 5.2). The difference between primary and secondary viraemia or bacteraemia is often not clear, but it is a useful distinction. Clearly, the stepwise spread of infection through the different compartments of the body takes time. Hence the incubation period before disease occurs is longer in these systemic infections than when multiplication and shedding is all accomplished locally at a body surface (see also pp. 68 and 164).

Circulating microorganisms generally localise in organs such as the liver and spleen. This is because they are phagocytosed by the macrophages of the reticuloendothelial system lining the sinusoidal blood vessels (see below). Infected leucocytes also tend to be arrested here if they show signs of damage, or surface changes. However, certain viruses (arthropod-borne viruses for instance) and rickettsias localise in capillary endothelial cells elsewhere in the body (see below).

Bacteria that do not generally cause a systemic infection can enter the blood in very small quantities as an accidental phenomenon, and this is important when it enables them to establish infection elsewhere in the body. Transient bacteraemias are probably not uncommon and under normal circumstances they are of no consequence, circulating bacteria being rapidly removed and inactivated by reticuloendothelial cells. But when host resistance is seriously impaired or when vulnerable tissues are exposed to the circulating bacteria, transient bacteraemias are a more serious matter. Certain bacteria, for instance, particularly those associated with the teeth (e.g. the *Streptococcus viridans* group) readily enter the blood, especially during dental extractions and even during tooth brushing or biting onto hard objects. If the heart valves are abnormal, the circulating bacteria settle on them to cause the disease infective endocarditis. Artificial joints are occasion-

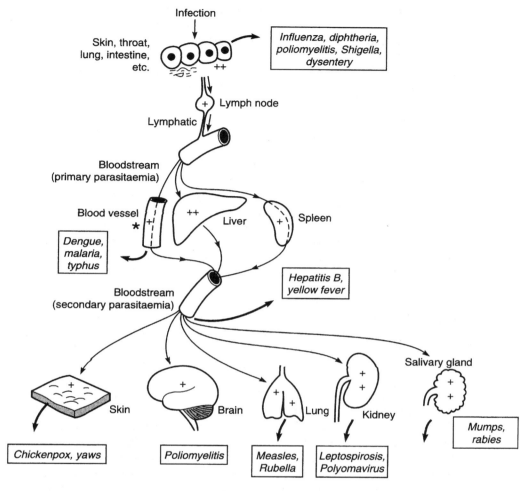

Fig. 5.2 The spread of infection through the body. + = sites of possible multiplication; large arrows = sites of possible shedding to the exterior; * indicates multiplication in bloodstream or vascular endothelium rather than in viscera.

ally infected in the same way with oral bacteria such as *Streptococcus sanguis*. Trauma to the growing ends of bones promotes the localisation of bacteria during transient bacteraemias and thus predisposes to osteomyelitis. Staphylococcal osteomyelitis therefore commonly affects the metaphyses of the long bones of the leg in children. Larger numbers of bacteria sometimes enter the blood during severe infections with bacteria such as pneumococci, meningococci or *Streptococcus pyogenes*, giving rise to the condition called septicaemia. The lung is a common source of bacteria, as in pneumococcal pneumonia, and in the old days of postpartum sepsis, streptococcal infection of the uterus

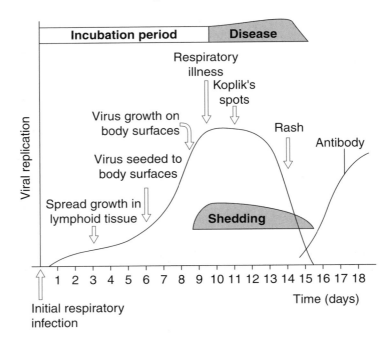

Fig. 5.3 The pathogenesis of measles. Virus in blood invades different body surfaces via blood vessels, and reaches surface epithelium, first in the respiratory tract where there are only 1–2 layers of epithelial cells, then in mucosae (Koplik's spots), and finally in the skin (rash). (Reproduced with permission from Mims *et al.* (1998). 'Medical Microbiology', 2nd edn, Mosby, London.)

often led to septicaemia. Finally, there are a number of specialised bacteria such as *Bacillus anthracis* and *Salmonella typhi* that regularly establish generalised infections. After entering the blood, often in large numbers, they establish focal infection in organs and cause serious disease (typhoid, anthrax). But there are few bacteria or fungi that regularly invade the blood, in contrast to the large numbers of viruses and rickettsias which nearly always do so. The very presence of considerable numbers of microorganisms in the blood causes disease if toxic materials are released during their metabolism and growth. Bacteria and fungi tend to release toxins when multiplying extracellularly, whereas viruses are more commonly being carried passively in the blood, and even when they multiply in cells, toxins are not released. Fungal or bacterial invasion of the blood is thus commonly associated with severe disease, whereas viraemia (e.g. hepatitis B carriers or HIV) is often silent. Cell-associated bacteraemia is sometimes silent, even when it continues for long periods. In lepromatous leprosy (see Ch. 9) there is a continuous bacteraemia, nearly all the circulating leprosy bacilli being inside blood monocytes. The bacteraemia persists for some months, but the bacilli remain inside cells, multiplying only very slowly (see Table 8.2), and there are no general or toxic signs.

The course of events after entry of microorganisms into the blood depends to some extent on the site in the body at which entry occurs. After entry into subepithelial blood vessels in the intestine the first slowing down of blood flow is in the sinusoids of the liver, and here microorganisms are exposed to macrophages (Kupffer cells) lining the sinusoids. Many bacterial products and at times intact bacteria enter the portal circulation and are removed by Kupffer cells in the liver, which can thus detoxify or disinfect portal blood before delivering it to the rest of the body. Microorganisms entering small blood vessels in the lung are carried straight to capillary beds in the systemic circulation, and those entering vessels in the systemic circulation are carried first to the pulmonary capillary bed and thence anywhere in the body. Clearly the lung is an important possible site for the localisation of circulating microorganisms, as well as an important possible source of microorganisms. Opportunities for localisation in other organs will depend to some extent on the blood flow, the kidney for instance, which receives about one-third of the cardiac output, having very good opportunities. Much more than this, localisation depends on the form in which the microorganism is carried in the blood, the activity of the reticuloendothelial system, and on the nature of the vascular bed in an organ. These will now be discussed in some detail.

Form in which microorganism is carried in the blood

Microorganisms may be carried free in the plasma, in the formed elements of the blood, or in both compartments (see Table 5.1).

Table 5.1.   Carriage of microorganisms in different compartments of blood[a]

| | Free in plasma | Leucocyte associated | | Erythrocyte associated | Platelet associated |
		Mononuclear cells	Polymorphs		
Viruses	Poliovirus Yellow fever Hepatitis B	Measles Epstein–Barr virus Herpes simplex Cytomegalovirus HIV		Colorado tick fever virus Human parvovirus B19[b]	Murine leukaemia virus LCM virus (see Glossary)
Rickettsias	All types				
Bacteria	Pneumococci *Leptospira* *B. anthracis* *Borrelia recurrentis*	*Mycobacterium leprae* *Listeria* *Brucella*	Pyogenic bacteria	*Bartonella bacilliformis*	
Protozoa	Trypanosomes	*Leishmania* *Toxoplasma gondii*		Malaria Babesia	

[a] Carriage in more than one compartment is possible, e.g. LCM virus is present in platelets, leucocytes and plasma of infected mice, and HIV in plasma and CD4+ cells.
[b] This virus binds to globoside, an erythrocyte protein, and this accounts for its tropism for the bone marrow.

Free in the plasma

Those carried free in the plasma include viruses such as poliomyelitis and yellow fever, bacteria such as *Bacillus anthracis* and the pneumococcus, and protozoa such as the trypanosomes. Their localisation in organs depends on their ability to adhere to or grow in vascular endothelial cells, and on phagocytosis by reticuloendothelial cells. They must also be resistant to any antimicrobial factors present in the plasma.

White cell associated

Certain microorganisms are carried either in or on white cells. The most important cells are lymphocytes and monocytes. These can be infected with viruses such as herpes simplex, HIV, cytomegalovirus, Epstein–Barr virus and measles, and monocytes are also infected with intracellular bacteria such as *Listeria*, tubercle bacilli and *Brucella*. Circulating monocytes are regularly infected by the protozoan parasite *Leishmania donovani* in the condition kala-azar, transmitted by blood-sucking sandflies. If the infected circulating cells remain healthy, they protect the microorganism from phagocytosis by reticuloendothelial cells and from antimicrobial factors in the plasma. They can also carry microorganisms with them on their migrations through tissues as they move in and out of the vascular system (see Fig. 6.5). Thus a mononuclear cell infected with measles virus in the subepithelial tissues of the respiratory tract can travel into the blood and then localise in the spleen and initiate infection in a splenic follicle (Fig. 5.3). Monocytes have also been implicated as the 'Trojan horse' for delivering microbes such as *Streptococcus suis* and HIV to the CNS. Microglial cells are of monocytic origin, and it is suggested that monocytes may carry infection into the CNS during the course of their normal movement, as part of the regular renewal and turnover of microglia.

Red cell associated

Some microorganisms travel in or on red blood cells. This gives them no opportunity to leave the vascular system, but those that are inside red blood cells are protected from phagocytosis by the reticuloendothelial system as long as the host cell remains normal. Viruses do not infect circulating erythrocytes, which are enucleated, metabolically impoverished cells, unsuitable for virus replication. Colorado tick fever virus (Table 5.1) is present inside circulating erythrocytes, but probably as a result of having infected precursor cells in the bone marrow, as does human parvovirus B19. Some viruses adsorb to the red cell surface, a phenomenon demonstrable *in vitro* and forming the basis for haemagglutination tests. Rubella virus haemagglutinates but is not significantly adsorbed to host erythrocytes in the infected individual. The most striking haemagglutinating viruses are those such as parainfluenza and influenza, infecting the respiratory tract. The haemagglutinating factor (haemagglutinin) is a protein subunit on the surface of the virus particle and is the mechanism by which virus adsorbs to specific receptors on susceptible respiratory epithelial cells. Red blood cells happen to have the same receptors, and therefore the viruses cause haemagglutination. Transient viraemias rarely occur in parainfluenza and influenza infections, and would presumably be partly red cell associated. Carriage on red blood cells is a feature of certain arthropod-borne virus infections. Rickettsia are often carried on red blood cells as well in the plasma.

The only bacterium ever found associated with human red cells is *Bartonella bacilliformis*. This occurs in Peru and causes oroya fever, a disease with acute haemolytic anaemia, transmitted by sandflies. Red blood cells are of particular importance in malaria. The malaria parasite lives in the red blood cell (see Table 2.1), breaking down and utilising the haemoglobin as it grows and divides, and up to 30 progeny parasites called merozoites are liberated from each red cell into the blood. Within a very short time the liberated merozoites have entered and infected another set of red blood cells by a process which seems similar to phagocytosis (see p. 114). During their brief extracellular period in the blood, merozoites are exposed to host antimicrobial forces, but the parasites in red blood cells might seem protected from the phagocytic reticuloendothelial cells, circulating antibodies and sensitised lymphocytes (see Ch. 6) that might otherwise inactivate or eliminate them. However, as the parasites mature certain plasmodial antigens are expressed on red cells. The red cells maintain the parasite in the blood and produce the sexual forms that must be ingested by the transmitting species of mosquitoes. The parasitised red cells undergo various changes before they are lysed. During passage through capillaries they are not deformed as easily as normal red cells and they also adhere more readily to vascular endothelium. As a result of this, and particularly in one type of malaria (*Plasmodium falciparum*), there is a piling up ('sludging') of red cells in the small vessels of various organs, leading to local anoxia, release of mediators such as tumour necrosis factor (TNF), and tissue damage.* In the brain this gives rise to the serious complication of cerebral malaria. Immunological events are of great importance in malaria (see Ch. 8) and the host–parasite relationship is complex. Each parasitic form is highly differentiated with many antigenic constituents. Parasitised red cells are lysed immunologically, and at times there is large-scale destruction of red

---

* In *P. falciparum* infection the parasitised red cells develop knobs, which bind the red cell to the wall of small blood vessels. In this site, where the parasite finally matures and ruptures, there are good opportunities for local tissue damage. Virulence of this type of malaria is also partly due to the fact that it can infect red cells of all ages (the life span of human red cells in blood is about 90 days), whereas the other types are restricted to young red cells.

cells, including normal ones, the suddenly released haemoglobin spilling over into the urine to give 'blackwater fever'.

Platelet associated

Platelets are phagocytic and, given the chance, are capable of ingesting microorganisms. They contain microbicidal granules and can generate free radicals, but their importance in infectious diseases is not clear. Viruses such as leukaemia viruses and LCM virus in carrier mice infect megakaryocytes, and therefore the circulating platelets derived from these cells are infected. The infected platelets, however, do not appear to be damaged. Microbial transport by platelets is likewise unimportant. Infected platelets could accumulate at a site of blood vessel injury, but they normally remain intravascular.

Reticuloendothelial system (RES)

Macrophages are present in all major compartments of the body, and those lining the sinusoids in the liver, spleen, bone marrow and adrenals monitor the blood, removing foreign particles, microorganisms or effete host cells. These macrophages constitute the reticuloendothelial system, and the liver macrophages (Kupffer cells) are quantitatively the most important component of the system.* Many studies have been made of the function of reticuloendothelial macrophages, as determined by the removal from the blood of intravenously injected dyes, marker particles such as carbon, viruses and bacteria. Microorganisms entering the circulation, as long as they are free in the plasma, are exposed to phagocytosis by these macrophages. Phagocytosis is influenced by a variety of factors. Larger particles are phagocytosed ('cleared' from the blood) more rapidly than small particles, and many larger viruses and bacteria are cleared completely after a single passage through the liver, so that more than 90% of intravenously injected particles disappear from the blood within a few minutes (see Fig. 5.4).†

---

* Macrophages lining the sinusoids in the spleen account for about one-tenth of the total clearance capacity of the RES. After a circulating microorganism or antigen has been phagocytosed by a splenic macrophage, antigenic materials are presented to neighbouring lymphocytes by antigen-presenting cells and an immune response is initiated. Phagocytosis by a liver macrophage, on the other hand, does not have this immunological consequence, except when there are mononuclear cell infiltrates (usually periportal) in the liver.

† Even large numbers of circulating bacteria can be dealt with very effectively by reticuloendothelial macrophages in the normal host. The antibacterial capacity of the peritoneal cavity, by comparison, is poor. If virulent encapsulated pneumococci are injected intravenously into mice about 100 000 bacteria are needed to cause death, but a single bacterium is lethal by the intraperitoneal route. Presumably this is partly because serum factors are important, and partly because there is less efficient phagocytosis in the peritoneal cavity than in a small blood vessel, and the bacteria therefore have more opportunity to multiply.

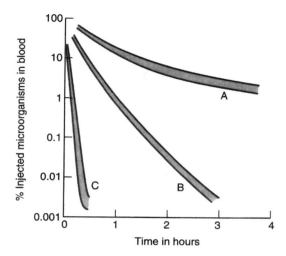

Fig. 5.4 Clearance of microorganisms from the blood. (A) encapsulated pneumococcus or small virus (e.g. T$_7$ bacteriophage, 30 nm diameter); (B) encapsulated pneumococcus coated with antibody in complement-depleted animal;
(C) encapsulated pneumococcus coated with antibody or nonencapsulated
bacterium (e.g. *Salmonella*) or large virus (e.g. *Vaccinia*, 250 nm diameter) in
normal animals.

Serum factors can be important, and specific antibodies to a microorganism and complement (pp. 177–8) promote rapid clearance by
opsonisation (see Fig. 5.4). Clearance also depends on the nature of the
microbial surface (see Ch. 4) and on the physiological state of the
macrophages (see 'macrophage activation', pp. 174–5). If microorganisms are cleared from the blood, the behaviour of the microorganism in the macrophages becomes of considerable importance.
Killing of the microorganism could mean termination of the infection,
whereas microbial persistence and growth in the macrophage could
lead to infection in the organ harbouring the macrophages, with
reseeding of progeny microorganisms into the blood.
  Because of their phagocytic activity reticuloendothelial macrophages
are inevitably involved in many systemic infections. Foci of infection in
liver, spleen and sometimes the bone marrow are features of brucellosis, leptospirosis and typhoid in man. The role of the RES can be illustrated by examples of the encounter of viruses with liver macrophages.
Viruses infecting the liver usually do so by way of the blood, and since
the macrophages (Kupffer cells) lining the liver sinusoids form a functionally complete barrier between blood and hepatic cells, no virus has
access to hepatic cells except through macrophages. There are also
distinct endothelial cells lining sinusoids, but the Kupffer cells, being
professional phagocytes, are particularly important. Uptake by these
cells is a necessary first step in infection of the liver. The types of

virus–macrophage interaction in the liver can be classified as follows (see Fig. 5.5).

(a) No uptake by macrophages. This has been described for a few viruses, such as LCM virus from congenitally infected mice, and favours persistence of the viraemia.
(b) Uptake and destruction in macrophages. This is the fate of most viruses circulating in the plasma. If viraemia is to be maintained, as much virus must enter the blood as is removed. Thus, in many arthropod-borne virus infections, there must be extensive seeding of virus into the blood to make up for clearance by macrophages and to maintain blood virus levels for ingestion by mosquitoes. Uptake of viruses by reticuloendothelial macrophages is rapid compared with the time taken for localisation in capillary vessels in other parts of the body. Hence, if neurotropic viruses such as polio are to reach the CNS from the blood, viraemia must be maintained for a period long enough to allow viral localisation in cerebral capillaries (see below). As long as virus is being cleared from the blood, a high rate of viral entry into the blood must be maintained. Experi-

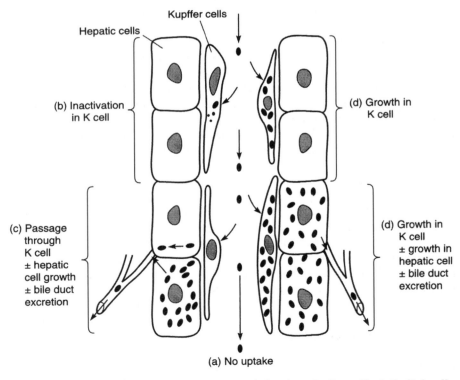

Fig. 5.5 Diagram to show types of virus behaviour in liver. Endothelial cells have been omitted, as their role is less clear.

mentally, viraemias can be prolonged and intensified by reducing the rate of virus clearance by reticuloendothelial cells. This is called reticuloendothelial 'blockade', and it used to be done, for instance, by injecting silica intravenously. This is taken up by reticuloendothelial cells and interferes with their phagocytic activity. These days, injection of macrophage-specific monoclonal antibodies, which lyse macrophages with the aid of endogenous complement, is the method of choice.

(c) Uptake and passive transfer from macrophages to hepatic cells. Circulating virus particles are taken up by Kupffer cells, and are passively carried across Kupffer and endothelial cells inside vesicles by transcytosis and presented to adjacent hepatic cells. If the virus cannot then grow in hepatic cells, things go no further, although sometimes the hepatic cells excrete virus into the bile. Viruses that infect hepatic cells, however, have an opportunity to cause hepatitis, in spite of their inability to infect macrophages. This is so for certain arthropod-borne viruses such as Rift Valley Fever virus (see pp. 225–6) and probably for hepatitis A and hepatitis B viruses. Excretion into bile is important in hepatitis A virus infection because it is the method of entry into the intestinal tract with shedding of virus in faeces. Hepatitis B on the other hand, is inactivated in bile.

(d) Uptake and growth in macrophages and/or endothelial cells. When this occurs the progeny virus particles are released in proximity to hepatic cells (Fig. 5.5d) and hepatitis again becomes a possibility. Liver infection in yellow fever is probably caused in this way.

This scheme gives a logical pathogenic background for the development of viral hepatitis, and the possibilities apply also to liver infection by the other microorganisms that grow inside cells. The infection may be restricted to macrophages (*Leishmania donovani*), or involve hepatic cells as with the exoerythrocytic stages of malaria, but microbial behaviour in the liver macrophage will exercise a determining influence on the infection, just as with virus infections. Microorganisms that can grow extracellularly, on the other hand, merely need to lodge in a liver sinusoid and grow. If taken up by a macrophage or a polymorph, their virulence is increased if they destroy the ingesting cell. This will result in an infectious focus containing necrotic cells, with a surrounding zone of inflammatory polymorphs, lymphocytes and macrophages. Such a pattern of liver involvement is seen with the hepatitis produced by *Entamoeba histolytica* or *Leptospira icterohaemorrhagica*.

Haematogenous spread and the nature of the vascular bed

If a circulating microorganism is to invade a tissue without sinusoids, it must first adhere to the endothelium of blood vessels in this tissue, preferably capillaries or venules, where the circulation is slowest and the vessel wall thinnest. The microorganism can then reach tissues by leaking through the vessel wall, being passively ferried across the vessel wall, or by growing through the vessel wall. These alternatives have been studied carefully only in the case of virus infections, especially in relation to the blood–brain junction (see Fig. 3.2). The anatomical nature of the obstacles separating blood from tissue is obviously important. It constitutes the blood–tissue barrier, but because it is not necessarily a barrier it will be referred to as the blood–tissue junction. Protection of tissues from invasion by circulating microorganisms depends to some extent on the anatomical nature of the blood–tissue junction. One of the most important barriers to the spread of virus into an organ is a layer of insusceptible cells that cannot be infected, either the capillary endothelium itself or other cells in extravascular tissues. Even if the vascular endothelium is infected, subsequent events may be determined by the topography of budding (see pp. 69–70). Viruses released exclusively from the apical surface of the cell would contribute to a viraemia, but unless the cell was destroyed, tissue invasion would depend on liberation of virus from the basal (tissue) side of the endothelial cell.

Viral invasion of a given cell depends on the presence of virus receptors on the cell (Ch. 2) and then on appropriate events taking place inside the cell (e.g. uncoating of the virus, presence of suitable transcription factors).

Capillaries in the CNS, connective tissue, skeletal and cardiac muscles are lined by a continuous layer of endothelium, whereas those in the renal glomerulus, pancreas, choroid plexus, ileum and colon have fenestrated gaps in the endothelium (Fig. 3.2). Once microorganisms have localised in the vessel wall, passage across the endothelium might be expected to be easier when there are fenestrated gaps. In all cases, however, there is a well-defined basement membrane which must also be negotiated if microorganisms are to reach extravascular tissues. The complexity of the basement membrane differs in different organs, but it is less well developed when capillaries are growing ('sprouting') during foetal development or during repair after injury. Circulating microorganisms also tend to localize more readily in abnormal, inflamed, or growing tissues. For instance circulating gonococci, staphylococci, etc. localise more frequently and cause septic arthritis in joints affected by rheumatoid arthritis or other pathological conditions, and circulating staphylococci localise in the growing ends of long bones in young people. Since the endothelial cells are not highly phagocytic, localisation in capillaries is slow compared with that in the reticuloendothelial cells lining sinusoids. Circulating microorganisms must circulate in the blood for long enough and in high enough concentration. Therefore the faster the clearance of viruses from the blood by reticuloendothelial cells, the less chance there is for localisation in capillaries. Removal of circulating viruses by reticuloendothelial cells, or their inactivation by serum antibody and complement, constitutes the most important barrier to invasion of those organs that have a capillary bed. In a capillary which is not 'sprouting' (see above) or inflamed, initial localisation of virus presumably depends on attachment to receptors on endothelial cells. Are there differences between capillaries from different organs? Now that endothelial cells from capillaries as well as from large vessels can be cultured *in vitro*, these matters are being investigated.

Removal of circulating bacteria, protozoa, etc. by reticuloendothelial cells tends to ensure their localisation in the liver, spleen and bone marrow. As with viruses, however, almost nothing is known about the factors governing localisation in particular capillary beds. The fact that circulating meningococci tend to localise in meninges, *Salmonella typhi* in the gall bladder and African trypanosomes in the cerebrospinal fluid (sleeping sickness) is of supreme importance, but the reasons for this localisation remain wrapped in mystery. Perhaps there are subtle differences in the nature of the capillary beds, or perhaps there is localisation in all parts of the body but only some sites provide suitable conditions for microbial growth. Positive activity on the part of a microorganism that is motile may determine localisation in a given capillary bed. For instance, the South American trypanosome causing Chagas' disease leaves the bloodstream by actively penetrating the capillary wall, nonflagellated end first. The heart, skeletal and smooth muscle are particularly affected, but the reasons for localisation here are not understood. The larger bacteria, fungi or protozoa may be large enough to be trapped mechanically in normal capillaries. Tissue invasion is made easier if there is preliminary multiplication or toxin production in the capillary lumen with damage to the vessel wall.

Central nervous system

Circulating viruses often localise more readily in the brains of immature animals than in those of adults. This is associated with sustained viraemia in the immature host, partly due to increased entry of virus into the blood from peripheral growth sites, and partly to less active reticuloendothelial clearance. Also it may be easier for viruses to traverse the blood–brain junction in the immature host because the basement membrane tends to be much thinner. Certain viruses grow through the walls of cerebral capillaries (herpes simplex, yellow fever, measles) and others leak through or are ferried across (polio). Many capillaries are enveloped by 'feet' from glial cells (Fig. 3.2) and the glial feet must be traversed before there can be infection of neurons. A few viruses infect only neurons (rabies) or glial cells (JC virus),* but most

---

* There is a rare neurological condition in man called progressive multifocal leucoencephalopathy. A papovavirus (JC virus) has been isolated from the brains of patients, and oligodendroglial cells in the brain are seen by electron microscopy to contain large numbers of virus particles. A related virus (BK virus) has been isolated from the urinary tract of pregnant women and immunosuppressed patients. Most adults have at some time been infected with these viruses because they have antibodies, but little is known about the primary infection and any associated illness is probably mild.

viruses that grow in the brain infect both types of cell. Viral encephalitis in man is rare, however, even though the potential viraemic invaders are numerous.

Bacterial meningitis due to meningococci, pneumococci, *Haemophilus influenzae* or *Mycobacterium tuberculosis*, is a tragic and often lethal disease, especially in developing countries. Almost nothing is known of the factors promoting bacterial localisation and invasion across meningeal blood vessels.*

Skeletal and cardiac muscle

Certain viruses infect skeletal or cardiac muscle after passage across the vessel walls, particularly coxsackie viruses, cardioviruses and certain arthropod-borne viruses. This occurs readily in the immature experimental host, and in man coxsackie virus infections can cause severe striated muscle or cardiac involvement. The trypanosomes causing Chaga's disease leave the blood and selectively parasitise the heart, skeletal and smooth muscle (see above), but it is not known how this is done. Circulating bacteria (e.g. the *Streptococcus viridans* group) may localise on abnormal heart valves and cause endocarditis.

The skin

The skin is involved in many systemic infections (Table 5.2). Rashes are produced after microorganisms or antigens have localised in skin blood vessels, sometimes spreading extravascularly. We do not know why the rash has such a characteristic distribution in many infectious diseases. Inflammation can certainly localise skin lesions, whether produced by sunburn, a tight garter, or pre-existing eczema, but the factors accounting for the characteristic distribution of the rash in chickenpox, smallpox (in the old days), or hand, foot and mouth disease (caused by certain coxsackie A viruses) remain unknown. The blood–skin junction consists of the endothelial cells forming a continuous lining to dermal capillaries, together with a basement membrane. There may be a fibroblastic cell applied to the basement membrane, and between the capillary and the epidermis lies the connective tissue matrix containing scattered fibroblasts and histiocytes. The skin has

---

* The meningococcus, for instance, is present in the oropharynx of 5–30% of normal people, but very occasionally it invades the blood, perhaps because of a genetically determined weakness in the capacity to form circulating bactericidal antibodies. It then has the opportunity to localise in meningeal blood vessels and infect the cerebrospinal fluid. Mechanisms of localisation are not known. In the case of *E. coli*, an occasional meningeal invader in the newborn, molecules on bacterial fimbriae bind with especial affinity to endothelial cells of meningeal blood vessels. For *H. influenzae* there is evidence from studies with baby rats that both the invasion of blood from the nasopharynx and meningeal invasion can be accomplished by a single individual bacterium, and these are possibly rare events.

Table 5.2.   Principal rashes in infectious disease in man

Microorganism	Disease	Features
Measles virus	Measles	Very characteristic maculopapular rash
Rubella virus	German measles	
Parvovirus	Erythema infectiosum	Maculopapular rashes not distinguishable clinically
Echoviruses 4, 6, 9, 16		
Coxsackie viruses A9, 16, 23	Not distinguishable	
Varicella-zoster virus	Chickenpox, zoster	
Coxsackie virus A16	Hand, foot and mouth disease	Vesicular rashes
Rickettsia prowazeki and others	Typhus	
Rickettsia rickettsiae and others	Spotted fever group of diseases	Macular or haemorrhagic rash
Streptococcus pyogenes	Scarlet fever	Erythematous rash caused by toxin
Treponema pallidum	Syphilis	Disseminated infectious rash seen in secondary stage, 2–3 months after infection
Treponema pertenue	Yaws	
Salmonella typhi	Enteric fever	Sparse rose spots containing bacteria
Salmonella paratyphi B		
Neisseria meningitidis	Spotted fever	Petechial or maculopapular lesions containing bacteria
Blastomyces dermatitidis	Blastomycosis	Papule or pustule develops into granuloma; lesions contain organisms
Leishmania tropica	Cutaneous leishmaniasis	Papules, usually ulcerating to form crusted sores; infectious
Hepatitis B and viral exanthems	Prodromal rashes	
Dermatophytes (skin fungi)	Dermatophytid or allergic rash	Generalised rash due to hypersensitivity to fungal or viral antigens
Streptococcus pyogenes	Impetigo[a]	Vesicles, forming crusts, especially in children
Staphylococcus pyogenes		

[a] These skin lesions are multiple but like those of erysipelas or warts are formed locally at the sites of infection, not after spread through the body.

its own immune cells, particularly Langerhans cells (see p. 151), many mast cells (see p. 161), and recirculatory T-cells are always present in the dermis.

The skin of man is mostly naked, and is an important thermoregulatory organ, under finely balanced nervous control. It is a turbulent, highly reactive tissue, and local inflammatory events are commonplace. At sites of inflammation, circulating microorganisms readily localise in small blood vessels and pass across the endothelium. The skin of most animals, in contrast, is largely covered with fur. Skin lesions are a feature of many infectious diseases of animals, but these lesions tend to be on exposed hairless areas where the skin has the human properties of thickness, sensitivity and vascular reactivity. Hence, although virus rashes very occasionally involve the general body surface of animals, it is udders, scrotums, ears, prepuces, teats, noses and paws that are more regular sites of lesions. For instance, the closely related diseases of measles, distemper and rinderpest can be compared. Cattle with rinderpest may show areas of red moist skin with occasional vesiculation on the udder, scrotum and inside the thighs. In dogs with distemper the exanthem often occurs on the abdomen and inner aspect of the thighs. Yet in human measles there is one of the most florid and characteristic rashes known, involving the general body surface. Even in susceptible monkeys, the same virus produces skin lesions sparingly and irregularly.

Macules and papules are formed when there is inflammation in the dermis, with or without a significant cellular infiltration, the infection generally being confined to the vascular bed or its immediate vicinity. Immunological factors (see Ch. 8) are often important in the production of inflammation. Measles virus, for instance, localises in skin blood vessels, but the maculopapular rash does not appear unless there is an adequate immune response. Virus by itself does little damage to the blood vessels or the skin, and the interaction of sensitised lymphocytes or antibodies with viral antigen is needed to generate the inflammatory response that causes the skin lesion. Rickettsia characteristically localise and grow in the endothelium of small blood vessels, and the striking rashes seen in typhus and Rocky Mountain Spotted Fever are a result of endothelial swelling, thrombosis, small infarcts and haemorrhages. The immune response adds to the pathological result. Vascular endothelium is an important site of replication and shedding of viruses and rickettsias that are transmitted by blood-sucking arthropods and which must therefore be shed into the blood. After replication in vascular endothelium, they may be shed not only back into the vessel lumen, but also from the external surface of the endothelial cell into extravascular tissues (see also p. 136). Certain arthropod-borne viruses replicate in muscle or other extravascular tissues, and can then reach the blood after passage through the lymphatic system.

Circulating immune complexes consisting of antibody plus microbial antigen also localise in dermal blood vessels, accounting for the trichophytid rashes of fungal infections and the prodromal rashes seen at the end of the incubation period in many exanthematous virus diseases. Antibodies to soluble viral antigens appear towards the end of the incubation period in people infected with hepatitis B virus and form soluble immune complexes. These localise in the skin causing fleeting rashes and pruritis, and rarely the more severe vascular lesions of periarteritis nodosa (see Ch. 8).

Certain microbial toxins enter the circulation, localise in skin blood vessels, and cause damage and inflammation without the need for an immune response. An erythrogenic toxin is liberated from strains of

Streptococcus pyogenes carrying the bacteriophage β, and the toxin enters the blood, localises in dermal vessels, and gives rise to the striking rash of scarlet fever.

Vesicles and pustules are formed when the microorganism leaves dermal blood vessels and is able to spread to the superficial layers of the skin. Inflammatory fluids accumulate to give vesicles, which are focal blisters of the superficial skin layers. Virus infections with vesicles include varicella, herpes simplex and certain coxsackie virus infections. The circulating virus localises in dermal blood vessels, grows through the endothelium (herpes, varicella) and spreads across dermal tissues to infect the epidermis and cause focal necrosis. Only viruses capable of extravascular spread and epidermal infection can cause vesicles. Inevitably there is an immunopathological contribution to the lesion, although a primary destructive action on epidermal cells gives a lesion without the need for the immune response, as with the oral lesions seen in animals as early as 2 days after infection with foot and mouth disease virus. A secondary infiltration of leucocytes into the virus-rich vesicle turns it into a pustule which later bursts, dries, scabs and heals. Such viruses are shed to the exterior from the skin lesion. Certain other microorganisms are shed to the exterior after extravasation from dermal blood vessels. They multiply in extravascular tissues and form inflammatory swellings in the skin, which then break down so that infectious material is discharged to the exterior. This occurs and gives rise to striking skin lesions in the secondary stages of syphilis and yaws (caused by the closely related bacteria *Treponema pallidum* and *pertenue*) and is also seen in a systemic fungus infection (blastomycosis) and a protozoal infection (cutaneous leishmaniasis). In patients with leprosy, *Mycobacterium leprae* circulating in the blood localises and multiplies in the skin, and for unknown reasons superficial peripheral nerves are often involved. The skin lesions do not break down, although large numbers of bacteria are shed from sites of growth on the nasal mucosa. Bacterial growth is favoured by the slightly lower temperature of the skin and nasal mucosa.

Almost all the factors that have been discussed in relation to skin localisation and skin lesions apply also to the mucosae of the mouth, throat, bladder, vagina, etc. In these sites the wet surface means that the vesicles will break down and form ulcers earlier than on the dry skin. Hence in measles the foci in the mouth break down and form small visible ulcers (Koplik's spots) a day or so before the skin lesions have appeared (Fig. 5.3). Similar considerations apply to the localisation of microorganisms and their antigens on the other surfaces of the body (see Fig. 2.1). In chickenpox and measles, circulating virus localises in subepithelial vessels in the respiratory tract, and after extravasation there is only a single layer of cells to grow through in the nearby epithelium before the discharge of virus to the exterior. Hence in these infections the secretions from the respiratory tract are infectious a few days before the skin rash appears and the disease becomes recognisable. Much less is known about the localisation of circulating microorganisms in the intestinal tract. Probably localisation here is not often of great importance, but this is a difficult surface of the body to study. In typhoid, secondary intestinal localisation of bacteria takes place following excretion of bacteria in bile, rather than from blood. Virus localisation in the intestinal tract is a feature in rinderpest in cattle but occurs only to a minor extent in measles. When the patient with measles suffers from protein deficiency, however, it is more important and helps cause the diarrhoea that makes measles a life-threatening infection in malnourished children (see p. 378).

The foetus

The blood–foetal junction in the placenta is an important pathway for infection of the foetus. The number of cells separating maternal from foetal blood depends not only on the species of animal, there being four cell sheets for instance in the horse and only one or two in man, but also on the stage of pregnancy. The junction usually becomes thinner, often with fewer cell layers, in later pregnancy. There are regular mechanical leaks in the placenta late in most human pregnancies, and up to 4.0 ml of blood is transferred across the placenta, but this appears to be principally in one direction, from foetus to mother. There is little evidence for the passive carriage of microorganisms across the placenta, and foetal infection takes place by either of two mechanisms. If a circulating microorganism, free or cell associated, localises in the maternal vessels it can multiply, cause damage, locally interrupt the integrity of the junction and thus infect the foetus. *Treponema pallidum* and *Toxoplasma gondii* presumably infect the human foetus in this way. Alternatively, a circulating microorganism can localise and grow across the placental junction. This occurs with rubella and cytomegalovirus infections of the human foetus. In both instances, a placental lesion or focus of infection occurs before foetal invasion. The microorganisms causing foetal damage are listed in Table 5.3 (see also p. 334). These, however, are special cases, and special microorganisms. Nearly always the foetus is protected from microbial as well as from biochemical and physical insults. The factors that localise micro-organisms in the placenta are not understood, but blood flow is slow in placental vessels, as in sinusoids, giving maximal opportunities for localisation. Once microorganisms are arrested in placental vessels, their growth may be favoured by particular substances that are present in the placenta. Erythritol promotes the growth of *Brucella abortus*, and its presence in the bovine placenta makes this a target organ in infected cows. Susceptibility of infected cattle to abortion thus has a biochemical basis. Microorganisms can damage the foetus without invading foetal tissues. If they localise extensively in placental vessels and cause primarily vascular damage this of course can lead to foetal anoxia, death and abortion. Also the toxic products of microbial

Table 5.3.   Principal microorganisms infecting the foetus

Microorganisms	Species	Effect
Viruses		
Rubella virus	Man	Abortion
		Stillbirth
		Malformations
Cytomegalovirus	Man	Malformations
HIV	Man	About 1 in 5 infants born to infected mothers are infected *in utero*
Hog cholera virus (vaccine strain)	Pigs	Malformations
Bluetongue virus (vaccine strain)	Sheep	Stillbirths, CNS disease
Equine rhinopneumonitis	Horse	Abortion
Bovine diarrhoea – mucosal disease virus	Cow	Cerebellar hypoplasia
Malignment catarrh virus	Wildebeest	Foetus unharmed
Bacteria		
Treponema pallidum	Man	Stillbirth, malformations
Listeria monocytogenes	Man	Meningoencephalitis
Vibrio foetus	Sheep, cattle	Abortion
Protozoa		
Toxoplasma gondii	Man	Stillbirth, CNS disease

growth in the placenta or elsewhere and probably cytokines can reach the foetus and cause damage. High fever and biochemical disturbances in a pregnant female can adversely affect the foetus.

Miscellaneous sites

There are certain other sites where circulating microorganisms selectively localise. In rats and other animals infected with *Leptospira*, circulating bacteria localise particularly in capillaries in the kidney and give rise to a chronic local lesion. Infectious bacteria are discharged in large numbers into the urine, which is therefore a source of human infection. Microorganisms that are discharged in the saliva (mumps and most herpes-type virus infections in man) must localise and grow in salivary glands. Those that are discharged in milk must localise and grow in mammary glands (the mammary tumour virus in mice and *Brucella*, tubercle bacilli, and Q fever rickettsia in cows). A few examples, such as *Haemophilus suis* in pigs, Ross River virus in man (Table A.5), and occasionally rubella virus, localise in joints. Almost any site in the body, from the feather follicles (Marek's disease) to testicles or epididymis (mumps in man, the relevant *Brucella* species in rams, boars, bulls) can at times be infected. Nothing is known of the mechanism of localisation in these organs.

Spread via other Pathways

Cerebrospinal fluid (CSF)

Microorganisms in the blood can reach the CSF by traversing the blood–CSF junction in the meninges or choroid plexus. Capillaries in the choroid plexus have fenestrated endothelium and are surrounded by a loose connective tissue stroma (Fig. 3.2). Inert virus-sized particles and bacteriophages leak into the CSF when very large amounts are injected into the blood. It is assumed that the viruses causing aseptic meningitis in man (polio-, echo-, coxsackie, lymphocytic choriomeningitis and mumps viruses) enter the CSF by leakage or growth across this junction (Fig. 5.6). Once in the CSF microorganisms are passively carried with the flow of fluid from ventricles to subarachnoid spaces and throughout the neuraxis within a short time. Invasion of the brain itself and spinal cord can now take place across the ependymal lining of the ventricles and spinal canal, or across the pia mater in the subarachnoid spaces. Nonviral microorganisms entering the CSF across the blood–CSF junction include the meningococcus, the tubercle bacillus, *Listeria monocytogenes, Haemophilus influenzae, Streptococcus pneumoniae*, and the fungus *Cryptococcus neoformans*.

Pleural and peritoneal cavities

Rapid spread of microorganisms from one visceral organ to another can take place via the peritoneal or pleural cavity. Entry into the peritoneal

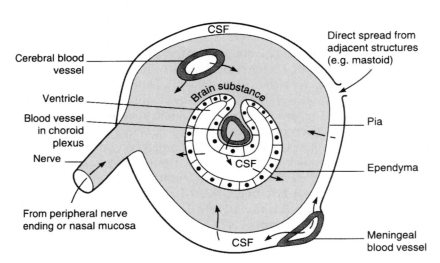

Fig. 5.6 Routes of microbial invasion of the central nervous system. CSF = cerebrospinal fluid.

cavity takes place from an injury or focus of infection in an abdominal organ. The peritoneal cavity, as if in expectation of such events, is lined by macrophages and contains an antimicrobial armoury, the omentum. The omentum, originating from fused folds of mesentery, contains mast cells and lymphocytes, macrophages and their precursors in a fatty connective tissue matrix. It is movable in the peritoneal cavity and becomes attached at sites of inflammation.* Microorganisms spread rapidly in the peritoneal cavity unless they are taken up and destroyed in macrophages or inflammatory polymorphs. Peritoneal contents drain into lymphatics opening onto the abdominal surface of the diaphragm, so that microorganisms or their toxins are delivered to retrosternal lymph nodes in the thorax, sometimes with slight leakage into the pleural cavity. Inflammatory responses in the peritoneum eventually result in fibrinous exudates and the adherence of neighbouring surfaces, which tends to prevent microbial spread.

Microbes entering the pleural cavity from chest wounds or from foci of infection in the underlying lung have a similar opportunity to spread rapidly. During pneumonia the overlying pleural surface first becomes inflamed, causing pleurisy, and later often infected. Pleurisy occurs in about 25% of cases of pneumococcal pneumonia. The pleural cavity, like the peritoneal cavity, is lined by macrophages.

Nerves

For many years peripheral nerves have been recognised as important pathways for the spread of certain viruses and toxins from peripheral parts of the body to the central nervous system (Fig. 5.6). Rabies, herpes simplex and related viruses travel along nerves at up to 10 mm h^{-1}, but the exact pathway in the nerve was for many years a matter of doubt and debate. Herpes simplex virus, following primary infection in the skin or the mouth, enters the sensory nerves and reaches the trigeminal ganglion (see Ch. 10). Here it remains in latent form until it is reactivated in later life by fever, emotional or other factors. The infection then travels down the nerve to reach the region of the mouth, where the skin is once again infected giving rise to a virus-rich cold sore. A similar sequence of events explains the occurrence of zoster long after infection with varicella virus. In cattle or pigs infected with pseudorabies, another herpes virus, the infection also travels up peripheral nerves to reach dorsal root ganglia, causing a spontaneous discharge of nerve impulses from affected sensory neurons, and giving rise to the signs of 'mad itch'. Another herpes virus (B virus) is often present in the saliva of apparently healthy rhesus monkeys, and people

---

* Because of its ability to attach to sites of inflammation and infection or to foreign bodies the omentum has been referred to as the 'abdominal policeman'.

bitten by infected monkeys develop a frequently fatal encephalitis, the virus reaching the brain by ascending peripheral nerves from the inoculation site. Rabies virus slowly reaches the CNS along peripheral nerves following a bite delivered by an infected fox, jackal, wolf, raccoon, skunk or vampire bat. It also travels centrifugally from the brain down peripheral nerves to reach the salivary glands and other organs. Poliovirus was long thought to reach the CNS via peripheral nerves, but this was a conclusion from studies with artificially neuro-adapted strains of virus. In natural infections, poliovirus traverses the blood–brain junction (Fig. 3.2). Peripheral nerves are affected in leprosy, the bacteria having a special affinity for Schwann cells, which are unable to control the multiplying bacteria. The molecular basis for this targeting of Schwann cells is being unravelled. This causes a very slow and insidious degeneration of the nerve, but it is certainly not a pathway for the spread of infection. Peripheral nerves are known to transport tetanus toxin to the CNS (see Ch. 8), and also prion agents (scrapie) in experimental infections of mice.

Possible pathways along nerves include sequential infection of Schwann cells, transit along the tissue spaces between nerve fibres, and carriage up the axon (Fig. 5.7). The last route is probably an important one, although at first sight it might seem less likely. There is a small but significant movement of marker proteins up normal axons from the periphery to the CNS, and in experimental herpes simplex and rabies infections virus particles have been seen in axons by electron microscopy. In experimental infections, herpes viruses can also travel in nerves by sequential infection of the Schwann cells associated with myelin sheaths, but this is not a natural route.

An alternative neural route of spread to the CNS is by the olfactory nerves. Axons of olfactory neurons terminate on the olfactory mucosa, the dendrites projecting beyond the mucosal surface giving a direct anatomical connection between the exterior and the olfactory bulbs in

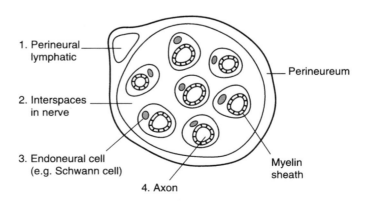

Fig. 5.7 Possible pathways of virus spread in peripheral nerves.

the brain. This route of infection, although at one time a popular postulate, is not often important. Aerosol infection with rabies virus (from the excreta of bats in caves in North America) presumably involves this route. When administered intranasally in experimental infections of mice, Semliki Forest virus rapidly enters the olfactory bulbs and thence into the rest of the brain. *Naegleria fowleri*, a free-living amoeba that can lurk in the sludge at the bottom of freshwater pools, causes a rare but often fatal meningitis in swimmers after infecting by the olfactory route. The meningococci that live commensally in the nasopharynx of 5–10% of normal people, and occasionally cause meningitis, were once thought to spread directly upwards from the nasal mucosa, along the perineural sheaths of the olfactory nerve, and through the cribriform plate to the CSF. More probably, the bacteria invade the blood, sometimes causing petechial rashes ('spotted fever'), and reach the meninges across the blood–CSF junction.

In summary, peripheral nerves are important pathways for the spread of tetanus toxin and a few viruses to the CNS, and for the passage of certain herpes viruses between the CNS and the surfaces of the body. Herpes and rabies viruses can travel both up and down peripheral nerves. The neural route is not generally used by bacteria or other microorganisms.

References

de Voe, I. W. (1982). The meningococcus and mechanisms of pathogenicity. *Microbiol. Rev.* **46**, 162–190.

Drutz, D. J. *et al.* (1972). The continuous bacteraemia of lepromatous leprosy. *N. Engl. J. Med.* **287**, 159–163.

Friedman, H. M., Macarek, E. J., MacGregor, R. A. *et al.* (1981). Virus infection of endothelial cells. *J. Infect. Dis.* **143**, 266.

Griffin, J. W. and Watson, D. F. (1988). Axonal transport in neurologic disease. *Ann. Neurol.* **23**, 3–13.

Johnson, R. T. (1982). 'Viral Infections of the Nervous System'. Raven Press, New York.

Mims, C. A. (1964). Aspects of the pathogenesis of virus diseases. *Bact. Rev.* **28**, 30.

Mims, C. A. (1966). The pathogenesis of rashes in virus diseases. *Bact. Rev.* **30**, 739.

Mims, C. A. (1968). The pathogenesis of virus infections of the foetus. *Prog. Med. Virol.* **10**, 194.

Mims, C. A. (1981). The pathogenetic basis of viral tropism. *Am. J. Pathol.* **135**, 447–455.

Moxon, R. E. and Murphy, P. A. (1978). *Haemophilus influenzae* bacteremia and meningitis resulting from survival of a single organism. *Proc. Natl Acad. Sci. U.S.A.* **75**, 1534–1536.

Pearce, J. H. *et al.* (1962). The chemical basis of the virulence of *Brucella abortus* II. Erythritol, a constituent of bovine foetal fluids which stimulates the growth of *Br. abortus* in bovine phagocytes. *Brit. J. Exp. Pathol.* **43**, 31–37.

Quagliarello, V. and Schell, W. M. (1992). Bacterial meningitis; pathogenesis, pathophysiology, and progress. *N. Engl. J. Med.* **327**, 864–872.

Rambukkana, A. (2000). How does Mycobacterium leprae target the peripheral nervous system? *Trends Microbiol.* **8**, 23–28.

Williams, A. E. and Blakemore, W. F. (1990). Pathogenesis of meningitis caused by *Streptococcus suis* Type 2. *J. Infect. Dis.* **162**, 474–481.

Williams, A. E. and Blakemore, W. F. (1990). Monocyte-mediated entry of pathogens into the central nervous system. *Neuropath. Appl. Neurobiol.* **16**, 377–392.

6

The Immune Response to Infection

The immune response is conveniently divided into the antibody and the cell-mediated component, the latter being transferable from one individual to another by lymphocytes but not by serum. Antibodies, since they can be tested and assayed without great difficulty, were the first to receive attention with the discovery of antibodies to tetanus and diphtheria toxins in the 1890s. Cell-mediated immunity (CMI) in the form of delayed hypersensitivity was described more than 50 years ago, and has received intensive study in the past 30 years. Specific antibodies and CMI are induced in all infections, but the magnitude and quality of these responses varies greatly in different infections. It is not often that the microbial antigens concerned have been individually defined or identified. More importantly, we have rarely identified the microbial antigens that induce protective immune responses.

Most antigens are proteins or proteins combined with other substances, but polysaccharides and other complex molecules also function as antigens. Substances called haptens, small molecules such as sugars, cannot by themselves stimulate antibody production, but do so when coupled to a protein. An antigen stimulates the production of antibodies that react specifically with that antigen. The reaction can be thought of as similar to that between lock and key, and it is specific in the sense that antibody produced against diphtheria toxin does not react with tetanus toxin. An antibody may, however, have weaker reactivity against antigens closely related to the one that stimulated its production. For instance, antibodies produced when human serum is injected into a rabbit will not react with the serum of cows, mice or chickens, but may give a weak reaction with the serum of the gorilla

149

and chimpanzee. The antibodies formed against a given antigen will include representatives from the four main immunoglobulin classes: IgG, IgA, IgE and IgM. A single antigen molecule may have several antigenic sites or epitopes, each of which stimulates the formation of a different antibody. Also, different immunoglobulin molecules vary in the firmness (avidity) with which they combine with the antigen, but little is known about antibody avidity in relation to infectious diseases.

The two arms of the immune response are expressed by different types of immunologically reactive lymphocytes, divided according to their origin into B (bursa in birds or bone marrow and foetal liver in mammals) and T (thymus) dependent cells. These two types of cells are both small to medium-sized lymphocytes, only distinguishable by specific cell surface molecules identified by immunological techniques. B cells are concerned with the antibody response and T cells with initiating the cell-mediated immune (CMI) response. B cells bear on their surface immunoglobulin molecules that act as receptors for antigen. Different B cells have different antigen-specific receptors (estimated to be of the order of 10^9 for each individual). There are about 10^5 receptors per cell; they are randomly generated by genetic recombination in the developing B cell, and when almost any antigen enters the body for the first time there will be a few B cells that react with it specifically. Following an encounter with antigen, B cells become activated and clonally expand to form a pool of memory cells or differentiate to form plasma cells, the antibody synthesising cells. B cells are located in various lymphoid tissues, notably spleen and lymph nodes and to a lesser extent in the blood.

T cells express on their surface the T-cell receptor (TCR), a structure not dissimilar to the immunoglobulin receptor, but which only recognises antigenic peptides associated with MHC molecules on cell surfaces. There are two main types of TCR, α/β and γ/δ,* which are present on distinct populations of T cells. As with B cells, T cells are clonally derived, each bearing a unique TCR derived by gene rearrangement during development. T cells are selected or educated to recognise foreign antigens in the thymus. A vast repertoire of TCRs are generated during thymic development, reactive against self-antigens as well as nonself or foreign antigens. Clearly, the host does not want T cells capable of damaging its own cells and tissues, and so removes these cells in the thymus by a process called clonal deletion (apoptotic death), referred to as negative selection. Equally, the host needs a mechanism for selecting those T cells destined to recognise foreign antigens and this is also achieved in the thymus by a process called positive selec-

---

* α/β and γ/δ denote the polypeptide chains composing the T cell receptor. Structurally, the TCR resembles the Fab region of immunoglobulin molecules containing similar constant and variable regions or domains. These domains form the basis of a diverse family of important immunological molecules belonging to the immunoglobulin superfamily. Included in this family are MHC molecules, Fc receptors, B7 molecules, CD2, CD3, CD4, CD8 and ICAM 1–3.

tion. Since all T cells must recognise self-MHC plus 'antigen' during development, it is still unclear why one population is deleted and the other selected. A possible explanation involves the avidity of the individual TCRs for self-MHC plus antigen: high-avidity interactions lead to negative signalling and cell death, whereas low-avidity binding leads to activation and an exit pass to the periphery. Two major classes of educated T cells leave the thymus, one expressing the CD8 glycoprotein (CD8 T cells) and the other expressing the CD4 glycoprotein (CD4 T cells). These cells patrol various lymphoid compartments, waiting for the opportunity to encounter foreign antigen presented by antigen presenting cells. When this happens, the reactive T cells proliferate and clonally expand, producing effector and memory cells.

The above is a simplified picture. Things are more complicated because, nearly always, appropriate responses are produced by cooperation between different types of cell. Dendritic cells and macrophages play a central role in the induction of immunological responses. Those in lymphoid tissues are strategically placed to encounter microbes or their antigens, and at the same time are in close proximity to lymphoid cells. Microbes and microbial antigens from sites of infection such as the body surfaces are 'focused' by afferent lymphatics into macrophages and dendritic cells in lymph nodes (see pp. 78–79), and when these materials enter the blood they are taken up by macrophages and dendritic cells resident in the spleen. These cells serve a vital immunological function. They act as antigen presenting cells whose function is to 'process' microbial and other antigens and present them to lymphocytes. An example is Langerhans cells* in the epidermis that send dendritic processes far into the surrounding epithelium. They sample their environment by endocytosis and macropinocytosis collecting antigens which are then transported into local lymph nodes.

This is separate from the antimicrobial function of macrophages described in Ch. 4 in which infectious agents are phagocytosed and killed. The all-embracing word macrophage can be misleading, because not all of them act as antigen-presenting cells in the induction of an immune response and it is clear that separate subpopulations of macrophages carry out the separate functions. For instance, most Kupffer cells are inefficient inducers of immune responses and therefore the uptake of microorganisms by these cells is generally non-productive from an immunological point of view.

Cell cooperation is an important feature in the induction and expression of the immune response. Virtually all effector responses are dependent on T-cell recognition of antigen associated with MHC class II molecules (see Glossary). These polymorphic membrane glycopro-

---

* Langerhans cells total 10^9 in man's skin, constituting 2–4% of all epidermal cells. They belong to the dendritic cell family which are the principal antigen presenting cells involved in the induction of the adaptive immune response. Dendritic cells are found in all tissues, with the exception of the brain and cornea.

teins are located on dendritic cells (including Langerhans cells), macrophages and B cells, all of which act as the 'professional' antigen presenting cells of the body, i.e. the main inducers of immunological responses. These cells function by endocytosing microbial antigens which become degraded in endosomes by lysosomal enzymes into short peptides (approximately 15 amino acids in length). These then associate with newly formed or recycled MHC class II molecules, which are presented on the cell membrane. This pathway of antigen presentation is sometimes referred to as the exogenous pathway (see Fig. 6.1). The peptide selectively binds to a groove on the MHC molecule, and it is

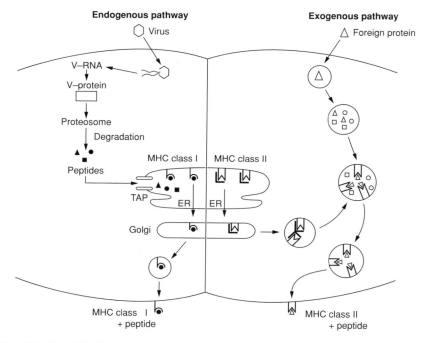

Fig. 6.1 Simplified scheme for the presentation of antigens via MHC class I and II molecules. In the endogenous pathway (left), an infecting virus will express viral RNA (v-RNA) and generate protein. The proteins are subjected to proteolysis by proteosomes to generate short peptides (●, ▲, ■) that are transported into the endoplasmic reticulum (ER) by peptide transporters associated with antigen presentation (TAP-1 and TAP-2). The peptides interact with MHC class I to form stable molecular complexes that become directed to the cell membrane and presented to CD8 T cells. The exogenous pathway (right) involves the uptake of antigen by endocytosis. Endosomes fuse with lysosomes leading to proteolysis. These early endosomes fuse with MHC class II-containing vesicles and selected peptides interact with the MHC molecules. These structures are displayed on the cell membrane and recognised by CD4 T cells. In the endoplasmic reticulum MHC class II molecules are protected from premature contact with other peptides and guided to the Golgi by the invariant chain (**L**). This is degraded when the endosome and vesicle fuse, thus allowing the peptides (○, △, □) to interact with the class II molecules.

this combination that is recognised by the TCR of CD4 T cells. These cells function by producing a variety of lymphokines involved in the activation and differentiation of other cells in the immune response, hence they are known as T-helper cells. Antibody responses are heavily dependent on T-helper cells for the generation of memory B cells and the presence of IgG, IgE and IgA, including high-affinity IgG antibodies in serum. The central role for T-helper cells in the immune response is summarised in Fig. 6.2.

T-helper cells can be further subdivided according to their function into two distinct populations of CD4 T cells, Th1 and Th2. These cells are distinguished from each other by the type of cytokines produced. Th1 cells are characterised by the expression of interleukin-2 (IL-2) and interferon-γ (IFN-γ) and fail to produce IL-4, IL-5 or IL-10. In contrast, Th2 cells produce IL-4, IL-5 and IL-10, but not IL-2 or IFN-γ. In terms of their function, Th1 cells are associated with delayed-type hypersensitivity (DTH) reactions resulting in the activation of macrophages and the production in mice of IgG2a antibodies. Th2 cells predominantly influence B-cell responses to produce IgE, IgA and IgG1 antibodies; these cells are not involved in DTH reactions. Depending on the nature of the antigen and the route of infection or immunisation, one particular Th subset will predominate. For example, microbial infection of skin will favour Th1 cells, where DTH reactions are important, whereas infections involving parasitic worms will favour Th2 cells, where IgE antibody is an important effector mechanism. T-cell cytokines are critical molecules in a number of immunological reactions. A summary of the cytokines and their actions is shown in Table 6.1.

CD8 T cells, also known as cytotoxic T cells, recognise foreign peptide in association with MHC class I molecules (found on virtually all cells of the body). In this instance peptides are generated from proteins derived within the cell (endogenously), for example, a protein from an infecting virus, but the pathway of antigen processing and presentation is different to that of the MHC class II system (see Fig. 6.1). The antigenic protein is degraded in the cytoplasm via an enzyme complex called a proteosome and peptide fragments (approximately nine amino acids in length) are actively transported into the endoplasmic reticulum where they encounter newly formed MHC class I molecules. The peptide–MHC complex is then transported to the cell membrane, where it is recognised by the TCR of CD8 T cells – these cells are often described as MHC class I restricted. The destruction of an infected cell by these cytotoxic T cells or the liberation of cytokines with antimicrobial action, is a major defence mechanism against intracellular microorganisms.*

* It is perhaps useful to attempt a rational explanation for these MHC requirements. A cytotoxic effector T cell, before releasing its powerful weaponry, needs to know that its physical contact with foreign antigen (peptide) is actually on the surface of a host cell. The recognition of antigen plus MHC class I (present on all cells) ensures that this is so. T-helper cells, on the other hand, need to know that peptide is being offered by a specialised cell that has been able to carry out suitable processing and presentation, and the MHC class II requirement for recognition ensures that this is so.

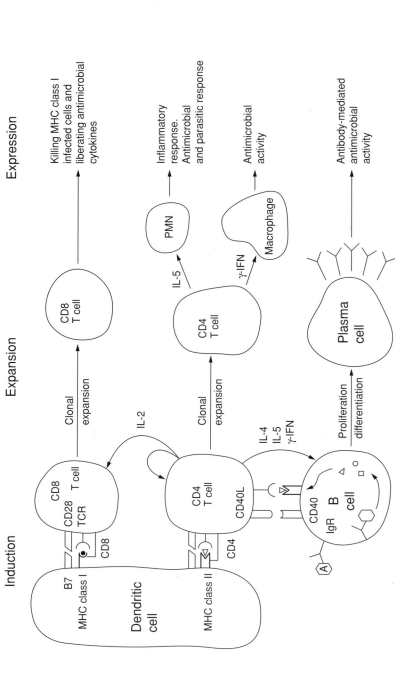

Fig. 6.2 Induction and expression of the immune response. Dendritic cells present antigen via MHC class I and class II molecules, resulting in the activation of CD8 and CD4 T cells. Key accessory molecules involved in activation are B7 and CD28, and also, via B cells, CD40 and CD40 ligand (CD40L). Once activated, T cells undergo proliferation and clonal expansion, driven by IL-2, producing memory cells and effector cells involved in antimicrobial immunity. Defence against pathogens is mediated by cytokines (e.g. TNF-α, IFN-γ) via activation of macrophages and polymorphonuclear (PMN) cells and also by CD8 T-cell-mediated cytolysis of infected cells. B cells recognise foreign antigens (A) by their immunoglobulin receptor (IgR). This complex becomes internalised by receptor-mediated endocytosis leading to antigen degradation (exogenous antigen presenting pathway). Processed antigen (○, △, □) is presented via MHC class II to previously activated CD4 T cells, which in turn induce proliferation of B cells and their differentiation to plasma cells. Again cytokines (e.g. IL-4, IL-5, IFN-γ) are instrumental in this process.

Table 6.1.   Key cytokines produced by T lymphocytes and macrophages in the immune response to microbial infection

Cytokine	Source	Target and action
IL-1	M	Co-stimulator of T cells. Activates macrophages. Inducer of fever
IL-2	L	Induces proliferation of T cells and activates natural killer (NK) cells. Induces antibody synthesis
IL-3	L	Growth and differentiation of precursor cells in bone marrow
IL-4	L	B-cell proliferation and differentiation
IL-5	L	Induces differentiation of B cells and activates eosinophils
IL-6	L,M	B- and T-cell growth and differentiation
IL-10	L,M	Activates B cells and inhibits macrophage function
IL-12	L,M	Activates NK cells and directs CD4 T cells to Th1 responses
IL-13	L	Induces proliferation of B cells and differentiation of T cells
IL-18	M	Induces proliferation of T cells
IFN-γ	L	Activates most lymphoid cells
TNF-α	L,M	Causes activation of macrophages. Induces inflammation and fever
TNF-β	L	Lymphotoxin. Inhibits B and T cells. Causes activation of macrophages
TGF-β	L,M	Inhibits B-cell growth and macrophage activation. Induces switch to IgA
GM-CSF	L,M	Induces production of granulocytes and macrophages

L = produced by T lymphocytes; M = produced by macrophages.

When an immune response is initiated, powerful forces are set in motion, which can be advantageous, but at times disastrous for the individual (see Ch. 8). So that each response can unfold in a more or less orderly fashion, it is controlled by a combination of stimulatory and inhibitory influences. The latter include antigen control and the activity of regulatory T cells producing immunosuppressive cytokines. Antigen itself acts as an important regulatory agent. Following its combination with antibody and uptake by phagocytic cells, it is catabolised and begins to disappear from the body. Since it is the driving force for an immune response, this response dies away as antigen disappears. Immune responses can therefore be regulated by controlling the concentration and location of antigen. A small amount of specific antigen or cross-reactive antigens from other sources is thought to be important for the maintenance of certain types of immunological memory. As already discussed above, cytokines are powerful regulators of the immune response (Table 6.1). Whereas some of these factors activate the immune system, others can exert inhibitory effects. For example, transforming growth factor-β (TGF-β) is a potent inhibitor of T- and B-cell proliferation. Other cytokines such as IFN-γ inhibit IL-4 activation of B cells, whereas IL-4 and IL-10 inhibit IFN-γ activation of macrophages and hence DTH reactions. T cells producing these cytokines can therefore be thought of as regulator or suppressor cells. Excessive production of any one of these cytokines may lead to an inappropriate balance between antibody and CMI responses, or to a more generalised immunosuppression affecting the immune response to other microorganisms (see Ch. 7).

In a naturally occurring infection, the infecting dose generally consists of only a small number of microorganisms, whose content of antigen is extremely small compared with that used by immunologists, and quite insufficient on its own to provoke a detectable immune response. But the microorganism then multiplies, and this leads to a progressive and extensive increase in antigenic mass. The classical primary and secondary immune responses merge into one (see Fig. 12.1). Antibodies of various types and reactivities are produced in all microbial infections, and are directed not only against antigens present in the microorganism itself but also against the soluble products of microbial growth, and in the case of viruses against the virus-coded enzymes and other proteins formed in the infected cell during replication. Of the antigens present in the microorganism itself, the most important ones in the encounter between microorganism and host are those on the surface, directly exposed to the immune responses of the host. Responses to internal antigenic components are generally less important, although they are often of great help in detecting past infection, may appear on infected cells as targets for cytotoxic T cells, and may play a part in immune complex disease (see Ch. 8).

There are three other important adjuncts to the immune response. These are complement, phagocytic cells (macrophages and polymorphs) and natural killer cells, which are described under separate headings below. Each is involved in various types of immune reactions.

Antibody Response

Types of immunoglobulin

By the time they reach adult life, all animals, including man, have been exposed to a wide variety of infectious agents and have produced antibodies (immunoglobulins) to most of them. Serum immunoglobulin levels reflect this extensive and universal natural process of immunisation. The different classes of immunoglobulin, with some of their properties, are shown in Table 6.2. All are glycoproteins. The major circulating type of antibody is immunoglobulin G (IgG). It has the basic four-chain immunoglobulin structure in the shape of a Y, as illustrated in Fig. 6.3, and a molecular weight of 150 000. The molecule is composed of two heavy and two light polypeptide chains held together by disulphide bonds. For a given IgG molecule the two light chains are

Table 6.2. Properties of immunoglobulin classes in man

Property	IgG	IgM	IgA[a]	IgE	IgD
Mol. wt	150 000	900 000	385 000 (170 000)	190 000	180 000
Heavy chain	γ	μ	α	ε	δ
Half-life (days)[b]	25	5	(6)[c]	2	2.8
Percentage of total immunoglobulin	80	6	(13)	0.002	0–1
Complement fixation	+	++	±	–	–
Transfer to offspring	Via placenta	No transfer	Via milk	No transfer	No transfer
Proportion in:					
blood	50–60%	90+%	0	V. low	90+%
extracellular fluids	40–50%	<10%	0	–	–
secretions	0[d]	0[d]	100%	High	0[d]
Functional significance	Major systemic immunoglobulin	Appears early in immune response Appears early in development	Present on mucosal surfaces	Allergenic responses, e.g. epithelial surfaces	Unknown; most of it present on surface of B cells

[a] Data for secretory IgA; serum IgA in parentheses. In human serum this antibody is mainly a monomer.
[b] Half-life generally shorter for smaller animals.
[c] Strictly speaking, the half-life of secretory IgA on mucosal surfaces is measured in minutes rather than days, because it is soon carried away in secretions or mucus.
[d] Can be increased in inflammation, IgA deficiency.

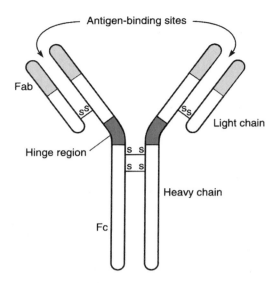

Region with variable amino acid sequence in heavy and light chains, conferring antigen specificity.
Region with Constant amino acid sequence.

Hinge region enables arms to swing out to 180° and bridge antigenic sites. Papain digestion of molecule yields two Fab (fragment antigen-binding) portions, and one Fc (fragment crystallisable) portion which confers biological activity on the molecule (placental passage, binding to phagocytes, etc.)

Fig. 6.3 Basic Y-shaped (four-chain) structure of immunoglobulin G molecule.

either kappa (κ) or lambda (λ), and both heavy chains are gamma (γ). The antigen-binding ends of the light and heavy chains have a unique amino acid sequence for a given antibody molecule and are responsible for its specificity, while the rest of the chains are identical throughout a given class of antibody. The molecule can be split into three parts by papain digestion. Two of these (Fab) represent the arms of the Y and contain the antigen-binding sites; the third part (Fc) has no antigen-binding sites, but carries the chemical groupings that activate complement and combine with receptors on the surface of polymorphs and macrophages (see below). This last activity of the Fc fragment mediates attachment of antibody-coated microorganisms to the phagocyte, giving the antibody opsonic activity. The Fc fragment also contains the groupings responsible for the transport of IgG across the placenta of some mammals. IgG can pass the placenta in primates, including man, but not in rodents, cows, sheep, or pigs. Most IgG antibody is in the blood, but it is also present in smaller concentrations in extravascular tissues including lymph, peritoneal, synovial and cerebrospinal fluids. Its concentration in tissue fluids is always increased as soon as there is inflammation, or when it is being synthesized locally. There are four subclasses of IgG in man, which differ in heavy chains and in biological properties such as placental passage, complement fixation and binding to phagocytes. The amounts present in serum are also different, but almost nothing is known of their relative importance in infectious diseases.

Serum IgM is a polymer of five subunits, each with the basic four-chain structure but with a different heavy chain (μ), and has a molecular weight of 900 000. Because it is such a large molecule, it is confined to the vascular system. Its biological importance is first that, molecule for molecule, it has five times the number of antigen-reactive sites as IgG. It therefore has high avidity and is particularly good at agglutinating microorganisms and their antigens. It also has five times the number of Fc sites and therefore at least five times the complement-activating capacity (see below). A mere 30 molecules of IgM attached to *E. coli* ensure its destruction by complement, whereas 20 times as many IgG molecules are required. Also, IgM is formed early in the immune response of the individual. An infectious disease can be regarded as a race between the replication and spread of the micro-organisms on the one hand, and the generation of an antimicrobial immune response on the other. A particularly powerful type of antibody that is produced a day or two earlier than other antibodies may often have a determining effect on the course of the infection, favouring earlier recovery and less severe pathological changes. As each immune response unfolds, the initially formed IgM antibodies are replaced by IgG antibodies, and IgM are thus only detectable during infection and for a short while after recovery. The presence of IgM antibodies to a microbial antigen therefore indicates either recent infection or persistent infection. A pregnant woman with a recent rubella-like illness would have rubella IgM antibodies if that illness was indeed rubella. Measles virus occasionally persists in the brain of children instead of being eliminated from the body after infection, and the progressive growth of virus in the brain causes a fatal disease called subacute sclerosing panencephalitis. The onset of disease may be 5–10 years after the original measles infection, but IgM antibodies to measles are still present because of the continued infection.

IgM antibodies are not only the first to be formed in a given immune response, but are also the first to be formed in evolution. They are the only antibodies found in a primitive vertebrate such as the lamprey. IgM antibodies are also the first to be found during the development of the individual. After the fifth to sixth month of development, the human foetus responds to infection by forming almost entirely IgM antibodies, and the presence of raised IgM antibodies in cord blood suggests intrauterine infection. The only maternal antibodies that can pass the placenta to reach the foetus are IgG in type, and thus the presence of IgM antibodies to rubella virus in a newborn baby's blood shows that the foetus was infected.

Secretory IgA is the principal immunoglobulin on mucosal surfaces and in milk (especially colostrum). It is a dimer, consisting of two subunits of the basic four-chain structure with heavy chains, and as the molecule passes across the mucosal epithelium, it acquires an additional 'secretory piece'. Secretory IgA has a molecular weight of 385 000. It does not activate complement (see Ch. 9); although monomeric IgA–antigen complexes do activate the alternative complement pathway. It has to function in the alimentary canal, and the secretory piece gives it a greater resistance to proteolytic enzymes than other types of antibody. In the submucosal tissues, the IgA molecule lacks a secretory piece, and enters the blood via lymphatics to give increased serum IgA levels in mucosal infections.

In the intestine, that seething cauldron of microbial activity, immune responses are of immense importance but poorly understood. On the one hand, commensal inhabitants are to be tolerated, but on the other hand, protection against pathogens is vital. Powerful immunological forces are present. The submucosa contains nearly 10^{11} antibody-producing cells, equivalent to half of the entire lymphoid system, and in man there are 20–30 IgA cells per IgG cell. Immune responses are probably generated against most intestinal antigens (see p. 28), and the sheer number of these antigens is formidable. It is a daunting prospect to unravel immune events and understand control mechanisms in this dark, mysterious part of the body. It has become clear that in some species most of the intestinal secretory IgA comes from bile. Although some of the IgA produced by submucosal plasma cells attaches to the secretory piece present on local epithelial cells and is then extruded into the gut lumen, most of it reaches the blood. In the liver, IgA attaches to the secretory piece which is present on the surface of hepatic cells, and is transported across these cells (see p. 134) to appear in bile. This is important in the rat, but perhaps less so in man. One consequence of the IgA circulation is that, when intestinal antigens reach subepithelial tissues, they can combine with specific IgA antibody, enter the blood as immune complexes and then be filtered out and excreted in bile as a result of IgA attachment to liver cells.

There is a separate circulatory system that involves the IgA producing cells themselves. After responding to intestinal antigens, some B cells enter lymphatics and the bloodstream, from whence they localise in salivary glands, lung, mammary glands and elsewhere in the intestine. Localisation at these sites is achieved by recognition of particular receptors on vascular endothelial cells called addressins (see later). In this way, specific immune responses are seeded out to other mucosal areas, where IgA antibody is produced and further responses to antigen can be made.

IgA antibodies are important in resistance to infections of the mucosal surfaces of the body, particularly the respiratory, intestinal and urinogenital tracts. Infections of these surfaces are likely to be prevented by vaccines that induce secretory IgA antibodies (see Ch. 12) rather than IgG or IgM antibodies. However, most patients with selective IgA deficiencies do not show undue susceptibility to infections of mucosal surfaces, probably because there are compensatory increases in the concentration of IgG and IgM antibodies on these surfaces.* Those that are more susceptible generally have associated deficiencies in certain IgG subclasses.

IgE is a minor immunoglobulin only accounting for 0.002% of the total serum immunoglobulins, and it is produced especially by plasma cells below the respiratory and intestinal epithelia. It has a marked ability to attach to mast cells, and includes the reagenic antibodies that are involved in anaphylactic reactions (see Ch. 8). When an antigen reacts with antibody attached to a mast cell, mediators of inflammation (serotonin, histamine, etc.) are released. Thus, if a microorganism, in spite of secretory IgA antibodies, infects an epithelial surface, plasma components and leucocytes will be focused on to the area as soon as microbial antigens interact with specific IgE on mast cells. IgE is considered to be important in immunity to helminths. Larval forms coated with IgE antibodies are recognised by eosinophils and destroyed.

In humans, intestinal antibody is measured in duodenal or jejunal aspirates, or in faeces ('coproantibody'). Antibody from the entire gut can be sampled by 'intestinal lavage', when an isotonic salt solution is drunk until there is a watery diarrhoea, one litre of which is collected, heat inactivated, filtered and concentrated.

IgD antibodies are for the most part present on the surface of B lymphocytes. The same cells also carry IgM antibody, and it might be expected that IgD serves as a receptor for antigen and is involved in the activation of B cells. However, its main function is not clear.

General features

The antibody response takes place mostly in lymphoid tissues (spleen, lymph nodes, etc.) and also in the submucosa of the respiratory and intestinal tracts. Submucosal lymphoid tissues receive microorganisms and their antigens directly from overlying epithelial cells, and lymphoid tissues in spleen and lymph nodes receive them via blood or lymphatics (see Ch. 5). Initial uptake and handling is by macrophages and dendritic cells, following which antigens are delivered to CD4 T cells (see above).

On first introduction of an antigen into the body, the antibody response takes several days to develop. Pre-existing antigen-sensitive

---

* Also they may show less deficiency in secretory IgA than in the serum IgA which is usually measured. In any case, the details differ in different species, and in sheep, for instance, IgG figures as prominently as IgA in the secretory immunoglobulins. Finally, it must be remembered that in the lower respiratory tract, at least, local CMI responses can be induced, and may contribute to resistance.

B lymphocytes encounter antigen via the immunoglobulin receptor. The antigen is internalised and processed via the exogenous pathway and presented in association with MHC class II molecules to activated T-helper cells. T-cell help is provided via CD40 activation and/or cytokine receptors on B cells, e.g. IL-4 receptor (see Fig. 6.2). The B cells then:

1. Divide repeatedly, forming a clone of cells with similar reactivity (clonal expansion), some of which remain after the response is over, as memory cells.
2. Differentiate, developing an endoplasmic reticulum studded with ribosomes, in preparation for protein synthesis and export. The cytoplasm of the cell therefore becomes larger and basophilic.
3. Synthesise specific antibody. The fully differentiated antibody-producing cell is a mature plasma cell. Each clone of cells forms immunoglobulin molecules of the same class and the same antigenic specificity.

Although the majority of antibody production occurs following T-cell help, B cells can also become activated directly by polymeric antigens (antigens with repeating epitopes) which cause cross-linking of specific immunoglobulin receptors. This is commonly seen with bacteria, but is also observed with viruses such as polyoma virus, rotavirus and vesicular stomatitis virus. T-cell-independent antibody responses are largely confined to the IgM isotype and have low affinity and short-lived memory. However, these responses can be protective and in the race to stem the dissemination of pathogens in the host such antibody responses may provide a key defence.*

In a natural infection the initial microbial inoculum is small, and the immune stimulus increases in magnitude following microbial replication. Small amounts of specific antibody are formed locally within a few days, but free antibody is not usually detectable in the serum until about a week after infection. As the response continues and especially when only small amounts of antigen are available, B cells producing high-affinity antibodies are more likely to be triggered, so that the average binding affinity of the antibody increases as much as 100-fold. The role of antibody in recovery from infection is discussed in Ch. 9, the relative importance of antibody and cell-mediated immunity depending on the microorganism. On re-exposure to microbial antigens later in life, there is an accelerated response in which larger amounts of mainly IgG antibodies are formed after only 1 or 2 days. The capacity to respond in this accelerated manner often persists for life, and depends on the presence of 'memory cells'.

---

* Remember that every infection is a race between the ability of the invading microbe to multiply and cause disease, and the ability of the host to mobilise specific and nonspecific defences – a delay of a day or so on the part of the host can be critical.

Antibodies to a given microbial antigen remain in the serum, often for many years. Since the half-life of IgG antibody in man is about 25 days, antibody-forming cells are continually active. In some instances (herpes viruses, tuberculosis) microorganisms remain in the body after the original infection, and can continuously stimulate the immune system. In other instances it seems clear that antibody levels are kept elevated partly by repeated re-exposure to the microbe, which gives subclinical re-infections and boosts the immune response. This is known to occur with whooping cough, measles and other infections. Sometimes, however, antibodies remain present in the serum for very long periods in the absence of persistent infection or re-exposure. For instance, five of six individuals who suffered an attack of yellow fever in an epidemic in Virginia, USA in 1855 were found to have circulating antibodies to yellow fever 75 years later. There had been no yellow fever since the time of the original epidemic. Similarly, evidence from isolated Eskimo communities in Alaska show that antibody to poliomyelitis virus persists for 40 years in the absence of possible re-exposure. It is now known that antigen can persist on the surface of follicular dendritic cells (another member of the dendritic cell family involved specifically with presenting antigen to B cells) in lymphoid follicles for prolonged periods. This provides a continual source of antigenic stimulation to promote B-cell survival and presumably maintains B-cell memory. Plasma cells have also been recorded to survive in the bone marrow for long periods, far in excess of what had previously been predicted for the half-life of these cells in lymph nodes and spleen.

As a general rule, the secretory IgA antibody response is short-lived compared with the serum IgG response.* Accordingly resistance to respiratory infection tends to be short lived. Repeated infection with common cold or influenza viruses often means infection with an antigenically distinct strain of virus, but re-infections with respiratory syncytial virus or with the same strain of parainfluenza virus, for instance, are common. Re-infection of the respiratory tract or other mucosal surfaces is more likely to lead to signs of disease, because of the short incubation period of this type of infection. After re-infection with a respiratory virus there can be clinical disease within a day or two, before the immune response has been boosted and can control the infection. This is in contrast to re-infection with say measles or typhoid; these are generalised infections, and the long incubation period gives ample opportunity for the immune response to be boosted and control the infection long before the stage of clinical disease (Fig. 6.4).

---

* One factor is that, although there are very large numbers of IgA-producing plasma cells in submucosal tissues, this immunoglobulin is exported to the outside world, whereas IgG accumulates in the blood as it is produced.

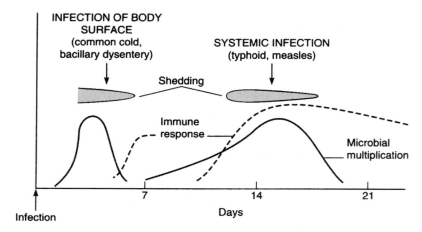

Fig. 6.4 Distinction between infections of body surfaces and systemic infections.

The newborn infant has acquired the IgG antibodies of the mother via the placenta and is protected against most of the infections that she has experienced. There is also transfer of secretory IgA antibodies initially via the milk, human colostrum containing 2–40 mg ml^{-1} IgA.* This maternal 'umbrella' of antibodies lasts for about 6 months in man, and the infant encounters many infectious agents while still partially protected. Under these circumstances the infectious agent multiplies, but only to a limited extent, stimulating an immune response without causing significant disease. The infant thus acquires active immunity while partially protected by maternal immunity. Very occasionally, a mother has not encountered a common microorganism and therefore has no immunity to transfer to her offspring. Certain virus infections, such as herpes simplex and rubella are especially severe in the totally unprotected small infant, causing systemic illness and often death. There are other major differences between the response to infectious agents of immature and adult individuals. They are due to age-related differences in the immune response, in the inflammatory response, in tissue susceptibility, etc. and are dealt with more fully in Ch. 11. As the child encounters the great variety of natural infections total serum antibody levels rise, reaching adult levels by about 5 years of age. Immunological reactivity reaches peak levels in the adolescent or

---

* Milk also contains other protective factors such as lactoferrin (see p. 387), lactoperoxidase, lysozyme and ill-defined lipids and glycoproteins with an antiviral activity. Lactoadherin, a 46K glycoprotein in human milk, binds to and inactivates rotaviruses. Oligosaccharides or glycolipids in milk can bind to pathogenic bacteria by resembling the natural receptors for these bacteria (see p. 14). In horses, cows, sheep, etc. the uptake of colostrum immediately after birth is vital for protection against certain infections. For instance, calves deprived of colostrum are likely to die of *E. coli* septicaemia within a few days of birth.

young adult, but falls off detectably in old individuals. This makes old people less resistant to primary infections, and less capable of keeping certain latent infections under control (see Ch. 10).

Protective action of antibodies

Antibodies are formed against a great variety of microbial components and products. The larger microorganisms have more components and products because they have more genes (see Table 12.1). The presence of antibody indicates present or past infection, but only some of the antibodies have a significant protective function. Protective antibodies generally combine with antigenic components on the surface of microorganisms and prevent them attaching to cells or body surfaces, prevent them from multiplying, and sometimes kill them. The antimicrobial actions of antibodies can be categorised as follows (in approximate order of importance):

1. Antibodies promote phagocytosis and subsequent digestion of microorganisms by acting as cytophilic antibodies or opsonins (see below).
2. Antibodies combining with the surface of microorganisms may prevent their attachment to susceptible cells or susceptible mucosal surfaces (streptococci, gonococci, rhinovirus) (see Table 2.1).
3. Antibodies to microbial toxins or impedins (see p. 98) neutralise the effects of these materials.
4. By combining with microbes or antigens and activating the complement system, antibodies induce inflammatory responses and bring fresh phagocytes and serum antibodies to the site of infection. This can have pathological as well as antimicrobial results (see Ch. 8).
5. Antibodies combining with the surface of bacteria, enveloped viruses, etc., may activate the complement sequence and cause lysis of the microorganism (e.g. *Vibrio cholerae*, *E. coli*, parainfluenza virus, *Mycoplasma pneumoniae*). Host cells bearing new antigens on their surface as a result of virus infection are lysed in the same way, often before virus replication is completed (see Ch. 9).
6. Antibodies enable certain leucocytes to kill infected host cells bearing viral or other foreign antigens on their surface. These include monocytes, polymorphs and natural killer (NK) cells, which act by recognising IgG antibody specifically attached to the target cell surface. Bacteria such as *Shigella* and meningococci can also be killed in this way. NK cells (see page 172) are present in blood and lymphoid tissues and bear receptors for the Fc region of IgG (FcγRIII or CD16). Antibody-dependent cell cytotoxicity (ADCC) of this type is more efficient per antibody molecule than complement-

dependent cell killing and is therefore more likely to be relevant *in vivo*.

7. Antibodies combining with the surface of microorganisms agglutinate them, reducing the number of separate infectious units and also, at least with the smaller microorganisms, making them more readily phagocytosed because the clump of particles is larger in size.

8. Antibodies attaching to the surface of motile microorganisms may render them nonmotile, perhaps improving the opportunities for phagocytosis.

9. Antibodies combining with extracellular microorganisms may inhibit their metabolism or growth (malaria, mycoplasmas).

10. Antibodies neutralise virus infectivity by a variety of mechanisms which nearly always affect a stage in the life cycle following attachment to the host cell (rhinovirus is the exception – see (2) above). This could be internalisation by the cell and/or other aspect of the uncoating process.

The ways in which antibodies can be detected or assayed in the laboratory are shown in Table 6.3.

Table 6.3. Tests for antibodies formed against microorganisms

Name of test	Nature of antigen	Positive test result	Microorganism (examples)
Haemagglutination inhibition	Haemagglutinin, forming part of surface of virus particle	Inhibition of erythrocyte agglutination	Rubella Influenza
Haemagglutination	Microbial antigen absorbed to surface of erythrocyte	Antibody to microbial antigen agglutinates erythrocytes	Hepatitis B
Precipitation or agglutination	Antigen on surface of microorganism Soluble microbial antigen	Antibody causes visible precipitation of microorganism or antigen	*Salmonella* (Widal test), *Brucella* Diphtheria toxin (Elek test)
Gel diffusion	Diffusible microbial antigen	Antibody reacts with antigen to form precipitation line in gel	Histoplasmosis, Hepatitis B
Complement fixation	Microbial antigen that reacts with antibody; resulting complex combines with ('fixes') complement	Complement depleted ('fixed')	Most microorganisms
Latex test	Microbial antigen adsorbed to latex particle	Antibody to microbial antigen agglutinates latex particles	Hepatitis B

Name of test	Nature of antigen	Positive test result	Microorganism (examples)
Neutralisation test	Viral surface antigen necessary for multiplication in experimental animal or cell culture	Antibody inhibits multiplication and prevents pathological lesions, death or cell damage	Most viruses
	Bacterial toxin	Biological effect of toxin inhibited	Diphtheria, etc.
Immobilisation test	Antigen on locomotor organ (flagellum, cilium)	Inhibition of mobility	*Treponema pallidum*
Immunofluorescence test (see pp. 124–125)	Antigen on microorganism or antigen formed in infected cell	Fluorescein-labelled antibody seen on microorganism or in infected cell by ultraviolet microscopy	*Treponema pallidum*, Respiratory syncytial virus, Toxoplasmosis
Radioimmunoassay	Microbial antigen	Radiolabelled antibody bound to microbial antigen	Hepatitis B
Enzyme-linked immunosorbent assay (ELISA)	Microbial antigen	Antibody linked to enzyme reacts with antigen. Specific binding revealed when enzyme causes colour change in substrate	Rubella, etc.; widely used
Capsular swelling	Capsules on surface of bacteria	Swelling of capsule	Pneumococci, *Klebsiella*
Western blot	Microbial antigen	Antigen separated by electrophoresis and identified by specific labelled antibodies	Most microorganisms

T-Cell-Mediated Immune Response

When T cells leave the thymus they enter into the peripheral circulation, touring the lymphoid system on the look out for foreign antigens. They are found in discrete areas in lymphoid organs, notably around the splenic arterioles and paracortical areas of lymph nodes; also in the blood and lymph. About 90% of the recirculation takes place from blood to lymph nodes via the postcapillary venules and then via lymphatics back to the blood (Fig. 6.5). This trafficking of lymphocytes is dependent upon the recognition of selective ligands on endothelial cells (called addressins) that act as postal codes enabling lymphocytes to identify the correct location. Key addressins associated with entry into lymph

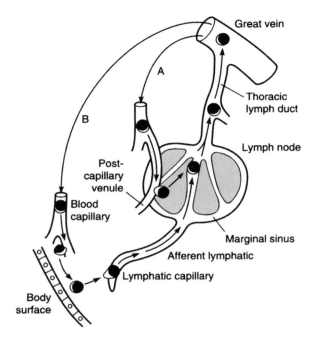

Fig. 6.5 Lymphocyte recirculation. Recirculating lymphocytes in man are mostly T lymphocytes; approximately 90% of recirculation is by Route A and 10% by Route B.

nodes are GlyCAM-1 (glycosylation-dependent cell adhesion molecule 1) and CD34. Recognition of these addressins by T cells involves L-selectin, which binds to GlyCAM-1, and LFA-1 (lymphocyte functional antigen 1), a member of the integrin family. The remaining 10% of cells leave capillaries in various parts of the body, moving through the tissues, entering lymphatics and passing through local lymph nodes. This last route is particularly important in the small intestine, where a different set of addressins operate characterised by MAdCAM-1 (mucosal addressin cell adhesion molecule 1) which interacts with lymphocytes expressing the integrin $\alpha 4/\beta 7$. A given T cell circulates about once in 24 hours in a man, and once in 2 hours in a mouse.

T cells become arrested in lymph nodes when they encounter antigen of the appropriate specificity presented on the surface of a dendritic cell. As mentioned above, dendritic cells acquire antigens at sites of infection where they become activated and migrate via the lymphatics to the local lymph node. In this environment dendritic cells differentiate, put out dendritic processes and express key accessory molecules important for interacting with and stimulating bound T cells. For example, a CD4 T cell will engage MHC class II plus foreign peptide with the α/β TCR and CD4 (which helps to stabilise the interaction) and also binds the accessory molecules CD40 and B7 on the dendritic cell; these react with the CD40 ligand and CD28, respectively, present on the T cell. A CD8 T cell would initially recognise MHC class I, but the accessory molecules would be the same as for CD4 T cells. These interactions are critical events in the evolution of the adaptive immune response as they lead to the activation and clonal expansion of both CD8 and CD4 T cell populations (see Fig. 6.2).

Dendritic cells continue to transport antigens from the site of infection, thereby servicing new waves of T cells that become attracted to the lymph node. After around 4–5 days* clonally selected T cells begin to leave the node via the efferent lymphatics and join the bloodstream. They now target the tissue where the infection is raging. Identification of and access to the site of infection from the bloodstream involves the T cells detecting a gradient of chemokines (see Glossary), such as IL-8 and RANTES. These chemokines are released from damaged tissues, and aid the exit of T cells via endothelial cells which must display the appropriate addressins. Once at the site of infection, T cells will encounter macrophages and polymorphs, part of the advanced guard of the innate immune defences (see below). Macrophages provide an additional stimulus to CD4 T cells following interaction with MHC class II, resulting in the release of a variety of cytokines with antimicrobial activity. A list of some of the common T-cell-derived cytokines and their properties is shown in Table 6.1. The infected tissue now undergoes further change with the recruitment of more monocytes from the blood. As they enter the tissue, they become activated by IFN-γ released from activated T cells, predominantly CD4 T cells. This in turn causes dramatic changes in macrophages resulting in antimicrobial activity (see Ch. 4). The tissue becomes swollen and is characterised by mononuclear cell infiltration, the hallmark of a delayed-type hypersensitivity reaction. These reactions are particularly important for controlling intracellular bacterial infections such as *Listeria monocytogenes* and *M. tuberculosis*.

Cytotoxic T cells (CD8 T cells) entering a site of infection will sample the MHC class I molecules on infected cells through their TCR. Specific recognition will lead to T-cell activation and the release of IFN-γ, but in addition they become cytolytic, killing infected cells by inducing apoptosis (see Glossary). Cytolysis involves intimate contact between the T cell and target cell, resulting in the T cell delivering a 'lethal hit' in the form of perforin, a molecule similar to C9 of the complement system which forms a 'plug' in cell membranes causing cell lysis (see p. 176) and

---

* The time taken for some responses to manifest depends upon the infecting organism and the immune competence of the host. In the majority of virus infections, cytotoxic T cells are first detected in lymph nodes or spleen 4–5 days after infection. A delayed hypersensitivity response to vaccinia virus is positive within 1 week, whereas the same response to infections such as tuberculosis, brucellosis and leishmaniasis, is not seen for several weeks.

granzymes (proteases). Perforin inserts into the membrane of the target cell enabling passage of the granzymes resulting in cell death. The cytotoxic cell then disengages and homes on to another target. This is an efficient and rapid killing mechanism, capable of destroying virus-infected cells in minutes, well before new virions are assembled and released. Perforin appears to be a crucial molecule in this process, since mice lacking the perforin gene are unable to eliminate infection caused by LCMV (lymphocytic choriomeningitis virus). Other mechanisms of cell killing performed by cytotoxic cells include Fas/Fas ligand interactions. Fas is a member of the TNF receptor family found on several cell types. These receptors have 'death' domains which, when activated by Fas ligand on CD8 T cells, lead to target cell death by apoptosis.

As the effector phase of the T-cell response unfolds, some of the clonally expanded cells become memory T cells. Memory T cells can be distinguished from naive T cells (yet to encounter antigen) by the presence of particular membrane markers. The commonest marker involves different isoforms of CD45. The high molecular weight isoform of CD45, CD45RA, is found on naïve T cells and the low molecular weight isoform, CD45RO, is found on memory T cells. It is still unclear whether memory cells arise directly from naïve cells or whether they arise from effector cells. The purpose of immunological memory is to provide a group of cells capable of a rapid response to pathogens on successive encounters. This is clearly the case in delayed-type hypersensitivity responses where mononuclear cell infiltration can be seen within 24–48 hours of antigen challenge.* This type of test can be used in the clinic to determine prior exposure to an infectious agent. For example, the tuberculin test, is of some practical value in determining previous exposure to tuberculosis. The test involves delivering mycobacterial antigens into the skin. Those with a positive response have at some time been infected or are at present infected with tuberculosis, with related mycobacteria, or with the attenuated mycobacteria in the BCG vaccine. Those with a negative response have never been infected, or have been infected but have recovered and eliminated bacteria from the body. The response may also be negative early after infection before T-cell immunity has had time to develop, or in acute disseminated infection where the T-cell response is feeble.

A question of major importance in immunology is how is memory maintained. Current ideas suggest that antigen must be the driving force for maintaining a memory cell pool. However, in the absence of any obvious source of antigen, there is evidence that T-cell memory still persists. Even here, it could be argued that there is cross-reactivity between other antigens that serve to stimulate the memory pool of cells. In order to overcome this argument, recent evidence has shown

---

* A positive response is seen by skin swelling at the site of injection (forearm in man, ear or footpad in mouse). The word 'delayed' is used to contrast it with antibody-mediated responses which appear within an hour.

that memory T cells can be maintained in mice lacking MHC class I and MHC class II molecules. This implies that mechanisms other than TCR recognition of antigen are involved in maintaining the pool of memory T cells.

We know less about the population of T cells that express the γ/δ TCR. They are distributed throughout the body and, in some species, notably cattle and sheep, can account for up to 60% of all T cells. These cells do not appear to recognise peptides presented by MHC molecules, but associate directly with various structures, such as stress proteins, non-classical MHC molecules and glycolipids. A subpopulation of γ/δ T cells exists at epithelial surfaces, intraepithelial lymphocytes, which are thought to play an early defensive role in pathogen-induced damage of the epithelium. We still have a lot to learn about the properties of these T cells.

T-cell-mediated immune responses are detected and quantified by the methods listed in Table 6.4.

Natural Killer Cells

Natural killer (NK) cells represent a first line of defence against intracellular pathogens. These large granular lymphocytes constitute a separate lineage of lymphocytes formed in the bone marrow and found patrolling the blood and lymphoid tissues. They were functionally characterised by their ability to kill certain tumour cell lines *in vitro*. Classical NK cells are distinct from B and T lymphocytes in that they do not have immunoglobulin or α/β and γ/δ TCR receptors. NK cells have receptors, Ly49 (mouse) and killer inhibitory receptors (man), that interact with MHC class I alleles producing a negative/inactivation signal, preventing NK cell killing. These receptors serve to protect normal cells from the attention of NK cells. Other receptors, not yet identified, are thought to recognise changes in glycosylation patterns on infected cells causing activation of NK cells and destruction of the target cell. These cells are efficient killers of virus-infected cells when activated by IL-2, IFN-α/β or IL-12. In particular, those virus infections (herpes viruses, adenovirus) resulting in a reduction of MHC class I expression on the target cell are readily killed. Such target cells are only poorly recognised by cytotoxic CD8 T cells, suggesting an important niche for NK cells in immune surveillance against such infections. It seems as if the NK cell evolved to counter the threat of those viruses that switch off MHC class I expression on infected cells and thus evade T-cell recognition and killing. Presumably, NK cells treat any cell with reduced MHC class I expression as though it were virus infected. NK cells appear within hours to days of a virus infection, long before specific T cells are detected. They are thought to limit the spread of infection by either killing virus-infected cells (this is by a perforin-

Table 6.4. Detection and assay of T-cell-mediated immunity

Response of sensitised lymphocytes to antigen	Test system
DNA synthesis and mitosis	Incorporation of tritiated thymidine into lymphocyte DNA
Liberation of cytokines	Measurement of growth and proliferation of specific cytokine-dependent cell lines, e.g. IL-2, IL-3
	Inhibition of virus replication, e.g. IFN
	Inhibition of cell growth (e.g. TGF-β) or cytolytic activity on virus-infected or tumour cells (e.g. TNF-α)
	Capture of cytokines by specific antibodies fixed to plastic surfaces and detection of cytokine binding by a second specific antibody using ELISA – a general assay
Cytotoxicity	Destruction of cells bearing, e.g. viral antigen on surface. Cell death can be measured by release of radioactive chromium
Induction of inflammation and mononuclear infiltration into tissues	'Delayed' swelling and induration (DTH response), after injection of antigen into skin
Antimicrobial activity (attributable to above activities)	Transfer of 'immune' T lymphocytes into animal infected 1–2 days earlier (before onset of T-cell response) and test for reduction in microbial titre in organs 24 h later
Clonal analysis of T-cell responses	Limiting dilution analysis – detects the frequency of antigen-specific T cells using the above procedures
	MHC class I-peptide tetramers–monomers of particular MHC class I are chemically linked together *in vitro* and mixed with the relevant peptide in order to mimic events on the surface of a cell. These tetramers can be fluorescently labelled and used like antibody molecules to stain cells carrying the appropriate TCR .
	Quantification of reactive T cells to a specific antigen (peptide) can be achieved by staining for IFN-γ as a functional marker of cell activation. This is now used as one of the principal methods for quantifying memory T-cell responses.

dependent mechanism) or by the production of cytokines such as IFN-γ and TNF-α, both potent activators of phagocytic cells, critical for the defence against intracellular bacteria such as *Listeria monocytogenes*. NK cells also express the low-affinity receptor for IgG (Fcγ RIII or CD16). This enables IgG-coated target cells to be recognised and rapidly killed in a process called antibody-dependent cell cytoxicity. NK cells are the major exponents of this process.

Macrophages, Polymorphs and Mast Cells

Macrophages, because of their phagocytic prowess and their location in many tissues, are inevitably important in the uptake of invading microorganisms, and they have important functions as phagocytes whether or not an immune response has been generated. They are involved in the initiation of immune responses to infection, as described above and are also important in the expression of the immune response seen at a later stage in the infection. In this they operate in close association with both antibodies and T cells (see below).

Polymorphs are also of extreme importance, operating in association with antibody and complement. They are mainly present in the blood, and do not continuously monitor the tissues and fluids* of the body. They are, however, rapidly delivered to tissues as soon as inflammatory responses are initiated (see Ch. 3). They are short-lived; during an infection macrophages are always having to deal with dead polymorphs containing microorganisms in various stages of destruction and digestion. Both polymorphs and macrophages bear Fc and C3b receptors on their surfaces which promote the phagocytosis of immune complexes or microorganisms coated with antibody (see below). By preparing microorganisms for phagocytosis in this way, specific antibodies and complement act as opsonins. Complement also often increases the virus-neutralising action of antibody, presumably by adding to the number of molecules coating the virus particle and further preventing its attachment to susceptible cells. When a microorganism is coated with antibody it undergoes a different fate after phagocytosis. *Toxoplasma gondii*, for instance, normally manages to enter macrophages without triggering an oxidative metabolic burst (see p. 91) but this antimicrobial response does occur when the parasite is coated with antibody, and is presumably triggered by Fc-mediated phagocytosis.† In the case of viruses, antibody can promote uptake and degradation by macrophages. When C3 is associated with antibody on the surface of a microorganism, it often increases the degree of opsonisation. IgM antibodies attached to *Pseudomonas* or other Gram-negative bacilli may even require complement before there is opsonisation. Sometimes, however, C3 is activated on the microbial surface by the alternative pathway (see below), and acts as an opsonin

---

* Sometimes, however, circulating polymorphs are arrested in capillaries, especially in the lung, and can then phagocytose microorganisms present in the blood.

† Interestingly enough, with viruses like dengue that can infect macrophages, small amounts of antibody actually enhance infection of these cells, presumably by enhancing the uptake or altering the intracellular fate of the virus. The Fc receptor becomes a Trojan horse (see pp. 287–288). Larger amounts of antibody prevent infection in the conventional fashion.

independently of antibody. This may be important early in pneumococcal infection, for instance, when there is not much antibody available. Opsonised phagocytosis is the principal method of control of infections with microorganisms such as the streptococcus, staphylococcus or encapsulated pneumococcus, the antibody response and complement acting in conjunction with phagocytic cells.

Macrophages also help give expression to the T-cell response, and this seems particularly important in the case of microorganisms such as mycobacteria, *Leishmania*, herpes viruses, brucellas, lymphogranuloma inguinale, etc. that survive and multiply within phagocytes and other cells. When sensitised T cells encounter specific antigen, they release a number of cytokines, as described above, with a profound effect on macrophages. Some induce inflammation and are chemotactic, bringing circulating macrophage precursors (monocytes) to the site of the reaction, and others inhibit their movement away from the site. Mere assembling of macrophages at a focus of infection is sometimes enough to control the infection but, especially for microorganisms that are not easily killed in macrophages, something more than this is needed. Thus there are other cytokines, especially IFN-γ, that 'activate' macrophages, causing them to develop increased phagocytic and digestive powers. The increased phagocytosis can be demonstrated directly by the uptake of particles or microorganisms, and is also evident by increased attachment and spreading on a glass surface, in what can be regarded as a heroic attempt to phagocytose the entire vessel in which the macrophages are contained. The increased digestive powers are associated with increased lysosomes and lysosomal enzyme content, and there is also an increased ability to generate oxygen radicals (see p. 92). As a result of these changes, macrophages show increased ability to destroy ingested microorganisms. For instance, in mice that have recently developed a T-cell response to tuberculosis, macrophages are activated and have an increased ability to ingest and destroy tubercle bacilli. Indeed, resistance to tuberculosis in man is largely attributable to the antibacterial activity of activated macrophages. Mouse macrophages activated in this way by tuberculosis also show increased ability to ingest and destroy certain unrelated intracellular bacteria such as *Listeria monocytogenes*, and protozoa such as *Leishmania* (see below). In other words the macrophage is activated by the cytokine following an immunologically specific interaction between lymphocyte and microbial antigen, but expresses this reactivity nonspecifically against a wider range of microorganisms. Some of the cytokines necessarily have a restricted local area of action, but activated macrophages are not confined to the immediate vicinity of the lymphocyte encounter with antigen. Macrophages elsewhere in the body are often affected, suggesting that the mediators (or the activated macrophages) spread throughout the body. Activation lasts only for a short time and is no longer detectable a week after termination of the infection. In persistent infections such as tuberculosis, macrophages can remain activated for longer periods because of the continued expression of the T-cell response.

Macrophages are also activated during the course of certain virus infections, and can express this reactivity against unrelated micro-organisms. For instance, when mice are infected with ectromelia (mousepox) virus and 6 days later injected intravenously with *Listeria*, the reticuloendothelial macrophages in the spleen show an increased ability to ingest and destroy these bacteria. Macrophages activated in infections by viruses that grow in macrophages may show increased resistance to the infecting virus, and sometimes they are also resistant to infection with unrelated viruses. Macrophage activation is impor-tant in protozoal infections, and specific antibody responses may add to the macrophage's antimicrobial capacity. In a resistant host, *Leishmania* parasites are destroyed after phagocytosis by activated macrophages and unrelated microorganisms such as *Listeria* are also killed. Nonactivated macrophages, in contrast, generally support the growth of both *Leishmania* and *Listeria*. Normal macrophages support the growth of *Toxoplasma gondii* (see p. 106), but after activation during the infection they increase H_2O_2 production (see p. 92) 25-fold and kill the parasite.

Like macrophages, mast cells are also strategically located in tissues throughout the body, acting as an early warning system for intrusive pathogens. Mast cells are important initiators of inflammation, where they are able to respond to a variety of mediators associated with bacteria, viruses, parasites, as well as complement components (C3a, C5a), cytokines (TNF-α, IL-12, stem cell factor) and IgE, a product of Th2-mediated immune responses. Mast cells respond to these stimuli by the rapid release from cytoplasmic granules of proinflammatory mediators (e.g. histamine, proteases) and cytokines (TNF-α, IL-6). These substances have powerful effects on tissues causing, e.g. bron-choconstriction, increased gastrointestinal motility, and increased vascular permeability leading to the accumulation of phagocytes at sites of infection. Mast cells are an important component in host defence against parasitic worms. They achieve this in various ways, for example, by promoting the expulsion of worms through inflammatory mediators causing rapid convulsive movements of the gut, i.e. 'throwing out' the parasite, or by recruiting eosinophils to mediate antibody-dependent cell cytotoxicity on IgE antibody-coated parasites (see Ch. 9). The importance of mast cells in protective immunity has been studied using mast cell-deficient mice, which show a reduced ability to eliminate endo- and ecto-parasites (e.g. biting insects), and to control certain forms of bacterial infection, e.g. bacterial peritonitis. Whereas mast cells do have phagocytic potential and can engulf bacteria, it is thought that protection against bacterial peritonitis occurs via production of TNF-α.

Although mast cells also mediate allergic disorders, some of which can be fatal, one must reconcile these pathological mechanisms with the potential benefits to animals and presumably early man in terms of the evolutionary pressures to acquire and retain defence mechanisms to counter the diversity of macroparasite infections.

Complement and Related Defence Molecules

Complement is a complex series of interrelated proteins present in normal serum. It functions by mediating and amplifying immune reactions. The first component (C1) is a complex of three proteins, C1q, C1r and C1s. It is activated in the classical complement pathway, after C1q combines with immunoglobulin (IgG or IgM) in immune complexes (antibody bound to antigen).* The immune complex may be free in the tissues or located on a cell surface following the reaction of specific antibody with a cell-surface antigen. The activated first component is an enzyme system, and acts on the next component to form a larger number of molecules of the second component's enzyme. This in turn activates larger amounts of the next component, and so on, producing a cascade reaction (Fig. 6.7). A single molecule of activated C1 generates thousands of molecules of the later components and the final response is thus greatly amplified. The later complement components have various biological activities, including inflammation and cell destruction, so that an immunologically specific reaction at the molecular level can lead to a relatively gross response in the tissues.

After activation of the C1 components, C4 and then C2 are activated to form a C3-convertase, and this in turn acts upon C3 to generate C3a, which has chemotactic and histamine-releasing activity. The residual C3b becomes bound to the antigen–antibody complex, and the whole complex can now attach to C3b receptors present on macrophages and polymorphs.† C5 is the next component to be activated, forming C5a with additional chemotactic and histamine-releasing activity. C5b remains with the complex and binds with C6 and C7, and finally with C8 and C9. The membrane attack complex is formed when the last component (C9) is polymerised to form a pore, and is inserted so as to

---

* Fc sites on the immunoglobulin are slightly altered as a result of the combination with antigen, and the altered Fc sites (near the hinge region, see p. 158) bind to the C1q fraction of C1. Each C1q must bind to at least two Fc sites and this means that there must be several IgG molecules close together on the immune complex. With IgM, several Fc sites are present on a single molecule, and IgM therefore activates complement much more efficiently. Although in Fig. 6.6 antibody is shown attached to the complex throughout the sequence, the amplification phenomenon leads to the formation of thousands of additional and separate molecules of the later components.

† The complex also attaches to C3b receptors on non-phagocytic cells (platelets and red cells) in some species and this is called immune adherence. In the blood it can lead to aggregation and lysis of platelets with release of vasoactive amines (see p. 282).

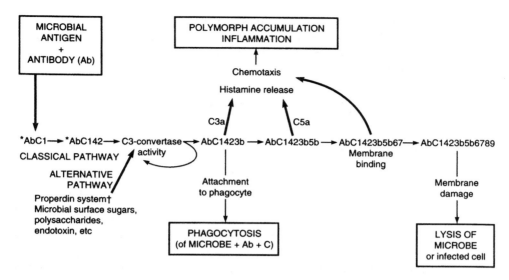

Fig. 6.6 Diagram to show complement activation sequence and antimicrobial actions. Unfortunately the components were numbered before this sequence of action was elucidated († see Glossary).

traverse the cell membrane. This allows a net influx of Na^+ and water, resulting in death of the cell.

Each activation must be terminated somehow, rather than snowball into generalised activation. The complement sequence is therefore controlled by a number of built-in safety devices in the form of inhibitors (regulatory proteins) and unstable links in the complement chain. The activated components have a short (msec) half-life and therefore cannot diffuse through the body and affect distant tissues.

Substances such as microbial polysaccharides and endotoxin can activate the complement system independently of antigen–antibody reactions and C1. These are directly involved in the activation of C3 and this process is referred to as the alternative complement pathway. Activation of the pathway involves the formation of a convertase between C3 and factor B (C3bBb – shares homology with C2bC4b convertase). This binds to bacterial surfaces where it activates more C3, resulting in more C3b deposited. This amplification system very quickly leads to the entire microbial surface being covered in C3b. The whole process is strongly regulated by two inhibitory proteins, factors H and I, which degrade C3b. The fact that the complement sequence can be activated without the need for an antigen–antibody reaction may be important in certain infectious diseases. The peptidoglycan of the cell wall of staphylococci, or the polysaccharides on the surface of the pneumococcus, for instance, could activate the alternative pathway very early in infection before specific antibodies have been formed, leading to antibacterial effects as described below.* The alternative pathway (so called because scientists discovered it after the 'classical' pathway) is probably an ancient defence system and appeared in evolution before the classical pathway.

Complement is capable of causing considerable inflammation and tissue damage, especially because of the amplification phenomenon. Once the sequence is activated, there are four principal antimicrobial functions, each of which is enhanced when both classical and alternative pathways are involved. (Fig. 6.6):

1. The inflammation induced at the site of reaction of antibodies with microbes or microbial antigens focuses leucocytes and plasma factors on to this site.
2. The chemotactic factors attract polymorphs to the site.
3. The C3b component bound to complexes attaches to C3b receptors on phagocytes and thus acts as an opsonin, promoting phagocytosis of microbes and microbial antigens.
4. Where antibody has reacted with the surface of certain microorganisms (Gram-negative bacilli, enveloped viruses, etc.) or with virus-infected cells, the later complement components are activated to form the membrane attack complex. Small pores, 9–10 nm in diameter, appear in the wall of Gram-negative bacilli, for instance, and lysozyme (present in serum) completes the destructive effect. Cells infected with budding viruses and bearing viral antigens on their surface (Fig. 9.2) can be destroyed by complement after reaction with specific antibody, even at an early stage in the infectious process (see also Ch. 9).

The binding or fixing of complement to immune complexes forms the basis of the complement fixation test. In the test for antibody, a known antigen is used in the reaction and vice versa. Complement is added to the reaction mixture and, if there has been a specific antigen–antibody interaction, this complement is fixed and is no longer detectable. The test for complement is by adding sheep red blood cells coated with specific antibody: if complement is present the cells are lysed, but if it has been used up (fixed) the cells are not lysed.

Another series of molecules acting as a first line of defence against microorganisms are the collectins, found in serum and various tissues. The name is derived from their structure, i.e. a <u>coll</u>agen 'stalk', a neck region and a globular carboxy-terminal C-type (calcium-dependent) <u>lectin</u>-binding domain. Included in this family are the mannose-binding protein (MBP), lung surfactant proteins A and D and serum bovine conglutinin. The structure of MBP is similar to C1q in that it

---

* On the other hand, Babesia activate the alternative pathway and depend on this for entry into susceptible erythrocytes, which bear C3b receptors.

resembles a 'bunch of tulips'. Furthermore, MBP is able to substitute for C1q in binding C1r and C1s in activating the classical pathway. These proteins recognise patterns of carbohydrates on the surface of bacteria, viruses and parasites (e.g. MBP binds mannose, fucose, *N*-acetyl glucosamine), where they mediate both complement-dependent and complement-independent protective responses. The lung surfactant proteins have been shown to function in defence against *Pneumocystis carinii* and *Cryptococcus neoformans*, two important respiratory pathogens in immunocompromised hosts, probably by aiding their phagocytosis by alveolar macrophages.

Conclusions Concerning the Immune Response to Microorganisms

Each T or B cell is committed to respond to a particular epitope. The initial encounter with this epitope, whether in lymphoid tissues or elsewhere in the body is a small-scale microscopical event. The purpose of the response, especially when the antigen is from an infecting microorganism, is to turn this microscopical event into a larger event as soon as possible, so that both antibody and T cells can be brought into action on a significant scale. Both types of immune reactive cell are small, and each must differentiate, generating the cytoplasmic machinery needed for synthesis of antibodies (B cell) or cytokines involved in the induction and expression of immunity (T cell). The stimulated cell also gives rise to a dividing population of cells with the same specific immune reactivity, and the response is thus magnified.

The two arms of the immune response to microorganisms are contrasted in Fig. 6.7. Antibody-forming cells (plasma cells) remain for the most part in lymphoid tissues, and antigens are brought to them via blood or lymph. The antibodies formed circulate through the body, acting at a distance from the plasma cells that are situated in lymphoid tissues. Antibodies needed on mucosal surfaces must pass through an epithelial cell layer onto these surfaces, and they are produced by plasma cells situated just below these surfaces. Antibodies bathe tissues and mucosal surfaces where they can react with microbes and microbial antigens in a mostly useful antimicrobial fashion. At the site of the antigen–antibody interaction in tissues, complement is activated and inflammatory responses are generated so that antibodies, phagocytes and more immune reactive cells are delivered to the scene of action.

In contrast, the ability to recognise and destroy infected host cells depends on the local action of individual sensitised T cells. The body's population of sensitised T cells must therefore be circulated throughout the tissues of the body like antibodies, and especially through the lymph nodes to which microbes and their antigens are brought from

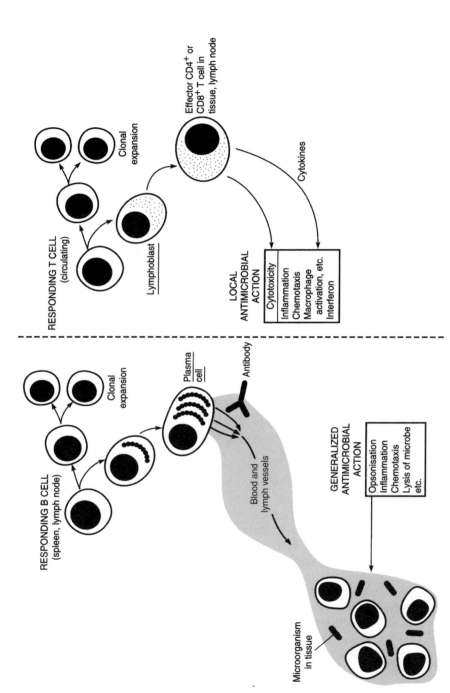

Fig. 6.7 Comparison of antimicrobial action of B and T lymphocytes. Antibody tends to act at a distance from the plasma cell, whereas effector T-cell responses generally require the local presence of the effector cell.

the tissues. In this way microbial antigens can be recognised wherever they are and the T-cell response initiated. The response, involving an accumulation of mostly lymphocytes and macrophages, can be generated locally in the tissue and also more centrally in lymph nodes or spleen. Microbial antigens are most commonly presented to the immune system at the periphery of the body. Langerhans cells in the skin (see p. 151), dendritic cells in submucosal lymphoid tissues and local lymph nodes are involved, and there is a tendency for T-cell responses to predominate. At a later stage in the response, central immune tissues in the spleen are also active. Sometimes, however, there is a reversal of this normal sequence, and antigens are presented directly to central immune tissues. There is then a tendency for the antibody response to be dominant. This is a generalisation, but it may have some bearing on problems of tolerance and suppression (see Ch. 7) and on the subject of T cells versus antibody in recovery from infection (see Ch. 9).

Immune reactions are specifically triggered by T and B cells by virtue of their ability to recognise antigen in a highly specific manner. Macrophages and the dendritic cell family, by processing and presenting antigens, exercise a controlling influence at this stage, and are in close physical association with T and B cells. Macrophages, polymorphs, NK cells and complement play an important part as effectors and amplifiers of the reaction in tissues.

The part played by antibodies, T cells, NK cells, polymorphs, macrophages and complement in recovery from microbial infections is discussed at greater length in Ch. 9.

References

Aderem, A. and Underhill, D. M. (1999). Mechanisms of phagocytosis in macrophages. *Annu. Rev. Immunol.* **17**, 593–623.

Arai, K., Lee, F., Miyajima, A., Miyatake, S., Arai, N. and Yokota, T. (1990). Cytokines: Coordinators of immune and inflammatory responses. *Annu. Rev. Biochem.* **59**, 783–836.

Banchereau, J. and Steinman, R. M. (1998). Dendritic cells and the control of immunity. *Nature* **392**, 245–252.

Bendelac, A., Rivera, M. N., Park, S-H. and Roark, J. H. (1997). Mouse CD1-specific NK1 T cells: Development, specificity and function. *Annu. Rev. Immunol.* **15**, 535–562.

Constant, S. L. and Bottomly, K. (1997). Induction of Th1 and Th2 CD4+T cell responses: The alternative approach. *Annu. Rev. Immunol.* **15**, 297–322.

Cyster, J. G. (1999). Chemokines and cell migration in secondary lymphoid organs. *Science* **286**, 2098–2102.

Doherty, P. C., Allan, W., Eichelberger, M. and Cording, S. R. (1992).

Roles of α/β and γ/δ T cell subsets in viral immunity. *Annu. Rev. Immunol.* **10**, 123–151.

Epstein, J., Eichbaum, Q., Sheriff, S. and Ezekowitz, R. A. B. (1996). The collectins in innate immunity. *Curr. Opin. Immunol.* **8**, 29–35.

Fearon, D. T. and Locksley, R. M. (1996). The instructive role of innate immunity in the acquired immune response. *Science* **272**, 50–53.

Galli, S. J., Maurer, M. and Lantz, C. S. (1999). Mast cells as sentinels of innate immunity. *Curr. Opin. Immunol.* **11**, 53–59.

Germain, R. N. (1994). MHC-dependent antigen processing and peptide presentation: Providing ligands for T lymphocyte activation. *Cell* **76**, 287–299.

Goldrath, A. W. and Bevan, M. J. (1999) Selecting and maintaining a diverse T-cell repertoire. *Nature* **402**, 255–262.

Gordon, S., Fraser, I., Nath, D., Hughes, D. and Clarke, D. (1992). Macrophages in tissues and *in vitro. Curr. Opin. Immunol.* **4**, 25–32.

Kaufmann, S. H. E. (1993). Immunity to intracellular bacteria. *Annu. Rev. Immunol.* **11**, 129–163.

Kerksiek, K. M. and Pamer, E. G. (1999). T cell responses to bacterial infection. *Curr. Opin. Immunol.* **11**, 400–405.

Lanier, L. L. (1998). NK cell receptors. *Annu. Rev. Immunol.* **16**, 359–393.

Nelson, P. J and Krensky, A. M. (1998). Chemokines, lymphocytes and viruses: What goes around, comes around. *Curr. Opin. Immunol.* **10**, 265–270.

Roitt, I., Brostoff, J. and Male, D. (2000). 'Immunology', 6th edn. Harcourt Brace, London.

Slifka, M. K. and Ahmed, R. (1998). Long lived plasma cells: a mechanism for maintaining persistent antibody production. *Curr. Opin. Immunol.* **10**, 252–258.

Springer, T. A. (1994). Traffic signals for lymphocyte recirculation and leukocyte emigration: The multistep paradigm. *Cell* **76**, 301–314.

Szomolanyi-Tsuda, E. and Welsh, R. M. (1998). T-cell-independent antiviral antibody responses. *Curr. Opin. Immunol.* **10**, 431–435.

Tough, D. F., Sun, S., Zhang, X. and Sprent, J. (1999). Stimulation of naive and memory T cells by cytokines. *Immunol. Rev.* **170**, 39–47.

Tomlinson, S. (1993). Complement defence mechanisms. *Curr. Opin. Immunol.* **5**, 83–89.

Underwood, B. J. and Schiff, J. M. (1986). Immunoglobulin A: Strategic defence initiative at the mucosal surface. *Annu. Rev. Immunol.* **4**, 389–417.

7

Microbial Strategies in Relation to the Immune Response

The very existence of successful infectious agents indicates that host defences do not constitute an impenetrable barrier for microorganisms. Infections are common. There are more than 400 distinct microorganisms that infect man alone, and all of us, by the time of death, have experienced at least 150 different infections. Many of these infections are asymptomatic, but a disease sometimes appears before the microorganism has been controlled and eliminated. In many instances, discussed in Ch. 10, the infection is not eliminated, but persists in the body. Persistence represents a failure of the host's antimicrobial forces – forces that can be regarded as having been designed to eliminate invading microorganisms from tissues. The infecting microorganism can then continue to cause pathological changes or continue to be shed from the body.

Once the epithelial surfaces have been penetrated, the major host defences are NK cells, complement, phagocytic cells, interferon and later antibody and T cells. These constitute a mighty hexad whose action is described in Chs 6 and 9. Generally speaking, if there is a way in which host defences can be successfully by-passed or overcome, then at least some microorganisms can be expected to have 'discovered' this. Microorganisms evolve very rapidly in relation to their host, so that most of the feasible antihost strategies are likely to have been tried out and exploited. The microbial devices for overcoming the phagocytic cell system and thus contributing to invasiveness, virulence or persistence have been discussed in Ch. 4. This chapter is largely devoted to an account of the microbial strategies that have been developed to over-come or by-pass the immune response. As new strategies are revealed one has the impression that microbes know more about the immune response than the immunologists! Strategies for interference with immune defences are summarised in Table 7.3.

Infection Completed before the Adaptive Immune Response Intervenes

Many infections that are restricted to body surfaces (see Table 3.1) have short incubation periods, with less than a week between initial infec-tion and disease onset. This is equally true for the common cold, influenza, viral or bacterial diarrhoea and gonorrhoea. It is the early, innate defences (Table 3.1) that control the infection, and it is all over before there has been time for the host to deploy specific antibody and T cells. The microbe thus evades the immune response by completing its growth and shedding within a few days. It is a 'hit and run' infection.

Induction of Immunological Tolerance

Tolerance is an immunologically specific reduction in the immune response to a given antigen. As discussed here, it is due to a primary lack of responsiveness, rather than to an active suppression of immune response, which is dealt with on pp. 190–191. If there is a feeble host immune response to the relevant antigens of a microorganism, the process of infection is facilitated and the possibility of persistence increased. This does not involve a general failure of the host immune response of the type discussed in Ch. 9, but a particular weakness in relation to an antigen or antigens of a given microorganism.* Some-times it is said that a particular microbial component is a 'poor antigen', an observation that suggests tolerance to this antigen. All microorganisms except the smallest viruses have numerous antigens

---

* Most tolerance is exerted at the level of the T cell, whether by failure to respond or following the generation of suppressor/regulatory T cells. B cells, however, are suscep-tible to 'tolerisation' by antigen during their development.

on their surfaces and, if infection with a given microorganism is to be favoured, then the immunological weakness must be in relation to the microbial antigens that are important for infectivity, invasiveness or persistence. Also, because in a given infection either antibody or T cells may be the most important antimicrobial force (see Ch. 9), there must be a weakness in that arm of the immune response to which the microorganism is most susceptible. Tolerance can involve either antibodies or T cells to some extent independently. Tolerance to infectious organisms is rarely absolute, with no trace of an immune response to an antigen, but even slight specific weakness (or slowness) in a host may favour a microorganism (see Ch. 11). There are a variety of ways in which tolerance, defined in this way, can arise.

Prenatal infection

There is commonly a degree of tolerance to a microorganism when infection occurs during foetal or early postnatal life. At one time it was thought that any antigen present in the foetus during development of the immune system was regarded as 'self', and that as a result there was no immune response to it. It is now clear that immune responses do occur under these circumstances, but they are often weak and fail to control an infection.* For instance, rubella virus infects the human foetus, causing congenital malformations, and although the foetus receives rubella antibodies (IgG) from the mother and makes its own IgM antibody response to the infection, the T-cell response is particularly poor, enabling the virus to persist during foetal life and for long periods after birth. In the mouse, LCM virus is transmitted vertically (see Glossary) via the egg, so that the foetus is infected from the earliest stages of development. The congenitally infected mouse nevertheless makes a feeble antibody response (and no antiviral T-cell response) to the virus, but this fails to control the infection, and virus persists in most parts of the body for the entire life of the animal. In contrast to this, when adult mice are infected for the first time with LCM virus, they develop both antibody and a vigorous CD8 T-cell response, and the T-cell response becomes a pathogenic force that can lead to tissue damage and death (see Ch. 8). A similar situation exists in bovine virus diarrhoea virus (BVDV) infection of cattle. BVDV also induces tolerance when infection occurs in utero between 100 and 150 days gestation. In this instance, no antibody or T-cell response is made to the virus, which persists in the adult in a nonpathogenic form. However, infection of adult cattle with a pathogenic virus of the same

---

* Tolerance to the vast majority of 'self' antigens is complete since these antigens persist in the body and maintain the tolerogenic signal. Therefore, for a microorganism to utilise this method of persistence, it must survive in the host following a prenatal infection without producing overt disease, i.e. exist in a nonpathogenic form.

strain leads to virus growth and disease in the absence of an immune response. If a different BVDV strain or another microorganism infects the tolerised cattle, then an immune response is generated to this.

The above are examples of antigen-specific interference with the function of the immune response. However, some antigens can act in a more 'nonspecific' way, yet have a dramatic effect on the development of T cells and subsequent virus infection. These are called 'superantigens' and they are produced by certain bacteria (staphylococci, streptococci; see Ch. 8) and viruses (e.g. retroviruses). They function by binding to MHC class II molecules and interact with the T-cell receptor via the Vβ chain (all T cells belonging to a particular Vβ family, whatever their antigen specificity, will interact). In certain bacterial infections the superantigen induces T-cell proliferation resulting in the release of large quantities of cytokines. However, if T cells encounter these antigens early in their development (i.e. in the thymus), they become deleted and hence no T cell with that Vβ chain is detected in the spleen or lymph node. This is seen in mice carrying the retrovirus MMTV (mouse mammary tumour virus).* This virus is integrated into the germ line and produces a superantigen that results in the clonal deletion of particular T-cell subsets. The superantigen also causes a clonal expansion of B cells, required by the virus for its growth and also as a vehicle to spread to the main target organ, the mammary gland. Interestingly, mice deprived of those T cells are not compromised in their ability to recognise and respond to other microorganisms, but they are resistant to re-infection by MMTV.

Desensitisation of immune cells by circulating antigens

Tolerance to a given microorganism can arise when large amounts of microbial antigen or antigen–antibody complexes are circulating in the body. For instance, patients suffering from disseminated coccidiomycosis or cryptococcosis, both fungal infections, show antibodies but little or no T-cell response to the microorganisms. This is referred to as a state of anergy. It seems to be due to excessive amounts of circulating fungal antigen, and the T-cell response to unrelated microorganisms is not affected. Those suffering from kala-azar (visceral leishmaniasis) or diffuse cutaneous leishmaniasis have a defective T-cell response to the protozoal antigens, again associated with the presence of circulating leishmania antigens and resulting in systemic spread and chronicity of the infection. Antibodies are formed, at least in kala-azar, a severe generalised form of leishmaniasis, but this is not enough by itself; if there is to be recovery and healing, a good T-cell response is also neces-

---

* These antigens were first identified by transplantation biologists and called MLS (minor lymphocyte stimulating) antigens.

sary, enabling sensitised lymphocytes to destroy host cells infected with *Leishmania* microorganisms. A possible mechanism for tolerisation (desensitisation) of specifically reactive circulating T cells by antigen is as follows. When they are circulating through the body, T cells fail to make their usual intimate association with dendritic cells, B cells, etc. Instead, the T-cell response becomes diluted among various tissues leading to a reduction in the critical cell mass required for activation. In this environment, T cells may encounter antigen on non-professional antigen-presenting cells that lack the necessary co-stimulator molecules for activating T cells, and instead may deliver a tolerogenic signal. Under these circumstances, CD4 and CD8 T cells, although not damaged, lose their ability to respond to the specific antigen. They are 'defused' or anergised. In a similar way, developing B cells can be rendered immunologically impotent by direct exposure to circulating antigen.

After infection with *Treponema pallidum*, immobilising and other antibodies (see Ch. 6) are formed* and there is an initial T-cell response, as detected by lymphocyte transformation *in vitro* in the presence of treponemes. This initial response disappears as the bacteria multiply and spread through the body, and lymphocytes from patients with early secondary syphilis fail to respond *in vitro* to *Treponema pallidum*. Later in the secondary stage, weeks or months after infection, lymphocyte reactivity reappears, delayed skin reactions are demonstrable, granulomata appear in lymph nodes and the infectious process is finally brought under control. It is not known why lymphocytes from patients with early secondary syphilis fail to respond to the infecting bacteria. Antigen-specific suppression (see pp. 192–193) is a possibility, or T cells become anergised due to high levels of circulating antigen, or alternatively, T cells become sequestered in particular tissues and do not appear in the peripheral blood where sampling for reactive T cells would normally occur.

Molecular mimicry

If a microbial antigen is very similar to normal host antigens, the immune response to this antigen may be weak or absent, giving a

---

* The antibodies that are formed are not protective, and the bacterial antigens (fragile outer membrane protein (OMP), and peptidoglycan) are only present in low density on the bacterial surface and are poorly immunogenic. Could antigenic variation (see pp. 206–211) be a feature of this chronic infection? The genome of *T. pallidum* has now been sequenced, and it has no less than 38 genes for flagella structure and function, and 22 for lipoproteins in its envelope. However, the function of nearly half the genes is still unknown. Nowadays the few patients who get syphilis are treated and immunologists have not had the opportunity to catch up with this disease, but the major antigens are now being defined, characterised by monoclonal antibodies, and produced by recombinant DNA technology.

degree of tolerance. An example of this is the scrapie agent which shares a similar amino acid sequence to a host protein (PrP), thus rendering the scrapie agent invisible to the immune system. The mimicking of host antigens by microbial antigens is referred to as molecular mimicry. The hyaluronic acid capsule of streptococci, for instance, appears to be identical to a major component of mammalian connective tissue. The commonest resident bacteria of the normal mouse intestine are *Bacteroides*, and these share antigens with mouse intestine. Cross-reactions are seen, even with foetal mouse intestine, which absorbs antibacterial antibodies from serum. Mice are known to be generally rather unresponsive to *Bacteroides* antigens, and it is tempting to suggest that this facilitates establishment of these bacteria as life-long intestinal commensals (see Ch. 2). Generally, however, there is little evidence that molecular mimicry is a cause of poor immune responses. On the contrary, there is good evidence that antibodies formed against microorganisms sometimes cross-react with host tissues and therefore cause disease. Two diseases that follow human streptococcal infection have this basis (see Ch. 8). Also, antigens in *Treponema pallidum* cross-react with components in normal tissues, and antigens in Epstein–Barr virus cross-react with human foetal thymus, but this does not prevent antibodies being formed and providing the basis for the Wasserman and the Paul Bunnell (heterophile antibody) tests, respectively. There are one or two infections in which antibodies react with normal uninfected host cells. For instance, in atypical pneumonia caused by *Mycoplasma pneumoniae*, antibodies to heart, lung, brain and red blood cells may be formed. The antibodies to red blood cells (called cold agglutinins) very occasionally cause haemolytic anaemia.

Viruses such as influenza, measles and mumps mature by budding from the surface of infected cells, and viral antigens are incorporated into the host cell membrane (see Ch. 9 and Fig. 9.2). The envelope that forms the outer membrane of the virus particle could therefore contain some of the host cell antigens. At first sight there would appear no better way of microorganisms acquiring host antigens. But in fact apart from HIV (and other retroviruses), all the proteins of the envelope have proved to be viral in origin. The envelope lipids, however, are derived from the host cell and their carbohydrate moieties serve as antigenic determinants. Does this make the immune response to the virus any weaker? On the contrary, the immune response reacts powerfully against the antigens on the infected cell, often destroying the cell and serving a useful antiviral purpose.

Molecular mimicry, in summary, sounds like a good idea from the point of view of the infecting microorganism, and there are observations suggesting that it occurs, but so far there is no very convincing evidence. Indeed, it seems likely that antigenic determinants of microbes may resemble those of the host purely by accident rather than by sinister microbial design, and sometimes the common sequences reflect basic biological functions, common to many living creatures. It was found that about one-third of 800 different mono-clonal antibodies to defined virus antigens cross-reacted with normal host tissue components. Computer searches for shared amino-acid sequences between viral polypeptides and host components such as myelin basic protein showed that shared stretches of 8–10 amino acids, which could give cross-reactive immune responses, were quite common. Only a few viral polypeptides and one or two host components have been tested, and cross-reactive responses would turn out to be very common indeed, if other host components and the polypeptides of other viruses, protozoa, bacteria, etc. were examined. Although these phenomena are examples of molecular mimicry, it would be unreason-able to suggest that they have any meaning in terms of microbial strategies. Rather the host, in responding to such an immense variety of different microbial antigens, is always in danger of responding acci-dentally, as it were, to its own tissues. The resulting autoimmune response, however, only rarely leads to harmful, immunopathological, results (see Ch. 8).

Conclusion about inducing tolerance

Usually, when there is a weak immune response to a microbial antigen, it is not known which of the above mechanisms is responsible. There is, for instance, a very weak antibody response to the microorganisms present in the normal intestinal tract of mice. These microorganisms have been present during the evolution of the host animal. They are symbiotic in the sense that they may supply nutrients to the host and tend to prevent infection with other more pathogenic microorganisms (see Ch. 2). Perhaps the immune response is poor because the bacteria share antigens with the mouse intestine, as mentioned above. Perhaps mice have a genetically determined immunological weakness as regards these microorganisms. Perhaps infection shortly after birth has induced a large degree of tolerance. Perhaps large amounts of antigen are constantly absorbed from the intestine, desensitising lymphocytes and inducing tolerance. At present we do not have enough evidence to decide between these possibilities. In man, the urinary tract is commonly infected with *E. coli*, and the frequency of different bacterial serotypes is in proportion to their frequency in the faecal flora. Strains rich in the polysaccharide K antigens, however, are more likely to invade the kidneys. Children with pyelonephritis due to *E. coli* show a correspondingly poor antibody response to these antigens, but the cause of the poor response is not known. Sometimes one suspects tolerance to a microorganism, but there is no evidence. For instance, when species of dermatophyte fungi of animal origin infect the skin of man, there is inflammation followed by healing and relative resistance to re-infection. But with species of fungi adapted to man there tends to be less inflammation, a more chronic infection and less resistance to re-infection. This sounds as if it could be due to a weak immune response to antigens of the human type of fungus.

Studies of autoimmunity have revealed that in normal people there are lymphocytes that respond to autoantigens. Autoimmune disease is avoided by suppressing these responses. This leads to the possibility that specific suppression is commoner than primary unresponsiveness (clonal deletion or clonal anergy). Indeed, the autoimmune phenomena seen in certain infectious diseases (see Ch. 8 and pp. 370–371) could be attributed to a breakdown in specific immune control mechanisms. Also, there are indications that, in certain persistent infections, a weak response to microbial antigens is due to antigen-specific suppression rather than to a shortage of responding cells. The following section deals with immunosuppression in infectious diseases.

Immunosuppression

General immunosuppression

A large variety of microorganisms cause immunosuppression in the infected host. This means that the host shows a depressed immune response to antigens unrelated to those of the infecting microorganism. Infectious agents that multiply in macrophages or lymphoid tissue (viruses, certain bacteria and protozoa) are especially likely to do this. For instance, during the acute stage of measles infection, patients with positive tuberculin skin tests become temporarily tuberculin-negative. The exact mechanism is not clear and may be complex. The virus grows in monocytes and could for instance induce production of immunosuppressive cytokines (e.g. IL-10) that inhibit the induction of Th1 responses (e.g. DTH responses), or affect the expression of MHC or accessory molecules on cells. Even after vaccination with live attenuated measles virus immune responses are depressed, and there is a reduction in cutaneous sensitivity to poison ivy,* lasting several weeks. Depressed T-cell or antibody responses to unrelated antigens have also been described in people with mumps, influenza, Epstein–Barr virus and cytomegalovirus infections. Immunosuppression is a feature in mice infected with cytomegalovirus, LCM virus, murine leukaemia virus or *Toxoplasma gondii*, and in cattle infected with rinderpest

---

* Poison ivy is a common plant on the east coast of North America. The leaves bear a substance (urushiol, a catechol derivative) of low molecular weight that sticks to the skin and in most people induces delayed hypersensitivity (CMI). On subsequent exposure the delayed hypersensitivity is expressed as contact dermatitis. The lesions are localised to the site of contact with the leaf and consist of erythema, papules and vesicles, later becoming scaly and thickened.

virus. Patients with certain types of malaria, trypanosomiasis, leish-maniasis and lepromatous leprosy show reduced responses to various unrelated antigens and vaccines.*

At present we do not know how such a variety of microorganisms inhibit immune responses. Interference with the immune functions of dendritic cells, macrophages and lymphocytes is important.† The reduced responses may partly be due to 'antigenic competition' rather than actual suppression of responses by the infecting microorganisms. Antigenic competition might be expected when an urgent, generalised response to an invading microorganism commandeers a large propor-tion of the available space, uncommitted cells, etc., in lymphoid tissues. When the spleen, for instance, is enlarged, a seething mass of cells responding with maximum immunological effort against the invader would leave less space and fewer cells for responses to unrelated antigens.

Interest in virus-induced immunosuppression received a great stim-ulus with the appearance of the acquired immunodeficiency syndrome (AIDS) in 1979. In this disease the infecting virus (HIV, human immunodeficiency virus) infects both CD4 T cells and macrophages. This results in serious loss of immune reactivity, and particularly of T-helper function.

In severely affected patients the immune deficit allows a variety of persistent yet normally harmless infections (*Pneumocystis carinii*, cytomegalovirus, tuberculosis, toxoplasmosis, candidiasis, etc.) to become active, and these, together with various other infections, even-tually prove fatal. HIV is responsible for the immunosuppression which gives the other microorganisms the opportunity to cause the lethal disease. (HIV also causes a late-onset, independently evolving disease of the brain.) The virus persists in the body, and patients remain infectious for life. Feline leukaemia virus (also a retrovirus) causes a similar condition in cats, and infected cats are more likely to die of secondary infection than of the leukaemia itself. People who are HIV positive but asymptomatic have virus-specific antibody and T-cell responses and low levels of infectious virus. This state continues for years, but symptoms of AIDS eventually appear, coinciding with a decline in immune responses and a concomitant increase in virus. This

---

* In the case of lepromatous leprosy, the signs of immunological disturbance include impaired delayed-type hypersensitivity responses to unrelated antigens, a polyclonal activation of B cells (see p. 199) and in addition there is a specific unresponsiveness to *Mycobacterium leprae*. Suppressor T cells obtained from skin lesions have been shown to inhibit the response of other T cells specifically to *M. leprae* antigens, perhaps by lysing cells that are presenting these antigens. The suppressor T cells are possibly induced by the terminal sugars on a leprosy-specific phenolic glycolipid.

† Interference with the recirculation and homing of T cells (see Ch. 6) would be a theo-retically attractive strategy for a microbe. Pertussis toxin (pertussigen) has this effect, but its relevance *in vivo* is not known.

could be regarded as a useful result from the point of view of the virus, favouring persistence and transmission to fresh hosts during the asymptomatic period, the more general and disastrous immunosuppression being an 'unfortunate' side effect.

Antigen-specific suppression

A general immunosuppression induced in the host is of no particular significance for an infectious agent if it merely promotes infection by unrelated microorganisms. If the immunosuppression is to be of value it must involve the response to the infecting microorganism itself, facilitating its spread, multiplication and persistence. A number of infectious agents, especially viruses, induce this type of immune suppression. In other words, their strategy is to suppress host immune responses specifically to their own antigens. The phenomenon occurs with other infectious agents, such as leprosy (see footnote, p. 191) and in tuberculosis, where there are reduced delayed-type hypersensitivity (DTH) and interleukin-2 (IL-2) responses to tubercular antigen (PPD) but normal responses to streptococcal antigens.

Antigen-specific suppression might enable a virus to persist indefinitely in the body (see Ch. 10). Also, many human viruses (measles, mumps) have an incubation period of 10 days or more, because they take time to spread in the body, multiply and be shed to infect a new host. Immune responses might normally be expected to curtail the infection before the full sequence of events had unfolded, and it would therefore be advantageous for such viruses to suppress responses to their own antigens. The argument applies even more to hepatitis B and rabies viruses, with incubation periods of several months.

The most audacious strategy would be to infect the very tissues in which the immune response is generated, and interfere with it. In other words, to evade immune defences and invade immune tissues. It turns out that infection of lymphoreticular tissues is very common in systemic and in persistent virus infections (see Table 7.1). There are several possible mechanisms for antigen-specific suppression. During a normal response there is careful control over the distribution and concentration of antigen (see p.150), especially in lymphoid tissue. Antigens are normally delivered to lymphocytes in minute quantities and in an appropriate setting after processing by dendritic cells and macrophages. If larger amounts of antigen are liberated locally in these tissues by an invading virus, a disordered response is to be expected, perhaps by tolerising T or B cells, as discussed on p.186. Another mechanism would be for the invading virus to infect preferentially the T or B cells that responded specifically to its own surface antigens, and either inactivate or destroy these cells. This would eliminate the clones of cells that might otherwise generate a specific antiviral response. Finally, the invading virus might generate antigen-

Table 7.1. Infection of lymphoreticular tissues by viruses and other infectious agents exhibiting systemic infection or persistence

Infectious agents	Host
Viruses	
Adenoviruses (L)	Man
Epstein–Barr virus (L)	Man
Kaposi's sarcoma herpesvirus (L)	Man
Cytomegalovirus (M)	Man, mouse.
Leukaemia virus (L,M)	Mouse, etc.
Visna virus (M)	Sheep
LCM virus (L,M,DC)	Mouse
Murine gammaherpesvirus (L)	Mouse
Thymic necrosis (L)	Mouse
Measles (L)	Man
Rubella (L,M)	Man
Lactic dehydrogenase virus (M)	Mouse
Infectious bursal disease virus (L)	Chicken
Aleutian disease virus (M,DC)	Mink
Equine infectious anaemia (M)	Horse
African swine fever virus (M)	Pig
HIV (L,M,DC)	Man
Other infectious agents	
(see also Table 4.2)	
Scrapie (DC)	Mouse, sheep
Theileria parva (L)[a]	Cattle

L = lymphocytes known to be infected; M = macrophages known to be infected; DC = dendritic cells known to be infected.
[a] A tick-borne protozoal parasite responsible for East Coast fever, an important cattle disease in East Africa. The parasite invades lymphocytes, becomes associated with the mitotic apparatus, and stimulates cell division. But the infected lymphocytes are later killed by other immune cells. Presumably the parasite betrays its presence by allowing its antigens to be displayed on the infected cell.

specific suppressor/regulatory cells (see p. 155, Ch. 6) or other suppressor factors. Antigen-specific suppressor T cells are induced during infection of experimental animals with herpes simplex and other viruses, but this may turn out to be a regular feature of most immune responses, one of the mechanisms by which the magnitude of the response is controlled.

Some of the above possibilities involve antigen-specific unresponsiveness rather than active immunosuppression, and another example of unresponsiveness is when the microorganism exploits 'holes' in the immunological repertoire of the host. Immune responses to given antigens, as noted earlier, are controlled by immune response genes. Successful microorganisms therefore would tend to develop surface antigens that are poorly seen and poorly responded to by the host.

Absence of a Suitable Target for the Immune Response

There are various ways in which intracellular microorganisms can avoid exposing themselves to immune forces. They evade host immune responses as long as they stay inside infected cells, and allow at the most a low density of microbial antigen to form on the cell surface. This is what happens in dorsal root ganglion cells persistently infected with herpes simplex or varicella virus (see Ch. 10), in circulating lymphocytes infected with Epstein–Barr virus, and in most of the cells of a mouse infected with persistent murine leukaemia virus. In Epstein–Barr virus-infected B cells, EBNA-1 (Epstein–Barr nuclear antigen) is responsible for maintaining the latent viral genome and, because it persists during virus latency, it is a potential target for cytotoxic CD8 T cells. However, this protein is able to resist proteolysis by the proteosome thus avoiding processing and presentation via MHC class I molecules. Malaria parasites are present in liver cells during the exoerythrocytic stages of infection and, during this silent latent period, the parasite avoids stimulating or presenting a target for the immune response. Even when the malaria parasites are growing in red blood cells and causing the disease, their very presence inside the red blood cells protects them from circulating antibodies. The merozoites that emerge from infected red cells are only briefly exposed to antibodies before entering fresh uninfected cells. Some of the parasite components, however, may be present on the surface of the infected red cell, which then becomes a less protected site.*

Some of the intracellular microorganisms that expose their antigens on the infected cell surface benefit from a host-mediated mechanism for disposal of those antigens. This involves antibodies and depends on the phenomenon called capping. Substances can move in the fluid matrix of the cell membrane and, when microbial antigens on the cell surface react with a specific antibody, the antigen–antibody complex moves to one pole of the cell (capping). Here the complex is either shed or taken into the cell by endocytosis. The antibodies that should have prepared the infected cell for immune destruction are diverted from this purpose and used to rid the cell of microbial antigens, making it less susceptible to immune lysis. Capping is frequently observed *in vitro* on cells infected with various enveloped viruses, notably measles and herpes simplex viruses. However, the significance of this process *in vivo* is unclear. In certain protozoa, capping by antibody can lead to the loss of the microbe's own surface antigens. It occurs with *Toxoplasma*

---

* Red blood cells infected with *Plasmodium falciparum* express on their surface a receptor for the cell surface ligand – ICAM-1, present on endothelial cells. Its significance for the parasite is not clear, but it would promote binding of the infected red cell to post-capillary venules, and could be a mechanism for the sludging of red cells in cerebral blood vessels in cerebral malaria.

gondii and with *Leishmania* but its importance in resistance to host immune defences is unknown.

Intracellular microorganisms also escape the action of antibodies if they spread directly from cell to cell without entering the extracellular fluids. This is seen when herpes simplex virus spreads progressively from cell to cell through cytoplasmic 'tunnels' in the presence of potent neutralising antibody. This virus also avoids detection by the immune system when travelling within axons of sensory nerves. Some enveloped viruses (corona and flaviviruses) avoid displaying their antigens on the cell surface by budding into cytoplasmic vesicles. Virions are then released directly into the external medium by fusion of the vesicle with the plasma membrane. Cells formed by division of an infected cell are also infected without virus entering extracellular fluid. For example, cells derived from the ovum of a mouse infected with leukaemia virus are all infected, whether or not virus is released from the cell, and in the newborn infant with rubella virus, a cell that was initially infected in the foetus has given rise to a group of infected progeny cells in spite of the presence of neutralising antibodies. The ability to stay inside cells certainly contributes to the success of persistent intracellular microorganisms, without in most cases being the sole factor.

An entirely different approach for a virus is the concealment of an antigenic site on the virion, behind a carbohydrate moiety. A single point mutation creates the glycosylation site and renders the site invisible to the immune system so that antibody is no longer made against it. The reverse mutation can restore the site. This situation occurs naturally in influenza A and rabies viruses, and presumably confers some selective advantage.

Since the only target for T cells is a peptide antigen complexed with a MHC protein, it follows that viruses can render infected cells invisible to T cells if they can downregulate the expression of MHC proteins. This is achieved in two ways by different adenoviruses. Adenovirus type 2 synthesises a protein which is anchored to internal cell membranes of the endoplasmic reticulum and binds to MHC class I proteins, so preventing them from being transported to the cell surface. This seems to be this protein's main function, for its deletion does not affect virus multiplication in cell culture, but in experimental infection of rats results in a more severe lung disease. Another adenovirus (type 12) somehow prevents the movement of mRNA encoding MHC class I protein from the nucleus with similar results. Some cytomegaloviruses and poxviruses decrease the expression of MHC class I on the infected cell. In the case of mouse cytomegalovirus an early viral protein(s) inhibits transport of peptide-loaded MHC class I complexes into the medial-Golgi compartment. The virus-induced inhibition of expression of MHC proteins is countered by interferon-γ (IFN-γ) which upregulates both MHC class I and II proteins. The activity of T cells depends largely on the density of MHC proteins on the surface of the infected cell, so the balance between their downregulation and upregulation is all important. In a similar way, cell surface molecules whose function permits the adhesion of T cells to their targets (ICAM-1, LFA-3, etc.) can be downregulated, as seen in some cases of Epstein–Barr virus-infected Burkitt's lymphoma cells.

Microbial Presence in Bodily Sites Inaccessible to the Immune Response

Many viruses persist in the infected host and are shed to the exterior via the saliva (herpes simplex virus, cytomegalovirus, rabies virus in vampire bats), milk (cytomegalovirus in man, mammary tumour virus in mice (MMTV)) or urine (polyomavirus in mice). The surface of the infected cell can be said to face the external world as represented by the lumen of the salivary gland, mammary gland or kidney tubule (Fig. 7.1). As long as virus particles and viral antigens are only formed on

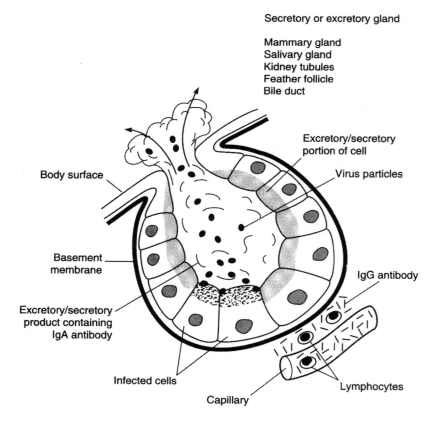

Fig. 7.1 Viral infections of cell surfaces facing the exterior.

the lumenal surface of the cell and there is little or no cell destruction, it is difficult for T cells* or antibodies to reach the site and eliminate the infection. This is the case for murine cytomegalovirus infection, where T cells efficiently control the productive infection in all tissues, except the salivary gland where a prolonged excretion of the virus occurs. Secretory IgA antibodies could react with viral antigens on the infected cell surface, but complement is unlikely to be activated and the cell would not be destroyed. IgA antibodies could also react with extracellular virus particles but would at the most render them non-infectious, again without acting on the source of the infection. The same considerations apply to epidermal infection with human wart viruses, or to infection of the epidermis lining the chicken's feather follicle with Marek's disease virus (see Glossary). In the case of a wart, neither virus nor viral antigens are manufactured in significant quantities until the infected epidermal cell is keratinised and about to be released from the body, and is physically far removed from host immune forces. The same is true for fungi that grow in the dead keratinised cells of skin and hair.

Bacteria that are present and multiply in the lumen of glands, tubes and tubules also enjoy some freedom from immune forces. In rats persistently infected with *Leptospira*, for instance, the bacteria multiply in the lumen of kidney tubules and are shed in urine. If the urine enters water in a river or puddle, it remains infectious and can cause leptospirosis in man. Commensal intestinal bacteria, unless they are very closely associated with the intestinal epithelium, enjoy similar freedom and it is therefore impossible to eliminate these bacteria by artificially inducing immune responses against them. Other bacteria, such as *Brucella abortus* in the cow, persistently infect mammary glands and are shed in the milk. In typhoid carriers, the bacteria colonise scarred avascular sections of the biliary or urinary tract, and are thence shed, often in large quantities, into the faeces or urine. Bacteria may also lurk in biliary or renal stones, and in a similar way the staphylococci in the devascularised bone of patients with chronic osteomyelitis are protected from host defences.

Induction of Inappropriate Antibody and T-cell Responses

Many types of antibody molecule are formed against a given antigen, reacting with different antigenic determinants (epitopes) on the molecule. For instance, studies with monoclonal antibodies (see Glossary)

---

* Although Fig. 7.1 shows a continuous layer of epithelial cells, infiltrating lymphocytes with possible antimicrobial potential are a normal feature of certain epithelial surfaces (e.g. intestinal).

have shown that there are many different epitopes on a protein such as the haemagglutinin that occupies most of the surface of influenza virus (see pp. 22–23), but only those located within five particular antigenic sites are capable of neutralising infectivity. Also, antibodies tend to have a range of avidities (see Glossary). If the antibodies formed against a given microorganism are of low avidity, or if they are mostly directed against unimportant antigenic determinants on the micro-organism, then they will only have a weak antimicrobial action and there are likely to be difficulties in controlling infection with that particular microorganism. For instance, there are several persistent (life-long) virus infections of animals in which antibodies are formed and react specifically with the surface of the infecting virus, but fail to render it noninfectious. The virus–antibody complexes are therefore infectious and they circulate in the blood. These viruses include LCM and leukaemia virus infections in mice, also Aleutian disease virus in mink. In the latter disease (see Glossary) there is a stupendous immune response on the part of the infected animal, with a fivefold increase in total IgG levels and viral antibody titres of 1/100 000. The antibody is not only of no antiviral value, but causes life-threatening immunopathological damage. Since there is also no effective T-cell response to these infections, and since these viruses grow in host cells without harming them, the infections persist for life.

Non-neutralising antibodies of this sort are particularly important if they combine with microbial antigen and block or sterically hinder the binding of any good-quality neutralising antibodies that may also be present. Antibodies to LCM virus formed in infected mice are known to have this property.

It is not known how commonly ineffective antibodies are induced in other microbial infections, but if they are induced, the antimicrobial task of the host is certainly made more difficult. The antibodies formed in patients with syphilis (a persistent infection) are only very feebly antimicrobial. Although they combine with the surface of the treponemes and perhaps aid phagocytosis (act as opsonins), they cause little neutralisation and, after 36 hours' treatment with antibody plus complement, most of the treponemes remain infectious.

From a microorganism's point of view it would also be an advantage to be able to induce the host to make the wrong type of immune response. There is a tendency for the antibody and the T-cell response to given antigens to vary inversely, and if for a given infection the host's major antimicrobial force was the T-cell response, the micro-organism could with advantage induce the formation of a strong anti-body response.* This could be done by inducing the formation of Th2

---

* The $BCRF_1$, gene of Epstein–Barr virus codes for an IL-10-like molecule that enhances humoral rather than cell-mediated responses, and the virus itself can infect the extra B cells that are formed.

rather than Th1 cell responses. Parasitic worms induce strong Th2 immune responses, benefiting the host by producing IgE antibodies important for the expulsion of the parasite. However, co-infection of parasitic worms with other microorganisms can inadvertently deviate normally protective Th1 responses against the microorganism to a poorly protective Th2 response. This can be considered a form of opportunism, which on the one hand can benefit a microbe, but at the same time may have adverse effects on the host, i.e. a more severe infection may ensue. Deviation of immune responses from strong Th1-mediated CMI responses can also occur when microorganisms infect the host through different routes. For example, herpes simplex virus infection of the epidermis promotes strong Th1 activity, measured as delayed type hypersensitivity, but a weaker antibody response. In contrast, infection via the bloodstream induces strong antibody responses, but negligible CD4 (Th1) cell responses. Once these pathways are set in motion it is difficult to reverse them, i.e. Th1-mediated CMI can no longer be induced in the host against the virus once it has been delivered in the bloodstream. Infections with Gram-negative bacteria such as *Salmonella typhi* are controlled by macrophages and T cells, and it has been suggested that endotoxin (see Ch. 8), which acts as a general B cell (antibody) stimulator, directs the host response in favour of antibody rather than T-cell activation of macrophages, to the benefit of the infecting bacteria. Patients with lepromatous leprosy or pulmonary tuberculosis show poor CMI responses to the invader, suggesting that the bacteria have induced an inappropriate immune response in a susceptible type of host.

Persistent protozoal infections such as malaria and African trypanosomiasis are characterised by the formation of very large amounts of antibody. But most of this appears to have little or no protective value in the host, although it sometimes shows some *in vitro* inhibition of parasite motility, viability, multiplication or metabolism. Although some of these antibodies are directed against microbial antigens, most are truly nonspecific in the sense that they do not react at all with any microbial antigens.* Some react with host tissues, such as the heterophile antibodies (see Glossary) and the antibodies to DNA, Schwann cells and cardiac myofibrils, that are seen in the various types of trypanosomiasis, and this raises the question of autoimmune damage (see Ch. 8). Similar antihost antibodies occur in certain virus infections such as those caused by Epstein–Barr and cytomegaloviruses. The basis for these irrelevant or excessive antibody responses is B-cell proliferation induced by the infection, often referred to as poly-

---

* Adults in West Africa chronically infected with malaria show seven times the normal (European) rate of IgG production, on a body weight basis. Much of this is associated with malaria infection because it is reduced by 30% after several years of prophylactic antimalarial therapy. Only 5% of the circulating IgG, however, is found to react with malarial antigens.

clonal activation. This is seen in malaria, lepromatous leprosy, and also in infection with *Mycoplasma pneumoniae*, *Trypanosoma* species, Epstein–Barr virus and many other microorganisms. It would make sense if it reflected microbial interference with host immune responses (see Ch. 10) but its significance in these important infectious diseases is still shrouded in mystery. For microbes such as MMTV and Epstein–Barr virus that grow in B cells, proliferation of these cells can be interpreted as a useful strategy (see p. 186).

Antibodies Mopped up by Soluble Microbial Antigens

The antimicrobial action of antibodies is to a large extent due to their attachment to the surface of microorganisms. Antibody on the microbial surface prevents entry into susceptible cells, promotes uptake by phagocytes, activates complement lysis of the microorganism, etc., as discussed in Ch. 6. One strategy that microorganisms could use to defend themselves against the antibody weapon would be to liberate their surface components in soluble form into tissue fluids. These surface components would combine with and 'neutralise' antibody before it reached the microorganism.

Soluble antigens are liberated into tissue fluids in most microbial infections, but it is not often that these are known to be surface antigens. Most normal tissue cells bud off tiny membrane-bound blebs of cytoplasm into surrounding fluids. Cells infected with budding viruses produce virus particles but they probably also liberate tiny blebs of cytoplasm whose limiting membrane contains viral antigens. The 20 nm particles present in the serum of patients and carriers with hepatitis B virus infection are produced in this way. There are up to 10^{13} particles per millilitre of serum.* Polysaccharide antigens from *Candida* contain mannan, and this material, which is present in the serum of patients suffering from candidiasis, inhibits lymphocyte proliferation in response to *Candida* antigens. Perhaps *Candida* polysaccharides are handled abnormally in these patients, allowing free mannan to stay in the circulation and interfere with macrophage responses. In certain bacterial infections surface polysaccharides are liberated. Bacterial polysaccharides are detectable in the serum in pneumococcal pneumonia, and in the serum and cerebrospinal fluid in fulminating meningococcal meningitis. The surface polysaccharide in *Pseudomonas aeruginosa* is also released from multiplying bacteria *in vitro*, and presumably *in vivo*. Even endotoxin is released in small amounts into the surrounding fluid by Gram-negative bacteria.

---

* They are not infectious. Could they be regarded as a viral 'device' to tolerise (pp. 186–187) host lymphocytes?

Antigens from *Trypanosoma cruzi, Candida albicans, Toxoplasma gondii, Plasmodium* spp. and *Babesia* spp. are present in serum during systemic infections. The phenomenon may prove to be a common one. But in spite of the theoretical advantages for the microorganism, it is not known whether the released surface components mop up enough antibody or inactivate enough T or B cells to be of significance in the infection.

Local Interference with Immune Forces

There are several ways in which microorganisms, without preventing the generation of immunity, interfere with the local antimicrobial action of immune forces. For instance, a few microorganisms induce the formation around themselves of a capsule or cyst. Cysts are formed in certain protozoal infections, but they occur inside cells, especially macrophages, and protect the microorganism from destruction by the host cell rather than from host immune forces. *Toxoplasma gondii*, for instance, infects man and a wide range of mammals and birds, forming cysts inside macrophages in the central nervous system, muscle and lung.* Shielded by the tough wall of the cyst, the microorganism multiplies without provoking a host reaction, and thousands of parasites may be present in a single cyst in the chronic stage of the disease. Cysts that restrict the access of antibody and phagocytic or immune cells are best seen with helminth parasites. The dog tapeworm, for instance, lives in the alimentary canal of dogs, and enormous numbers of eggs are present in faeces. When humans ingest these eggs, the parasites develop and travel from the intestine to reach the lung or liver, where hydatid cysts are formed. A cyst consists of larval worms inside a firm capsule made up of parasite and host components. The cyst gradually grows larger, in spite of the antibody and cell-mediated immune response of the host, often reaching the size of a coconut, and may cause serious disease. The natural hosts for the cysts are grazing animals such as sheep, which ingest the eggs; after the sheep has been killed and eaten by a predatory carnivore such as the dog, the larval worms in the cyst grow to form adult worms in the intestine.

Virulent staphylococci produce a coagulase and this acts on plasma components at the site of infection and leads to the deposition of a layer of host fibrin around the bacteria. It is possible that this restricts the local access of host cells, and that the bacteria are also disguised immunologically, so that they are less readily identified as targets for

---

* *Toxoplasma gondii* is primarily a parasite of members of the cat family, in whom it infects intestinal epithelium and is shed in faeces. Human infection is world-wide, and common in the UK, but nearly always symptomless.

immune responses. There are other examples from bacterial infections. The K antigens on the surface of *E. coli* are closely associated with the pathogenicity of these bacteria. They can mediate attachment of *E. coli* to intestinal epithelial cells and certain K antigens can increase the ability of *E. coli* to grow in the kidney or other sites. First, the K antigen is a polysaccharide and makes phagocytosis a more difficult task for host cells (see Ch. 4). Also, some of them are poor immunogens, possibly resembling host polysaccharides. Finally, the K antigen interferes with the alternative pathway of complement activation, so that the bacteria escape this early antibacterial defence mechanism.

One theoretically simple strategy for bacteria would be to produce an enzyme that destroys antibodies, and it has been shown that pathogenic strains of the gonococcus, a human pathogen, liberate a protease that specifically cleaves human IgA1 subtype antibodies. The protease acts at the site of a Pro-Thr peptide bond in the hinge region of the heavy chain (see Fig. 6.3).* The significance of this enzyme *in vivo* is not yet clear because antigen-combining sites on the cleaved fragments would remain intact, but such enzymes are obviously likely to be important, and might help account for the apparent indifference of the gonococcus to the host's antibody response. A similar enzyme is produced by the meningococcus, a frequent resident in the nasopharynx, by many strains of *Haemophilus influenzae* and *Streptococcus pneumoniae*, and also by *Streptococcus sanguis*, one of the common commensal bacteria of the mouth. In each case the bacteria producing the protease are normally exposed to secretory IgA antibody. The staphylococci provide a more plausible example of the local interference with the action of antibody by bacteria. Virulent staphylococci have a factor called protein A in the cell wall which is excreted extracellularly, and this inhibits the phagocytosis of antibody-coated bacteria by attaching to the Fc portion of the antibody molecule. *Pseudomonas aeruginosa* produces an elastase that inactivates the C3b and C5a components of complement, and thus tends to inhibit opsonisation and the generation of chemotactic and other inflammatory responses. Perhaps this contributes to invasiveness;† *Pseudomonas* infections tend to show minimal inflammatory responses. Factors that inhibit the action of complement or opsonising antibodies are perhaps also produced by other bacteria, but even when such factors are discovered and defined, it can be difficult to decide how important they are in the actual process of infection. *Entamoeba*

---

* Human IgA2 antibodies lack a Pro-Thr in their hinge region and are therefore resistant to the action of the bacterial proteases. This IgA subtype is mainly associated with secretions.

† The *Pseudomonas* elastase also inactivates lysozyme and cleaves type I collagen. As with many other bacterial products, its role in pathogenicity may be clarified when elastase-negative strains are produced by genetic engineering and tested for virulence.

histolytica, for example, produces proteases that degrade C3a and C5a and the amoebic adhesin prevents assembly of C8, C9 into the membrane attack complex.

Gram-negative bacteria have a very complex cell wall, consisting primarily of a membrane-like arrangement of phospholipids, lipopoly-saccharide (LPS or endotoxin; Figs 4.4 and 8.15) and protein. In virulent strains of bacteria the polysaccharide chains project from the general bacterial surface, carrying on their tips the important O antigens of the cell wall. The O antigens are the key targets for the action of host antibody and complement, but when this reaction takes place on the end of the polysaccharide chains, a significant distance external to the general bacterial cell surface, complement fails to have its normal lytic effect. Such bacterial strains are virulent because of this resistance to host immune forces. Their colonies happen to have a smooth appearance on agar surfaces and they are therefore called 'smooth' strains. If the projecting polysaccharide chains are shortened or removed, antibodies react with O antigens on the general bacterial surface or very close to it, and complement can then lyse the bacteria. Strains without the projecting polysaccharides are therefore nonvirulent and they have a 'rough' colonial morphology. Gram-negative bacteria therefore can be thought of as protecting themselves from the damaging consequences of antibody and complement reactions by having antigens that project a short distance out from the bacterial cell surface.

Recent studies have highlighted the importance of LPS in determining the *in vivo* phenotype of gonococci. Gonococci in urethral exudates resist complement-mediated killing by human serum. In most cases this resistance is lost after one subculture in laboratory media. A terminal Galβ1–4GlcNAc site on a conserved 4.5 kDa LPS surface component is sialylated by host-derived cytidine 5'-monophos-phate-*N*-acetyl neuraminic acid (CMP-NANA). The transfer is catal-ysed by a gonococcal sialyltransferase. Not only does this prevent lysis by complement but confers on the organism the ability to avoid contact with and ingestion by phagocytes, and masks LPS from reacting with specific antibodies. It also, paradoxically, prevents the uptake of the organisms by endothelial cells. This latter property is probably of importance in the extracellular phase of transmission. Presumably, in order for re-infection to occur, then the organisms must undergo desia-lylation, presumably by host enzymes. Meningococci – Groups A, B and C – show similar properties to the gonococcus except that for Groups B and C the sialylation requires endogenous CMP-NANA. In one Group B meningococcal epidemic, almost all cases had sialylated LPS and were virulent in a mouse model of infection; in contrast, most carrier isolates were not sialylated and were not virulent in mice.

Certain bacteria activate complement via the alternative pathway (without antibody) and this promotes opsonisation and thus increases host resistance early in infection (see p. 177). Group A streptococci would do so were it not for their special outer covering. These bacteria are poorly opsonised by complement alone, but when the M protein on the pili is gently removed with trypsin to expose peptidoglycan, the alternative pathway is activated, the bacteria bind complement and are then efficiently opsonised and phagocytosed. Encapsulated strains of *Staphylococcus aureus* activate the alternative pathway and bind complement, but C3 is somehow hidden by capsular material and the bacteria are not opsonised.

The DNA viruses have evolved a number of strategies to combat the local actions of the immune response. These include viral genes that mimic: (a) cytokines and chemokines, (b) receptors for cytokines and chemokines, (c) MHC molecules, (d) antiapoptotic factors and cell cycle proteins, and (e) complement regulatory proteins. These genes have been acquired by viruses during their evolution by acts of molecular piracy from the cell. The viral genes show close homology with cellular genes and perform similar functions. Thus many act as decoy proteins subverting the action of the normal cellular counterpart. Vaccinia virus disrupts cytokine and chemokine responses through production of soluble receptors against TNF, IFN-γ, IFN-α/β (see below), IL-1β and through the production of chemokine binding proteins (a new set of regulatory molecules so far only found in vaccinia and herpes viruses). A similar armory is found in the gammaherpes viruses which possess cellular homologues of IL-10 (Epstein–Barr virus), IL-6, MIP-1α/β (KSH virus) and IL-8 receptor (MHV-68, KSH virus). These viral cytokines and chemokines (virokines) act on the relevant target cell in a way analogous to their cellular counterpart. This could involve recruiting target cells to sites of infection in order to infect them. Lymphocytes are the principal site of latency in gammaherpes virus infection (see Ch. 10). To ensure target cells survive to support a latent infection, members of the gammaherpes viruses also carry the anti-apoptotic genes, vBcl-2 and vFLIP. Viral Bcl-2 has been shown to protect cells from death induced by TNF and presumably serves a similar function *in vivo* by preventing premature B-cell death, thus insuring virus survival in the B-cell compartment.

Surviving the attentions of NK cells and cytotoxic T cells is an important goal for all viruses. The herpes viruses and adenoviruses have solved this problem by interfering with the expression of MHC molecules on the cell surface. To inhibit cytotoxic T-cell recognition of MHC class I molecules, cytomegalovirus and adenovirus produce proteins that either anchor MHC class I in the endoplasmic reticulum or block their progression through the Golgi complex. This prevents the MHC molecules reaching the cell surface and being recognised by T cells. Herpes simplex virus and cytomegalovirus also inhibit peptide transport into the endoplasmic reticulum by producing a protein that binds to the TAP molecules (see Fig. 6.1). Reducing MHC class I expression is a key strategy for outwitting cytotoxic T cells; however, this strategy only serves to alert NK cells to these target cells (see Chapter 6). To overcome this and outwit NK cells, human and murine cytomegalovirus encode a MHC class I-like protein on the surface of infected cells which can interact with NK cells instructing them to ignore the infected cell.

The fact that pox viruses and herpes viruses encode proteins designed to disrupt the complement system suggests that viruses see this system as a threat. Pox viruses are able to inactivate C4b through production of C4B-BP-like molecules and herpes simplex virus gC on the surface of the virus or the infected cell binds C3b and thereby inhibits the complement cascade. Herpes simplex virus and human cytomegalovirus induce virus-coded Fc receptors on virions and on the surface of infected cells. This could be useful for the virus by binding IgG nonspecifically to the cell surface and thus protecting it from immune lysis. Fc receptors are present on staphylococci (protein A and G, see p. 202), on certain types of streptococci and on trypanosomes, and could protect these microorganisms in a similar fashion. The Fc and C3 receptors present on the various forms of *Schistosoma mansoni* conceivably aid the survival of this large parasite in the host.

Reduced Interferon Induction or Responsiveness

The interferons are cytokines and play a major role in the innate defence against virus infection by augmenting NK cell and macrophage activity, and promoting an antiviral state in cells, a form of intracellular immunity (see Ch. 9). This gives the IFNs an important role in early defence, before the adaptive immune response has been generated. Their production is stimulated by foreign macromolecules (Ch. 9); IFN-α and IFN-β are induced in all cell types in response to double-stranded RNA,* and IFN-γ is produced following the activation of T cells (Ch. 6).

Viruses are generally sensitive to interferon and they can evade this host defence mechanism if they fail to induce interferon in the host or if they are resistant to the action of interferon. In the latter category is vaccinia virus which synthesises no less than three products which counter the effects of interferon. It secretes soluble receptors which bind to and inactivate IFN-α/β and IFN-γ, and makes two other proteins which act intracellularly. One of these prevents the interferon-induced phosphorylation of eIF2α, so inhibiting the initiation of translation, and the other blocks the activation of a dsRNA-dependent protein kinase (PKR – an interferon-induced enzyme) by producing a protein that competes for dsRNA. Adenoviruses (E1A proteins) and hepatitis B virus (polymerase protein) block IFN-induced signalling.

---

* dsRNA is formed by all viruses, and is indeed the chemical signature of a virus infection.

Other viruses, including HIV, adenovirus, and Epstein–Barr virus produce small RNA molecules that bind to PKR, blocking its activation by dsRNA.

There are a few persistent virus infections in which interferon is not induced. Mice persistently infected with LCM or leukaemia virus do not produce detectable interferon in spite of the continued multiplication of virus. This is also true of mouse cells infected *in vitro* with these viruses, although virus multiplication is readily inhibited when interferon is added. The infected mice form interferon normally when infected with other viruses, so the defect is specifically in relation to these particular viruses. Mink infected with Aleutian disease virus fail to respond with interferon production, although they give normal responses to other interferon inducers. Here there is an additional feature, because the infecting virus appears to be insensitive to the action of interferon. Presumably the cells of mice have difficulty in recognising LCM or leukaemia virus nucleic acid as foreign, but nothing is known about this. Some persistent viruses such as human adenoviruses tend to be insensitive to interferon (see above), and others (e.g. hepatitis B) are poor inducers of interferon. The fact that many viruses have evolved mechanisms for evading interferons (inducing less, or becoming insensitive to its action) suggests in itself that interferons are an important part of host defences. Mice with experimentally disrupted IFN-α/β receptor genes are highly susceptible to many viruses (e.g. vesicular stomatitis virus (VSV), Semliki Forest virus, herpesviruses).

Antigenic Variation

One way in which microorganisms can avoid the antimicrobial consequences of the immune response is by periodically changing their antigens. They present a moving target to the immune response.

Antigenic variation within the infected individual

This happens in a few bacterial and protozoal infections where it is an important factor promoting their persistence in the body. The spirochaetal microorganism *Borrelia recurrentis* is transmitted from person to person by the body louse and causes relapsing fever. After infection the bacteria multiply and cause a febrile illness, until the onset of the immune response a week or so later. Bacteria then disappear from the blood because of antibody-mediated lysis or agglutination, and the fever falls. But antigenically distinct mutant bacteria then arise in the infected individual so that 4–10 days later bacteria reappear in blood and there is another febrile episode, until this in turn is terminated by the appearance of a new set of specific antibodies. Ensuing attacks become progressively less severe, but there may be up to ten of them before final recovery. The disease is called relapsing fever because of the repeated febrile episodes, each caused by a newly emerging antigenic variant of the infecting bacterium.

Some protozoa have a similar antigenic versatility. Sleeping sickness is a disease of man in Africa caused by parasitic protozoa of the *Trypanosoma brucei* group, and is spread by biting (tsetse) flies. The infection spreads systemically to the lymph nodes and blood, and is characterised by recurrent fever and headache. In the later stages the central nervous system is involved to give chronic meningoencephalitis, occasionally with the condition of lethargy from which the disease gets its name.* The surface coat of the trypanosome is 12–15 µm thick, and is composed of carbohydrate and a glycoprotein of mol. wt 65 000. During the infection, an antigenically new coat is produced spontaneously in about 1 in 10 000 trypanosomes, to give a series of antigenic variants. These arise by changes in gene expression rather than by mutation, and a single clone of trypanosomes can express hundreds of variants. The systemic stage of the infection consists of a series of parasitaemic waves, each wave being antigenically different from preceding and successive waves. During this time the immune system is constantly trying to catch up, as it were, with the trypanosomes. A large part (about 10%) of the genome of the trypanosome is taken up with the different surface coat genes, but this is a worthwhile investment for the parasite, allowing it to stay for long periods in the blood, and also offering a dismal outlook for a vaccine.

Parasitic worms have even greater opportunities for this type of hide-and-seek with the immune response. Schistosomiasis is a common disease of man in Egypt and elsewhere in Africa, caused by trematodes (flukes) of the genus *Schistosoma*. There is a larval stage of the parasite in the blood, and the adult worm lives in the veins around the bladder and rectum, causing frequent, painful and bloody urination. The adult worm liberates antigens into the blood, and although the antibodies formed are effective against new larval invaders, they have no effect on the adult because it is safely covered with a layer of host antigens. This ensures that the adult worms remain few in number, and prevents overcrowding in the host.

Antigen variation in *Neisseria gonorrhoeae* contributes to the pathogenicity of this resourceful parasite. During the initial stages of infection, adherence to epithelial cells of the cervix or urethra is mediated by pili (see p. 14), but equally efficient attachment to phagocytes would be undesirable. Hence rapid switching on and off of the genes controlling pili are necessary at different stages of the infection. Changes are also seen in the expression of the outer-membrane proteins of the

---

* But many patients are said to suffer from insomnia.

bacteria. Finely tuned control of expression of the genes for pili and outer-membrane proteins, giving changes in adherence to different host cells, in resistance to cervical proteolytic enzymes, in cytotoxicity, etc., are presumably necessary for successful infection, spread through the body, growth and shedding of these bacteria.

The different strains of gonococci circulating in the community also show great antigenic variation in pili and in outer-membrane proteins, which helps account for the multiple attacks of gonorrhoea that can occur in an individual (see p. 60 footnote). *Pilin*, the protein subunit of the pili, consists of constant, variable and hypervariable regions (analogous with immunoglobulin molecules) and genetic rearrangements and recombinations occurring in the repertoire of pilin genes forms the basis for the antigenic variation.

The above microorganisms are complex enough to be capable of undergoing a series of antigenic variations during the course of a single infection. The lentiviruses (visna in sheep, equine infectious anaemia in horses and HIV in man) cause persistent infections of long duration and show antigenic variation within a given infected individual. In HIV, replication errors, like those of other RNA viruses, are not checked because there are no proofreading mechanisms. Consequently, mutations can arise in key antigenic molecules such as gP120 leading to evasion of antibody defences. The HIV gP120 mutations are more extensive than say those in the H and N of influenza virus, probably because there is more opportunity for these to occur; HIV replicates more widely in the body than influenza virus and for a longer period. Mutations can also appear in viral proteins recognised by CD8 T cells, resulting in (a) failure of key antigenic peptides to bind to MHC class I molecules, or (b) loss of a critical amino acid involved in binding peptide to the T-cell receptor. This could influence the existing repertoire of memory T cells leading to a failure in immune surveillance and favouring the emergence of new antigenic variants within the host. For many viruses antigenic variation occurs within one individual, but the variants are rarely seen in a second individual.

Antigenic variation at the population level

When antigenic variants are transmitted, however, they can accumulate as the virus spreads in the host community, so that eventually a strain is formed that differs antigenically from the original to such an extent that it can come back and re-infect the population. This occurs especially with infections limited to mucosal surfaces, where resistance is often of limited duration (see Ch. 6) and there is a strong selective advantage for virus strains with altered antigenicity. The time between initial infection and shedding is only a few days, and an antigenically altered virus variant can infect, replicate and be shed from the body before a significant local secondary immune response is generated. In contrast to this, re-infection with viruses such as rubella, measles, or mumps, which cause systemic infection, is less likely. The incubation period is 2–3 times as long as in a respiratory virus infection, and the secondary immune response has time to come into action and prevent the spread of infection through the body (Fig. 6.4). The growth of virus in the skin and respiratory tract that occurs late in the incubation period is therefore prevented, and there is no shedding of virus to the exterior. Partly for these reasons, systemic viruses such as rubella, measles, poliomyelitis, or mumps tend to be of uniform character (monotypic) antigenically, and all known isolates world-wide are neutralised by antiserum produced against any other isolate (see footnote, p. 336).

The significance of antigenic variation is well illustrated by influenza viruses. Both influenza A and influenza B viruses evolve continuously, undergoing small antigenic changes due to point mutations, deletions and insertions, as the virus spreads through the community. This operates on a world-wide scale and is called antigenic drift. As new variants appear they replace the previous variant, so that within about 4 years a given individual can be re-infected with an antigenic variant that has been gradually generated by infection of other individuals. This results in local epidemics. Foot and mouth disease virus also evolves by antigenic drift. Influenza A virus, however, shows in addition 'antigenic shift', which results in the creation of new pandemic strains. These probably occur by genetic recombination between human and animal virus strains in a doubly infected host. It is thought that the animal virus 'reservoir' consists of populations of susceptible birds (mainly sea birds and ducks), which are known to harbour their own strains of influenza A virus. Recombinants between different type A influenza viruses are readily formed because the influenza virus genome consists of eight segments which can reassort independently of each other. To be of pandemic potential, antigenic shift must involve the RNAs encoding one or both of the surface components of the virus, either haemagglutinin (H) or neuraminidase (N), but any of the RNAs can be shifted. Very occasionally the recombinant virus shows major antigenic differences from previous human influenza A virus strains, having H or N antigens of bird origin, and is at the same time capable of infecting and being efficiently transmitted in man. Initial infection of man with the new strain perhaps takes place in parts of the world where people live in close association with domestic birds. The entire world's population, with no previous immunological experience of such a virus, is completely susceptible. In modern times pandemic strains of influenza A virus have arisen to give major global outbreaks in 1918, 1957 (Asian 'flu) and 1968 (Hong Kong 'flu). All have originated in Asia. Influenza occurs only in the winter and the new shift variants spread from the Southern Hemisphere winter in June northwards to peak around December in the Northern Hemisphere. Inevitably some infected individuals on one side of the

Table 7.2. Pandemic strains of influenza virus in modern times

Time of pandemic	Strain designation
1918–1919[a]	H1N1
1957–1958 ('Asian flu')	H2N2
1968–1969 ('Hong Kong flu')	H3N2

H = haemagglutinin; N = neuraminidase.

[a] Recently, RNA sequences were identified from the HA of the 1918–1919 virus. This was done from an Alaskan victim of the epidemic, using paraffin-embedded lung sections and the actual lung of a corpse frozen in the permafrost. So far this has not told us why the strain was so lethal, with its peak mortality in 15–45-year-olds.

world will carry the infection to the opposite hemisphere, but efficient person-to-person spread and an epidemic will have to wait until the winter. Pandemic strains are designated according to the H or N antigens (see Table 7.2).* Although antigenic shift causes a dramatic increase in the amount of influenza, it is a comparatively rare event, and antigenic drift is in fact responsible for a far greater number of influenza epidemics. Monitoring of the antigenic evolution of influenza A and B viruses is carried out by local, national and international laboratories coordinated by the World Health Organisation (WHO). It is the responsibility of the WHO to advise the vaccine manufacturers when a significantly different strain of influenza has emerged.

As a viral adaptation for the overcoming of host immunity, antigenic variation is more likely to be important in longer-lived species such as the horse or man where there is a need for multiple re-infection during an individual's lifetime if the virus is to remain in circulation, and if the virus does not have the ability to become latent (Ch. 10). In shorter-lived animals such as chickens, mice or rabbits, on the other hand, populations renew themselves rapidly, and fresh sets of uninfected individuals appear fast enough to maintain the infectious cycle. Human respiratory viruses are among the most successful animal viruses in the world. Some show regular antigenic variation and in others local immune responses are weak. Because of assured increases in human numbers and density, these viruses are perhaps entering their golden age, with an almost unlimited supply of susceptible hosts in the foreseeable future, and poor chances of control by vaccination (see Ch. 12).

The bacteria responsible for superficial infections also tend to show something similar to immunological drift, with the appearance of new variants or subtypes that can re-infect the individual. Staphylococci and streptococci, for instance, exist in a great variety of antigenic types, and this, although it may have some other biological signifi-

*In Hong Kong in 1997 a new virus strain (H5N1), caused 17 cases of human influenza, six of them fatal. Luckily the virus, which was acquired from chickens, did not spread effectively from person to person.

cance, can perhaps be regarded as antigenic drift. Some of the 150 surface proteins of *Mycoplasma pneumoniae* show great antigenic variation. Among the intestinal bacteria, *E. coli* shows a similar antigenic variety but it is not clear that this has the immunological significance suggested.* Just as with virus infections, the bacteria that cause systemic infections are relatively conservative, antigenically speaking, and tend to be monotypic in type, as for instance with plague, tuberculosis, syphilis, typhoid, etc.

Microorganisms that Avoid Induction of an Immune Response

There is an intriguing group of virus-sized microorganisms of uncertain nature that multiply, persist and spread in the infected host, giving pathological changes only after very lengthy incubation periods consisting of a large fraction of the host's life span. The microorganism multiplies in the brain and causes a neurological disease that is nearly always fatal (if the host lives long enough). These 'slow' infections are scrapie (sheep), transmissible mink encephalopathy (mink), and kuru and Cruetzfeld–Jacob disease (CJD) in man; they make up the 'transmissible viral dementias' or 'spongiform encephalopathies' from the description of the pathology produced in the brain (see p. 351). Recently scrapie has spread to cattle where it is known as BSE or bovine spongiform encephalopathy, and BSE has infected about 40 people who had eaten contaminated beef, causing a 'new variant' type of CJD. Although the agent responsible has been adapted to mice and hamsters and behaves experimentally as an infectious microorganism, its mode of replication is not understood. Resistance to irradiation indicates that the infectious principle is smaller than that of any known virus, and no DNA or RNA has been isolated. This led to the revolutionary hypothesis that the infectious material is in fact protein, and the agent has been named a prion. The prion protein, coded by the host and of known sequence, is slightly altered in affected tissues. The exact mode of replication is not clear.

However, during infection with the best studied of these, scrapie, not the slightest flicker of an immune response has been detected and the infection neither induces nor is susceptible to the action of interferon. These microorganisms (if they can be called that) therefore 'multiply' in an unrestricted and inexorable fashion in the host. Presence of scrapie is inferred when a disease with characteristic pathology appears after injection of test material into animals, and the incuba-

---

* Bacterial mutability is high when there are defects in DNA repair. This, together with the acquisition of DNA from other bacteria, makes them 'hypermutable'. Some of the clinically isolated *E. coli* strains for instance are hypermutable, and those that thrive on salted or acidic foods and have antibiotic resistance have obvious advantages.

Table 7.3. Microbial interference with or avoidance of immune defences

Type of interference/avoidance		Mechanism	Example	Status
Hit and run raid		Infection, replication and shedding before adaptive immune defences come into play	Common cold, Rotavirus, gonorrhoea	++
Induction of:	Ineffective antibody	Antibody of poor specificity or affinity fails to neutralise or opsonise	LCM virus *Treponema pallidum*	+ +
	Blocking antibody	Ineffective antibody bound to microbe blocks action of 'good' antibody or immune cells	Disseminated gonorrhoea?	±
	Enhancing antibody	Antibody bound to microbe enhances infection of phagocyte by attaching to Fc receptor	Dengue virus	++
	No antibody	No neutralising antibody	African swine fever virus	++
Destruction of antibody		Liberation of IgA protease	Gonococcus *Haemophilus influenzae* Streptococci	+ + +
Switch-on of T cells or of B cells nonspecifically, nonproductively		Polyclonal activation of T cells (by superantigen) of B cells	Staphylococcal toxins (see Ch. 8) Epstein–Barr virus *Mycoplasma pneumoniae*	+ ± ±
Antigenic variation		Microbial antigens vary within individual host	Trypanosomiasis Relapsing fever HIV	++ ++ ++
		Microbial antigens vary within host population	Influenza virus Streptococci	++ +
Infection in bodily site inaccessible to antibody and immune cells		Persistent infection of glands etc. inaccessible to circulating antibody and immune cells (see Fig. 7.1)	Cytomegalovirus Rabies virus Marek's disease virus	++ ++ ++
'Silent' infection of host cell without making it vulnerable to immune lysis		Failure to display microbial antigen on infected cell surface	Herpes simplex virus Epstein–Barr virus	++ ++
		Loss of microbial antigen by capping	Measles virus	+

Mechanism	Description	Examples	
Fc receptors present on microbe or induced on infected host cell	IgG antibodies nonspecifically bind to microbe or infected cell in 'upside down' position and block immune lysis/opsonisation, etc.	Staphylococci (protein A)	+
		Certain streptococci	+
		Herpes simplex virus	?
		Cytomegalovirus	?
Induction of antigen-specific immune suppression	Microbial invasion of lymphoid tissue leads to suppressor T-cell induction or clonal deletion or clonal anergy of T or B cells	LCM virus	+
		BVDV	++
		Lepromatous leprosy	?
		Hepatitis B virus	?
Antibodies mopped up by microbial antigens	Microbial surface antigens in extracellular fluids combine with and 'divert' antibodies	Hepatitis B virus	?
		Pneumococcal infection	?
Molecular mimicry	Microbial antigens mimic host antigens, leading to poor antibody response	Mycoplasma pneumoniae	?
Lack of recognition by T cells	Interference with the expression of MHC molecules	Adenovirus	+
		Cytomegalovirus	+
		Vaccinia	+
		Herpes simplex virus	+
Inhibition of cytokines, chemokines and complement proteins	Synthesis of 'imitation' cytokines or of soluble receptor (for IL-1, IL-8, TNF, IFN-γ, C3b) by viruses	Vaccinia virus	+
		Kaposi's sarcoma herpes virus	+
		Herpes simplex (C3b rec)	+
Prevention of interferon action	Synthesis of intracellular protein inhibitors	Vaccinia virus	+
		Adenovirus	+
	Induce little or no interferon	LCM virus	±
Concealment of antigenic site	Mutation to provide a nearby glycosylation site	Influenza A virus	++
		Rabies virus	?

tion period is long. Multiplication is slow and the most rapidly detectable member of the group is scrapie when injected intracerebrally into hamsters, the disease appearing within 4 months; kuru injected into monkeys may take several years, and kuru in humans exposed by cannibalism used to take up to 20 years. Infection cannot be detected serologically because there is no immune response. It seems likely there are more of these microorganisms waiting to be discovered, but they pose formidable problems for the investigator.

A summary of microbial interference with or avoidance of immune defences is set out in Table 7.3.

References

Bloom, B. R., Modlin, R. L. and Salgame, P. (1992). Stigma variations: observations on suppressor T cells and leprosy. *Annu. Rev. Immunol.* **10**, 453–488.

Clements, J. E., Gdovin, S. L., Montelaro, R. C. and Narayan, O. (1988). Antigenic variation in lentiviral disease. *Annu. Rev. Immunol.* **6**, 139–159.

Cohen, M. S. and Sparling, P. F. (1992). Mucosal infection with Neisseria gonorrhoeae. Bacterial adaptation and mucosal defences. *J. Clin. Invest.* **89**, 1699–1705.

Donelson, J. E., Hill, K. L. and El-Sayed, N. M. (1998). Multiple mechanisms of immue evasion by African trypanosomes. *Molec. Biochem. Parasitol.* **91**, 51–66.

Farrell, H. E., Degli-Esposti, M. A. and Davis-Poynter, N. J. (1999). Cytomegalovirus evasion of natural killer cell responses. *Immunol. Rev.* **168**, 187–198.

Gregory, C. D., Murray, R. J., Edwards, C. F. and Rickinson, A. B. (1988). Down-regulation of cell-adhesion molecules LFA-3 and ICAM-1 in Epstein–Barr virus-positive Burkitt's lymphoma underlies tumour cell escape from virus-specific T cell surveillance. *J. Exp. Med.* **167**, 1811–1824.

Kalvakolanu, D. V. (1999). Virus interception of cytokine-regulated pathways. *Trends Microbiol.* **7**, 166–171.

Karp, C. L. (1999). Measles: immunosuppression, interleukin 12 and complement receptors. *Immunol. Rev.* **168**, 91–102.

Luther, S. A. and Acha-Orbea, H. (1997). Mouse mammary tumor virus: immunological interplays between virus and host. *Adv. Immunol.* **65**, 139–243.

McMichael, A. (1997). How viruses hide from T cells. *Trends Microbiol.* **5**, 211–214.

McMichael, A. J. and Phillips, R. E. (1997). Escape of human immunodeficiency virus from immune control. *Annu. Rev. Immunol.* **15**, 271–296.

Mahr, J. A. and Gooding, L. R. (1999). Immune evasion by adenoviruses. *Immunol. Rev.* **168**, 121–130.

Maizels, R. M., Bundy, D. A. P., Selkirk, M. E., Smith, D. F. and Anderson, R. M. (1993). Immunological modulation and evasion by helminth parasites in human populations. *Nature* **365**, 797–805.

Male, C. J. (1979). Immunoglobulin A protease production by Haemophilus influenzae and *Streptococcus pneumoniae*. *Infect. Immun.* **26**, 254–261.

Mandrell, R. E., McLeod Griffis, J., Smith, H. and Cole, J. E. (1993). Distribution of a lipooligosaccharide-specific sialyltransferase in pathogenic and non-pathogenic *Neisseria*. *Micro. Pathogen.* **14**, 315–327.

Miller, D. M. and Sedmak, D. D. (1999). Viral effects on antigen processing. *Curr. Opin. Immunol.* **11**, 94–99.

Pererson, P. K. *et al.* (1979). Inhibition of the alternative complement pathway opsonisation by Group A streptococcal M protein. *J. Infect. Dis.* **139**, 575–585.

Pincus, S. H., Rosa, P. A., Spangrude, G. J. and Heinemann, J. A. (1992). The interplay of microbes and their hosts. *Immunol. Today* **13**, 471–473.

Seiler, K. P. and Weis, J. J. (1996). Immunity to Lyme disease: Protection, pathology and persistence. *Curr. Opin. Immunol.* **8**, 503–509.

Skehel, J. J., Stevens, D. J., Daniels, R. S., Douglas, A. R., Knossow, M., Wilson, I. A. and Wiley, D. C. (1984). A carbohydrate side chain on hemagglutinin of Hong Kong influenza virus inhibits recognition by a monoclonal antibody. *Proc. Natl Acad. Sci. U.S.A.* **81**, 1779–1783.

Smith, G. A., Symons, J. A., Khanna, A., Vanderplasschen, A. and Alcami, A. (1997). Vaccinia virus immune evasion. *Immunol. Rev.* **159**, 137–154.

Webster, R. G., Bean, W. J., Gorman, O. T., Chambers, T. M. and Kawaoka, Y. (1992). Evolution and ecology of influenza viruses. *Microbiol. Rev.* **56**, 152–179.

8

Mechanisms of Cell and Tissue Damage

The impact on the host of microbial damage depends very much on the tissue involved. Damage to muscle in the shoulder or stomach wall, for instance, may not be serious, but in the heart the very existence of the host depends on a strong muscle contraction continuing to occur every second or so, and here the effect of minor functional changes may be catastrophic. The central nervous system (CNS) is particularly vulnerable to slight damage. The passage of nerve impulses requires normal function in the neuronal cell membrane, and viruses especially have important effects on cell membranes. Also a degree of cellular or tissue oedema that is tolerable in most tissues may have serious consequences if it occurs in the brain, enclosed in that more or less rigid box, the skull. Therefore, encephalitis and meningitis tend to cause more severe illness than might be expected from the histological changes themselves. Oedema is a serious matter also in the lung. Oedema fluid or inflammatory cell exudates appear first in the space between the alveolar capillary and the alveolar wall, decreasing the efficiency of gaseous exchanges. Respiratory function is more drastically impaired when fluid or cells accumulate in the alveolar air space.* The effect of tissue damage is much less in the case of organs, such as the liver, pancreas or kidney, which have considerable functional reserves. More

---

* In the normal lung, bronchioles and alveoli have an immense capacity to absorb surplus fluid, as indicated by the observation that 21 litres of fluid can be given intratracheally to a horse over the course of 3.5 h with no ill effect. There are nearly 10^9 alveoli in man, richly supplied with lymphatics, and with a combined absorptive area of 90 m^2.

than two-thirds of the liver must be removed before there are signs of liver dysfunction.

Cell damage has profound effects if it is the endothelial cells of small blood vessels that are involved. The resulting circulatory changes may lead to anoxia or necrosis in the tissues supplied by these vessels. Here too, the site of vascular lesions may be critical, effects on organs such as the brain or heart having a greater impact on the host, as discussed above. Rickettsiae characteristically grow in vascular endothelium and this is an important mechanism of disease production. By a combination of direct and immunopathological factors there is endothelial swelling, thrombosis, infarcts, haemorrhage and tissue anoxia. This is especially notable in the skin, and forms the basis for the striking rashes in typhus and the spotted fevers. These skin rashes, although important for the physician, are less important for the patient than similar lesions in the central nervous system or heart. It is damage to cerebral vessels that accounts for the cerebral disturbances in typhus; involvement of pulmonary vessels causes pneumonitis, and involvement of myocardial vessels causes myocardial oedema. In Q fever, rickettsiae sometimes localise in the endocardium, and this causes serious complications.

Sometimes an infectious agent damages an organ, and loss of function in this organ leads to a series of secondary disease features. The signs of liver dysfunction are an accepted result of infections of the liver, just as paralysis or coma is an accepted result of infection of the central nervous system. Diabetes may turn out to be caused by infection of the islets of Langerhans in the pancreas. Coxsackie and other virus infections of the islets of Langerhans can certainly cause diabetes in experimental animals, and coxsackie viruses have been associated with juvenile diabetes in man.

There are many diseases of unknown aetiology for which an infectious origin has been suggested. Sometimes it is fairly well established that an infectious agent can at least be one of the causes of the disease, but in most instances it is no more than a hypothesis, with little or no good evidence. For conditions as common and as serious as multiple sclerosis, cancer and rheumatoid arthritis it would be of immense importance if a microorganism were incriminated, since this would give the opportunity to prevent the disease by vaccination. Accordingly, there is a temptation to accept or publicise new reports even though the evidence is weak or the observations poorly controlled. As if to warn us about this and remind us of possibilities from environmental toxins, Parkinson's disease, a chronic neurological condition in which there is a loss of neurons in a sharply defined region of the brain (substantia nigra), can be caused by exposure to the chemical MPTP. One example which raises the possibility that subtle CNS disturbances may be caused by viruses is experimental infection with Borna disease virus. This virus was used to infect tree shrews (*Tupaia glis*) which are primitive primates. There is little overt disease, but afterwards the male is no longer able to enact the ritual courtship behaviour, which (as students well know), is an essential preliminary to mating in all primates, and the frustrated male usually ends up bitten by the female. Thus it can be said that infection with Borna disease virus renders the male psychologically sterile. Presumably the virus in some way alters the functioning of neurons concerned in this particular pathway. All other behavioural and physiological aspects appear normal. Borna disease virus is not known to occur in man, but speculation about an analogous human situation is fuelled by the finding of Borna disease virus-specific antibodies in patients with psychiatric/behavioural disorders. In an entirely different clinical context, infection of a particular strain of rats with Borna disease virus causes immense obesity, the underlying physiological basis of which is not understood. Since the aetiology of such diseases raises interesting problems in pathogenesis, the present state of affairs is summarised in Table 8.1, which includes some of the human diseases whose infectious origin is probable, possible, conceivable, or inconceivable.

Causal connections between infection and disease states are particularly difficult to establish when the disease appears a long time after infection. It was not too difficult to prove and accept that the encephalitis that occasionally occurs during or immediately after measles was due to measles virus. But it was hard to accept that a very rare type of encephalitis (subacute sclerosing panencephalitis or SSPE), occurring up to 10 years after apparently complete recovery from measles, was also due to measles virus and this was only established after careful studies and the eventual difficult isolation of a mutant form of measles virus from brain cells. 'Slow' infections, in which the first signs of disease appear a long time after infection, are now an accepted part of our outlook. The disease kuru occurred in New Guinea and was transmitted from person to person by cannibalism. The incubation period in man appears to be 12–15 years, and the disease was caused by an unconventional infectious agent that grew in the brain. This was established when the same disease appeared in monkeys several years after the injection of material from the brain of Kuru patients (see pp. 351–353). A similar agent called scrapie (see Ch. 7) infects sheep, mice and other animals and also has an incubation period representing a large portion of the life span of the host. In both Kuru and SSPE the agent was eventually shown to be present in the brains of patients. So far this has only been demonstrated indirectly as, despite strenuous efforts, the causative agent has yet to be isolated. If in a slow infection, the microorganism that initiated the pathological process is no longer present by the time the disease becomes manifest, then the problem of establishing a causal relationship will be much greater. This may possibly turn out to be true for diseases like multiple sclerosis and rheumatoid arthritis. Liver cancer in humans and certain leukaemias in mice, cats, humans and cattle can be caused by slow-type virus infections. Cancer or leukaemia appears as a late and occa-

Table 8.1. Microorganisms as causes of human diseases of unknown aetiology

Disease	Features	Microorganism	Pathogenic mechanism	Comments	Status of infectious aetiology
Juvenile diabetes	Onset early in life; sensitive to insulin	Coxsackie B viruses	Infection and damage of islets of Langerhans; secondary immune phenomena	Accounts for some cases.	+
		Mumps		No direct evidence	±
		Rubella		Late result congenital rubella	+
Crohn's disease	Granulomatous inflammation of intestine	Mycobacteria Viruses	Not clear; secondary immune phenomena	No good evidence	−
Ulcerative colitis	Inflammation of colon	Viruses	Not clear; secondary immune phenomena	No good evidence	−
Multiple sclerosis	Demyelinating disease of central nervous system. Waxes and wanes	A variety of enveloped viruses?	Autoimmunity triggered by presentation of brain autoantigens in the envelope of a succession of different viruses	No direct evidence	−
Rheumatoid arthritis	Chronic inflammation and damage to joints	Mycoplasmas	?	Cause arthritis in animals but no evidence for man	−
		Viruses (Epstein–Barr, rubella, parvovirus B 19)	?	No good evidence	−
Paget's disease of bone	Localised deformation of bone	Measles virus	Persistent infection of osteoclasts		±

Disease	Features	Microorganism	Pathogenic mechanism	Comments	Status of infectious aetiology
Duodenal ulcer, gastritis	Ulceration, inflammation	Helicobacter pylori [a]	Bacterial cytotoxins?	Treated with antacids and antibiotics	+
Ankylosing spondylitis	Chronic arthritis of spine	Klebsiella spp.	Immune response to bacterial antigen cross-reacts with joint antigen, giving autoimmune damage	Strong association with HLA B27 genotype	+
Chronic fatigue syndrome	Tiredness, muscle weakness, lasting months or years	Epstein–Barr virus? HHV6, etc.	Unknown Upset of hypothalemic-adrenal axis (p. 380)?	Some cases (see p. 358)	
Alzheimer's disease	Presenile (<55 years) dementia	'Slow virus'?	Infectious agent replicates slowly in brain, destroying cells	Some cases?	±
Senile dementia	Loss of neurons; very common at 65+ years			No evidence	–
Old age	Not a disease but early death offers reliable prophylaxis. Degenerative changes	?	?	No evidence. Universal infection plus very long incubation period could give onset of 'disease' with ageing	–

Cancer		Agent	Mechanism		Comment
Carcinomas	Nasopharyngeal carcinoma	Epstein–Barr virus	Transformation of epithelial cell	+	Susceptibility gene in Chinese people
	Cervical/penile carcinoma	Papillomaviruses	Transformation of epithelial cell	+	Associated with sexual promiscuity
	Carcinoma of liver	Hepatitis B virus	Transformation of hepatic cell	+	Liver cancer especially common in those with persistent hepatitis B infection[b]
	Skin cancer (basal cell carcinoma)	Papillomaviruses	Ultraviolet light as co-carcinogen	+	Evidence in animals but so far not in humans
	Stomach cancer	Helicobacter pylori	Chronic inflammation?	±	Association (in small proportion of cases) is with chronic gastritis and ulcer (? role of host genes, diet, cofactors)
Lymphomas	Burkitt's lymphoma	Epstein–Barr virus	Transformation of B lymphocyte plus cofactor (? malaria)	+	Evidence compelling but not conclusive
	Hodgkin's disease	Epstein–Barr virus	Transformation of B lymphocyte	–	No direct evidence
Leukaemias		Retroviruses	Transformation of white cell precursor	+	Cause leukaemia in animals, and certain T-cell leukaemias in humans (HTLV 1 and 2)

[a]An ancient human parasite (see p. 26) detected in 2000-year-old corpses from Chile.
[b]In a study of 22 797 civil servants in Taiwan, 1.2% of hepatitis B carriers developed liver cancer compared with 0.005% of noncarriers.

sional sequel to infection. The virus, its antigens or fragments of its nucleic acid are detectable in malignant cells.

One important factor that often controls the speed of an infectious process and the type of host response, is the rate of multiplication of a microorganism.* Different infectious agents show doubling times varying from 20 min to 2 weeks, and some of these are listed in Table 8.2. Often the rate of multiplication in the infected host, in the presence of antimicrobial and other limiting factors, and when many bacteria are obliged to multiply inside phagocytic cells, is much less than the optimal rate in artificial culture. Clearly a microorganism with a doubling time of a day or two will tend to cause a more slowly evolving infection and disease than one that doubles in an hour or less.

It is uncommon for an infectious agent to cause exactly the same disease in all those infected. Its nature and severity will depend on infecting dose and route, and on the host's age, sex, nutritional status, genetic background, and so on (see Ch. 11). Many infections are asymptomatic in more than 90% of individuals, clinically characterised disease occurring in only an occasional unfortunate host, as 'the tip of the iceberg'.† Asymptomatically infected individuals are important because they are not identified, move normally in the community, and play an important part in transmission.

Table 8.2.   Growth rates of microorganisms expressed as doubling times

Microorganisms	Situation	Mean doubling time
Most viruses	In cell	<1 h
E. coli, staphylococci, streptococci etc.	*In vitro*	20–30 min
Salmonella typhimurium	Mouse spleen	5–12 h
	In vitro	30 min
Tubercle bacillus	*In vitro*	24 h
	In vivo	Many days
Fungi		
Candida albicans	*In vitro* (37°C)	30 min
Dermatophytes	*In vitro* (28°C)	1–24 h
Treponema pallidum	*In vivo* (rabbit)[a]	30 h
Scrapie group	Mouse brain[a]	4–7 days
Leprosy bacillus	*In vivo*[a]	2 weeks
Plasmodium falciparum	*In vivo* or *in vitro*[b]	8 h

[a] Cannot be cultivated *in vitro*.
[b] Erythrocyte or hepatic cell.

* Every infection is a *race* between the spread and multiplication of the microbe and the generation of an antimicrobial response by the host. A day or two's delay in this response may let the microbe reach the critical levels of growth that give tissue damage and disease.

† The incidence of clinical disease varies from zero (*P. carinii*), through 1–2% (poliomyelitis and Epstein–Barr virus infections in small children) to virtually 100% (measles and HIV).

This chapter deals with demonstrable cell and tissue damage or dysfunction in infectious diseases. But one of the earliest indications of illness is malaise, 'not feeling very well'. This is distinct from fever or a specific complaint such as a sore throat and, although it is difficult to define and impossible to measure, we all know the feeling. It can precede the onset of more specific signs and symptoms, or accompany them. Sometimes it is the only indication that an infection is taking place. Almost nothing is known of the basis for this feeling. 'Toxins', of course, have been invoked and the earliest response to pyrogens (see pp. 329–331) before body temperature has actually risen, may play a part. Interferons may have something to do with it because pure preparations of human α or β interferons cause malaise and often headaches, and muscle aches after injection into normal individuals. Soluble mediators of immune and inflammatory responses, such as interleukin-1 (IL-1; see Glossary) or other cytokines doubtless also play a part. Several cytokines induce release of prostaglandin E2 which, in addition to its effect on fever, reduces the pain threshold in neurons, and this could account for aches and pains. In some infectious diseases weakness and debility are prominent during convalescence. This can be especially notable following influenza and hepatitis, but its basis is as mysterious as in the case of malaise.

Infection with no Cell or Tissue Damage

The infections that matter are those causing pathological changes and disease. Before giving an account of the mechanisms by which these changes are produced, it is important to remember that many infectious agents cause little or no damage in the host. Indeed, it is of some advantage to the microorganism to cause minimal host damage, as discussed in Ch. 1. Virus infections as often as not fall into this category. Thus, although infection with rabies or measles viruses nearly always causes disease, there are many enterovirus, reovirus and myxovirus infections that are regularly asymptomatic. Even viruses that are named for their association with disease (poliomyelitis, influenza, Japanese encephalitis) often give an antibody response as the only sign of infection in the host. Tissue damage is too slight to cause detectable illness. There is also a tendency for persistent viruses to cause no more than minor or delayed cellular damage during their persistence in the body, even if the same virus has a more cytopathic effect during an acute infection, e.g. adenoviruses, herpes simplex (see Ch. 10). A few viruses are remarkable because they cause no pathological changes at all in the cell, even during a productive infection in which infectious virus particles are produced. For instance, mouse cells infected with LCM (see Glossary) or leukaemia virus show no pathological changes. A mouse congenitally infected with LCM virus shows a high degree of immune tolerance, and all tissues in the body are infected. Throughout the life of the animal, virus and viral antigens are produced in the cerebellum, liver, retina, etc. without discernible effect on cell function. But sometimes there are important functional changes in infected cells which lead to a pathological result. For example, the virus infects growth-hormone-producing cells in the anterior pituitary. Although the cells appear perfectly healthy, the output of growth hormone is reduced, and as a result of this, suckling mice fail to gain weight normally and are runted.

When bacteria invade tissues, they almost inevitably cause some damage, and this is also true for fungi and protozoa. The extent of direct damage, however, is sometimes slight. This is true for *Treponema pallidum*, perhaps because the lipopolysaccharide–protein components that might have induced inflammatory responses, are not exposed on the surface of the bacteria. It produces no toxins, does not cause fever, and attaches to cells *in vitro* without harmful effects. Leprosy and tubercle bacilli eventually damage and kill the macrophages in which they replicate, but pathological changes are to a large extent caused by indirect mechanisms (see below). In patients with untreated lepromatous leprosy, the bacteria in the skin invade blood vessels, and large numbers of bacteria, many of them free, may be found in the blood. In spite of the continued presence of up to 10^5 bacteria ml^{-1} of blood there are no signs or symptoms of septicaemia or toxaemia. *Mycobacterium leprae* can be regarded as a very successful parasite that induces very little host response in these patients, even when the bloodstream is invaded. The resident bacteria inhabiting the skin and intestines of man and animals do not invade tissues and are normally harmless; indeed, as discussed in Ch. 1, they may benefit the host. Bacteria such as meningococci and pneumococci, whose names imply pathogenicity, spend most of their time as harmless inhabitants of the normal human nasopharynx: only occasionally do they have the opportunity to invade tissues and give rise to meningitis or pneumonia.

Direct Damage by Microorganisms

Cell and tissue damage are sometimes due to the direct local action of the microorganism. However, it is not at all clear how viruses cause the death of cells. Many virus infections result in a shutdown of RNA synthesis (transcription), protein synthesis (translation) and DNA synthesis in the host cell, but usually these are too slow to account for the death of the cell. After all, cells like neurons never synthesise DNA, and the half-life of most proteins and even RNAs is at least several hours. A possible alternative mechanism is the alteration of the differential permeability of the plasma membrane. This is important as the cell has a high internal K^+ concentration and low Na^+ concentration, while the reverse is true of body fluids. Viruses do alter membrane permeability, but the unresolved question is whether or not this is responsible for the death of the cell or whether it is merely an after effect.

It now appears that in many virus infections (including HIV, adenoviruses, herpesviruses, influenza virus, and picornaviruses) the cells commit suicide by a mechanism called 'programmed cell death' or 'apoptosis'. This is the natural process by which the body controls cell numbers and rids itself of superfluous or redundant cells during development. A familiar example is a tadpole 'losing' its tail. Cells do not disintegrate but round up, and are then removed by phagocytes. Apoptosis in virus infections can be regarded as a host strategy for destroying infected cells.* The chromatin condenses round the edge of the nucleus and a cellular endonuclease cleaves the DNA into 180–200 base pair fragments. The cell membrane forms blebs but stays intact while the cell as a whole breaks up into smaller bodies. The suicide process is more controlled, almost more dignified, than mere disintegration and necrosis. In the latter there is early loss of membrane integrity, spillage of cell contents and random break up of DNA.

There are two more characteristic types of morphological change produced by certain viruses, and these were recognised by histologists more than 50 years ago. The first are inclusion bodies, parts of the cell with altered staining behaviour which develop during infection. They often represent either cell organelles or virus factories in which viral proteins and/or nucleic acids are being synthesised and assembled. Herpes group viruses form intranuclear inclusions, rabies and poxviruses intracytoplasmic inclusions, and measles virus both intranuclear and intracytoplasmic inclusions. The second characteristic morphological change caused by viruses is the formation of multinucleate giant cells. This occurs, for instance when human immunodeficiency virus (HIV) 'fusion' proteins (gp120–gp41) present in nascent virus particles budding from an infected cell attach to CD4 receptors in the plasma membranes of neighbouring cells; membranes then fuse and multinucleate cells are formed. It also happens in measles and certain herpes virus infections.

Before leaving the subject of direct damage by viruses, one supreme example will be given. Here the direct damage is of such a magnitude that the susceptible host dies a mere 6 h after infection. If Rift Valley Fever virus, an arthropod-borne virus infecting cattle, sheep and man in Africa, is injected in very large doses intravenously into mice, the injected virus passes straight through the Kupffer cells and endothe-

---

* Although viruses often prevent apoptosis while they replicate in the cell, it can be useful to make the cell disintegrate at a later stage. Adenoviruses have a 'death protein' (E3 11.6K), a nuclear envelope transmembrane protein, that acts a few days after infection, breaking up the nucleus and allowing the progeny virus to escape.

lial cells lining liver sinusoids (see Ch. 5) and infects nearly all hepatic cells. Hepatic cells show nuclear inclusions within an hour, and necrosis by 4 h. As the single cycle of growth in hepatic cells is completed, massive liver necrosis takes place, and mice die only 6 h after initial infection. The host defences in the form of local lymph nodes, local tissue phagocytes, etc. are completely overcome by the intravenous route of injection, and by the inability of Kupffer cells to prevent infection of hepatic cells. Direct damage by the replicating virus destroys hepatic cells long before immune or interferon responses have an opportunity to control the infection. This is the summit of virulence. The experimental situation is artificial, but it illustrates direct and lethal damage to host tissues after all host defence mechanisms have been overwhelmed.

Most rickettsiae and *Chlamydia* damage the cells in which they replicate, and it is possible that some of this damage is due to the action of toxic microbial products. This action, however, is confined to the infected cell and toxic microbial products are not liberated to damage other cells. Mycoplasma (see Table A.3) can grow in special cell-free media, but in the infected individual they generally multiply while attached to the surface of host cells. As studied in culture and on the respiratory epithelium, they 'burrow' down between cells, inhibit the beat of cilia and cause cell necrosis and detachment. The mechanism is not clear. If a complete lawn of mycoplasma covers the surface of the host cell, some effect on the health of the cell is to be expected, but it is possible that toxic materials are produced or are present on the surface of the mycoplasma.

Bacteria generally damage the cells in which they replicate, and these are mostly phagocytic cells (see Ch. 4). *Listeria*, *Brucella* and *Mycobacteria* are specialists at intracellular growth, and the infected phagocyte is slowly destroyed as increasing numbers of bacteria are produced in it. Bacteria such as staphylococci and streptococci grow primarily in extracellular fluids, but they are ingested by phagocytic cells, and virulent strains of bacteria in particular have the ability to destroy the phagocyte in which they find themselves, even growing in the phagocytes, as described in Ch. 4. Many bacteria cause extensive tissue damage by the liberation of toxins into extracellular fluids. Various toxins have been identified and characterised. Most act locally, but a few cause pathological changes after spreading systemically through the body.

Dental caries provides an interesting example of direct pathological action. Colonisation of the tooth surface by *Streptococcus mutans* leads to plaque formation, and the bacteria held in the plaque utilise dietary sugar and produce acid (see p. 40). Locally produced acid decalcifies the tooth to give caries. Caries, arguably the commonest infectious disease of Western man, might logically be controlled by removing plaque, withholding dietary sugar, or vaccinating against *Streptococcus mutans*. However, fluoride in the water supply or in toothpaste has been the method of choice, and has been very successful. It acts by making teeth more resistant to acid.

Microbial Toxins

This is a huge and growing part of our subject and we need to define the term toxin, a task which is more difficult than one might think. An attempt was made by Bonventre who in 1970 defined toxins as a 'special class' of poisons which differ from, for example, cyanide or mercury by virtue of their microbial origin, protein structure, high molecular weight, and antigenicity. This view is too embracing, because it includes proteins of doubtful significance in disease, and also too restrictive, because it excludes nonprotein toxic complexes such as endotoxin. Another suggestion is that toxin must include all naturally occurring substances (of plant, animal, bacterial or whatever origin) which, when introduced into a foreign host, are adverse to the well-being or life of the victim. This, too, is unsatisfactory because some substances – potent toxins within the scope of this definition – are being used in some contexts as therapeutic agents! Perhaps it is point-less to strive for an all-embracing definition, although the obvious differences between bacterial and fungal toxins warrant the continued use of the appropriate prefix. For example, bacterial toxins are usually of high molecular weight and hence antigenic, whereas fungal toxins tend to be low molecular weight and not antigenic.

The problem of definition is compounded because there are substances (aggressins) which help to establish an infective focus as well as those whose action is uniquely or largely responsible for the disease syndrome. Also there are substances known to be produced by bacteria *in vitro*, whose properties on a priori grounds make them potential determinants of disease, but which have not been shown to play a role *in vivo*.

In the last few years a huge effort has been devoted to understanding the genetic basis of toxin acquisition, expression, assembly and secre-tion of toxins, the resolution of the three-dimensional structure of toxins, and their biochemical modes of action. As a result we now know a great deal about the spread of some virulence determinants in bacte-rial populations via bacteriophages and other transmissible genetic elements, the conditions under which toxins are expressed both *in vitro* and *in vivo*, how to disassemble complex protein toxins and form chimeric derivatives of known and potential use as therapeutic agents, and how to use some of the deadliest poisons known to man in treating certain physiological disorders. Elucidation of biochemical modes of action has resulted in toxins being used increasingly as important tools for the dissection of cell biological processes. Also, some new insights as to the role(s) of toxins in disease causation have been developed. The latter is the result of using isogenic *tox*(-) mutants *in vivo*, using more relevant biological test systems, and concentrating more on the effects of sublethal doses of toxin and less on the effects of injecting a toxin bolus into some animal. It is beyond the scope of this book to attempt to cover all these subjects, so only an outline treatment will be given with some examples. Fortunately, for a fuller treatment one can refer the reader to the recent excellent text on bacterial protein toxins by Alouf and Freer (1999).

Protein toxins

These are either secreted by, or released upon lysis from both Gram-positive and Gram-negative bacteria, and historically referred to as exotoxins. They are proteins, some of which are enzymes. When liberated locally they can cause local cell and tissue damage. Those that damage phagocytic cells and are therefore particularly useful to the microorganism have been described in Ch. 4. Those that promote the spread of bacteria in tissues have been referred to in Ch. 5. A description of some protein toxins, or families of toxins follows.

Toxins which act extracellularly

Helicobacter pylori is a specific human pathogen affecting billions of people world-wide. It is transmitted via the oro-faecal route and colonises the seemingly inhospitable niche of the stomach. Some 20% of infected patients can develop ulcers or stomach cancer. An essential virulence factor of *H. pylori* is a potent urease which is synthesised in vast quantity by the organism, and (at least in culture) released by autolysis and efficiently absorbed on to the surface of viable organisms. As noted in Ch. 2, it is important in local neutralisation of stomach acidity thereby allowing *H. pylori* to penetrate the protective mucus layer overlying the lining of the stomach where the organism attaches to gastric epithelial cells. However, urease is now considered by some as a toxin which acts outside cells, since NH_3, the product of urease activity, is toxic to cells.

Toxins which affect extracellular 'structural' elements

Proteases and *hyaluronidases*, which help the spread of bacteria through tissues have already been mentioned in Ch. 5. Here we consider toxins which act on extracellular substances and are responsible for many of the main features of the diseases caused by the infecting organism. *Pseudomonas aeruginosa* elastase, and one of at least six proteases of *Legionella pneumophila*, both induce fibrinopurulent exudation in the rat lung (a model for *P. aeruginosa*-induced pneumonia in human cystic fibrosis) and the guinea-pig lung (a model for legionnaires' disease), respectively. These characteristics almost certainly arise from the release of oligopeptides from extracellular matrix components of the host which are chemotactic for leucocytes and fibroblasts. The *L. pneumophila* protease is the same major secretory protein (the zinc metalloprotease) already considered in Ch. 4 in relation to survival within macrophages.

Staphylococcal exfoliatin (epidermolysin) is important in staphylococcal 'scalded skin syndrome' (SSSS), a disease of newborn babies. The disease is characterised by a region of erythema which usually begins around the mouth and, in 1–2 days, extends over the whole body. During this period, small yellowish exudative lesions often appear. The most striking feature of the disease, however, is that the epidermis, although apparently healthy, can be displaced and wrinkled like the skin of a ripe peach by the slightest pressure. Soon large areas of epidermis become lifted by a layer of serous fluid and peel at the slightest touch. Large areas of the body rapidly become denuded in this way and the symptoms resemble those of massive scalding. The toxin causes cleavage of desmosomes (specialised cell membrane thickenings through which cells are attached to each other) in the stratum granulosum. However, despite numerous attempts to characterise the biological activity of exfoliatin, the genetically predicted serine protease and/or lipase activity has never been demonstrated.

Vibrio cholerae ZOT (zonula occludens toxin) alters the permeability of junctional complexes in rabbit gut epithelia. In human cholera patients there is evidence of widening of the zonula adherens (Fig. 8.1). Again, like staphylococcal exfoliatin, we do not know the precise mode of action of the toxin responsible for such alterations in epithelial junctional complexes.

In the two preceding examples, it has not been formally proved that these cell–cell splitting toxins act from the outside. In contrast, *Clostridium difficile* toxins A and B which also affect epithelial tight junctions are first internalised and, by virtue of their ability to inhibit Rho (an intracellular target; see Fig. 4.1), cause the collapse of tight junctions thereby increasing the ease with which inflammatory cells arrive at the site of *C. difficile* infection, a highly characteristic feature of such infections.

Toxins which act on cell membranes

Some enterotoxigenic *E. coli* elaborate families of low-molecular-weight heat-stable (ST) peptides as well as heat-labile (LT; cholera-like) toxins. STs bind to a receptor which then activates a tightly coupled membrane-bound guanylate cyclase in gut cells, resulting in the transmission of a signal to the inside of the cell, thereby elevating cGMP, or some other second message. As described later in the section on diarrhoea, this gives rise to efflux of ions, and hence water, from enterocytes.

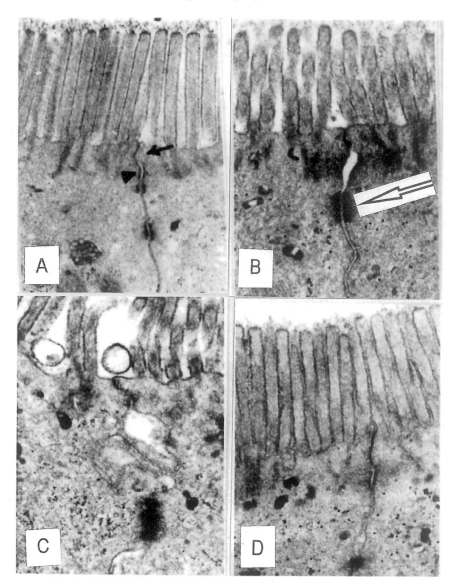

Fig. 8.1 Apical junctional complexes in duodenal biopsies from (A) a control and patients with cholera (B–D). (A) Control upper third of villus showing tightly apposed cell membranes in zonula occludens region (arrow) and widened space between cell membranes in zonula adherens (arrowhead). (B) Upper third of villus from patient with cholera. Note widening of zonula adherens and intact desmosome (arrow). (C) Saccular dilatation of zonula adherens in upper third of villus. (D) Lower third of villus from same sample as (B) showing no abnormalities. Crypt from same sample as B showed normal junctional complexes (not shown). Reproduced from *Gastroenterology* 1995, **109**, 422–430, Fig. 3, with kind permission of Professor M. M. Mathan, and the publisher W.B. Saunders.

Toxins which damage membranes

Some toxins destroy membranes by virtue of their proteolytic activities, and some by their ability to degrade lipid components, while others are pore-forming or detergent-like in their mode of action.

Proteases

In addition to their action on protein components of lung connective tissue referred to above, *Pseudomonas aeruginosa* elastase and the zinc metalloprotease of *Legionella pneumophila* are believed to destroy cell membranes by their proteolytic activity. This is the probable reason for the haemorrhage associated with lung infections caused by these pathogens, i.e. effects on type I alveolar epithelial and endothelial cells.

Phospholipases

Clostridium perfringens α-toxin

A large number of bacterial enzymes are phospholipases, some of which, but by no means all, are important toxins. The best example is the α-toxin of *Clostridium perfringens*, the organism most commonly associated with gas gangrene. It is strictly anaerobic and occurs as a normal inhabitant in the large intestines of man and animals; its spores are ubiquitous in soil, dust and air. *C. perfringens* does not multiply in healthy tissues, but grows rapidly when it reaches devitalised and therefore anaerobic tissues. This could be after contamination of a natural wound with soil or dust, particularly on battlefields or in automobile accidents, or after contamination of a surgical operation site with clostridia from the patient's own bowels or skin. After abortions, particularly in the old days before antibiotics, intestinal clostridia often gained access to necrotic or devitalised tissues in the uterus and set up life-threatening infections. Invasion of the blood was common and soon resulted in death, the clostridia localising and growing in internal organs such as the liver after death. *C. perfringens* has various enzymes that enable it to break down connective tissue materials, including collagen and hyaluronidase, thereby facilitating spread of the infection along tissue planes. Most of these enzymes are toxic to host cells and tissues, but α-toxin is easily the most important one. It is dermonecrotic, haemolytic (a feature seen mainly in tissues close to the focus of infection but sometimes responsible for large-scale intravascular haemolysis in infected patients), causes turbidity in lipoprotein-rich solutions and is lethal. While it is still true that these activities are all due to one molecular species, they are not (as was once thought) different expressions of the one enzymic activity.

Historically, *C. perfringens* α-toxin was the first bacterial toxin to be characterised as an enzyme: it is a zincmetallophospholipase C (PLC) which removes the head group, phosphoryl choline, from phosphatidyl choline and from sphingomyelin. It is of undoubted importance in gas gangrene. Toxoid prepared by formalin-treated toxin will protect sheep against infection caused by *C. perfringens*. However, one might ask why all enzymes with such biochemical specificity are not equally toxic or important in determining virulence; there are several reasons which can be put forward. We know that there are at least two functional domains in *C. perfringens* α-toxin. Comparison of *C. perfringens* α-toxin with the phosphatidyl choline-preferring, nontoxic zincmetallophospholipase of *Bacillus cereus*, reveals that two-thirds of the N-terminal sequence of *C. perfringens* α-toxin shows homology with the entire sequence of *B. cereus* PLC; this portion of *C. perfringens* α-toxin retains its PLC activity, but not its haemolytic and lethal activities. The C-terminal part is not haemolytic, not enzymatically active and not cytotoxic for mouse lymphocytes, but is necessary for conferring toxicity on the N-terminal part of the protein. In fact, the C-terminus is a potent immunogen that will solidly protect mice – and hopefully man – against experimental infection with *C. perfringens*. Surprisingly, the C-terminal domain of the nontoxic *C. bifermentans* enzyme shows sequence similarity with that of its *C. perfringens* α-toxin counterpart. The relative nontoxicity of this enzyme is ascribed to its comparatively much lower turnover rate, i.e. it is a much less efficient enzyme.

While haemolysis does occur in experimental gas gangrene (evidenced by haematuria), there is little evidence of massive haemolysis in naturally occurring cases of gangrene. It is now considered more likely that the basis of toxicity is not cytolysis, but rather the consequence of the ability of the α-toxin, in sublytic doses, to cause profound metabolic changes arising from release of phospholipid derivatives. For example production of inositol triphosphate (IP_3), a potent secondary messenger, would affect many cell functions. The activation of the arachidonic acid cascade would result in the production of leukotrienes (increasing vascular permeability), prostaglandins and thromboxanes (causing inflammation, muscle contraction and platelet aggregation). This toxin also upregulates expression of endothelial leucocyte adhesion molecule-1 (ELAM-1), intercellular adhesion molecule-1 (ICAM-1) and neutrophil chemoattractant-activator IL-8, thereby impairing delivery of phagocytes to the site of infection.

There are other pathogenic clostridia that cause gas gangrene and produce similar toxins. Infected tissues show inflammation, oedema and necrosis, not necessarily with the formation of gas, and the illness can be mild or very severe according to the extent of bacterial spread, and the nature and quantity of toxins that are formed and absorbed. Since the bacteria grow and produce their toxins only in devitalised tissues, the most important form of treatment is to remove such tissues. Clostridia are strictly anaerobic, and exposure of the patient to hyperbaric oxygen (pure oxygen at 2–3 atmospheres in a pressure chamber) has been found useful in addition to chemotherapy.

Staphylococcal β-toxin (haemolysin) is known to be produced *in vivo*. In Ch. 4, studies with isogenic mutants were described which indicate that it is important in killing neutrophils. It probably has the narrowest substrate specificity among the phospholipases, and is a hot–cold haemolysin: lysis of erythrocytes occurs only on cooling after incubation at 37°C. The phenomenon, although of doubtful significance *in vivo*, has attracted attention and generated speculation about its mechanism. Perhaps the most likely explanation is that, when cooled below their phase-transition temperature, the remaining phospho-lipids undergo quasi-crystalline formation, thereby generating intramembranous stresses incompatible with structural integrity.

Pore-forming toxins

Cholesterol-binding cytolysins (CBCs)

These proteins, more commonly known as 'SH-activated cytolysins', are made by some 23 taxonomically different species of Gram-positive bacteria, not all of which are pathogens. They are lethal, cardiotoxic, antigenically related, and their lytic and lethal activities are blocked by cholesterol. Recent work requires that we abandon certain percep-tions about these toxins which are enshrined in the older nomencla-ture. For example, *purified* toxins are not O_2-labile, and are not activated by sulfhydryl compounds, and do not depend on a cysteine residue for activity. Alouf has suggested the generic term used as our heading, since the one common feature which correctly applies to all members of this group is the ability of cholesterol to irreversibly inhibit the lytic and lethal properties of these toxins. Interaction with choles-terol is thought to be the key primary event in their interaction with susceptible membranes, which leads to the impairment of the latter; cholesterol plays no further part in the subsequent damage process. However, the role of cholesterol has been interpreted in terms of medi-ating the oligomerisation process (illustrated in Fig. 8.2) which leads to membrane damage. Four examples of CBCs from pathogenic species are considered briefly below. Despite the similarities which warrant their inclusion in the same toxin group, they play entirely different roles in disease causation by the organisms expressing these toxins.

Streptolysin O (SLO)

Some streptococci produce two haemolysins. The one considered here was originally thought to be oxygen-labile and designated streptolysin O (SLO), the other oxygen-stable and designated streptolysin S (SLS). In Ch. 4 (Table 4.1) the *in vitro* cytolytic properties of SLO were listed. However, the situation *in vivo* is far from clear. An injection of a bolus of SLO is lethal almost certainly due to its cardiotoxicity. More recently, SLO has been implicated (alone or in combination with other streptococcal toxins) in tissue damage. The accumulation of polymor-phonuclear leucocytes (PMNs) in lung and soft tissue in cases of strep-tococcal toxic shock syndrome, has been attributed to SLO. SLO-induced increases in proinflammatory cytokines IL-1β and tumour necrosis factor α (TNF-α) and several leukotrienes have also

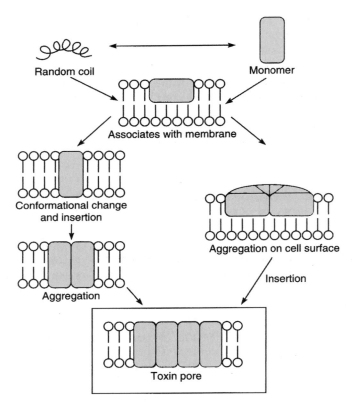

Fig. 8.2 Pore formation by pore-forming toxins. Newly synthesised proteins are soluble. On interaction with cell membranes they undergo conformational changes which allow reorganisation on and insertion into target cell membranes. Cholesterol is involved as primary receptor or mediator of aggregation for the CBC group. Others have specific receptors, but staphylococcal δ-toxin does not. (Reproduced with permission of authors (T. J. Mitchell *et al.*) and publisher (Gustav Fischer Verlag, Stuttgart, Germany) from Fig. 1, in 'Molecular studies of pneumolysin, the thiol-activated toxin of *Streptococcus pneumoniae* as an aid to vaccine design', Fifth European Workshop, Veldhoven; (B. Witholt *et al.* (eds)), *Zentralbl. Bakteriol.* 1992, Suppl. 23, p. 431.)

been demonstrated. SLO-deficient mutants of *S. pyogenes* induced less of these substances and SLO-deficient group A streptococci are less virulent in chick embryos. However, the precise role of SLO in the pathogenesis of infections caused by *S. pyogenes* and their non-suppurative sequlae (e.g. rheumatic fever) is still not clear due almost certainly to the lack of suitable animal models.

Perfringolysin O (PFO; Clostridium perfringens θ toxin)
This toxin is the first of the group for which a three-dimensional crystallographic structure has been obtained: it is a four-domain molecule. Elucidation of its structure has spawned a great deal of biochemical

and biophysical activity. However, the role of PFO in disease causation is still not entirely clear, probably due to the fact that it is only one of a very large number of known or potential virulence determinants produced by *C. perfringens*. The major features of the pathogenesis of *C. perfringens*-mediated gas gangrene is unquestionably explicable in terms of *C. perfringens* α-toxin described above. Experimental evidence suggests that PFO plays only a minor role in the pathogenesis of gas gangrene by upregulating ICAM-1 (see above for α-toxin) thereby slightly contributing to the inhibition of PMN migration to the site of infection.

Listeriolysin O (LLO; listeriolysin)
In contrast to the first two examples, LLO is the most important virulence determinant of *Listeria monocytogenes*. We have already met listeriolysin in Ch. 4: it plays an important part in mediating the escape of *L. monocytogenes* from intraphagocytic vacuoles. Exogenous addition of LLO will, like all other members of this group, rapidly kill cells by rupturing the cytoplasmic membrane. However, rupture of the phagocytic vacuole with release of organisms does not result in immediate death of the cell. This has been explained in terms of the initial acidification of the vacuole with concomitant activation of LLO; the subsequent rise in pH deactivates LLO. LLO is also a potent trigger of host cell-signalling molecules. Many of the responses elicited by LLO are believed to be the result of the activation of cytosolic NF-κB (host-cell stress-inducible transcription factor) and its translocation into the nucleus where it acts as a transcriptional activator of different genes involved in the immune response. The pore-forming activity of LLO results in the release of antigens from the vacuole and stimulation of CD8+ cytotoxic T cells which are known to afford protection against *L. monocytogenes*. This helps explain why patients recovering from *L. monocytogenes* infection have high levels of anti-LLO which are themselves non-protective in experimental infections.

Pneumolysin (PLY)
This protein is produced by the pathogen *Streptococcus pneumoniae* (pneumococcus) which causes bacteraemia, pneumonia, meningitis, and otitis media in humans. PLY is different from all other members of this group in that it is not actively secreted by the pathogen, but remains in the cytoplasm until released by lysis of the pneumococcus. Recent work has shown this toxin is, like PFO, a four-domain molecule. It possesses properties which could never have been predicted or deduced from classical studies of its haemolytic activity. Comparative studies in mice with both wild-type and a pneumolysin-negative mutant (PLN-A) of serotype-2 pneumococci demonstrated that pneumolysin was important in the induction of inflammation in the lung (not cell wall components as had long been believed), conferring ability to replicate in the lung and invade into the bloodstream, and altering alveolar permeability. It inhibits cilial beat in respiratory mucosa. In

experimental meningitis in guinea pigs, in contrast to the mouse lung model, PLY was not responsible for the potent inflammatory response, but did cause an increase in protein content of cerebrospinal fluid (CSF) presumably reflecting its ability to alter cell barriers. PLY has also been implicated in causing sensorineural deafness associated with meningitis caused by the pneumococcus (Fig. 8.3). PLY also activates the complement cascade thereby diverting complement from bacteria. From the foregoing, it is clear that PLY is an important virulence determinant of the pneumococcus.

Attempts to develop protective antipneumococcal vaccines have hitherto been based on the type-specific capsular polysaccharides. Unfortunately, there are at least 90 known types and current vaccine preparations comprise a blend of polysaccharides from some 23 types. Currently, efforts are being made to develop a broadly effective vaccine based on genetically engineered PLYs which are sufficiently non-toxic but immunogenic. Watch this space.

RTX toxins

This group of toxins has been designated RTX (<u>r</u>epeats in <u>tox</u>in) toxins by virtue of a common structural feature – the presence of an array of a nine amino acid repeat (*ca.* 10–40) to which Ca^{2+} binds thereby activating the toxins which form membrane pores of varying sizes. They constitute the largest group of bacterial pore-forming toxins and are widespread among Gram-negative pathogens. In general, the role of RTXs in disease is not clear but three examples are given where RTXs are important. *E. coli* α-haemolysin, regarded as the prototype of this group, is important in extraintestinal infections caused by this organism; the toxin is active against a broad range of mammalian cells. Leukotoxin from *Pasteurella haemolytica* exhibits narrow target cell and host specificities; it specifically kills ruminant leucocytes and is important in bovine pneumonic pasteurellosis. The third example is the 'invasive' adenylate cyclase toxin of *Bordetella pertussis*. This toxin is unique among this group in that it is a large bifunctional toxin: it has both haemolytic and adenylate cyclase activities hence the designations AC-Hly, AC toxin, CyaA. It is one of several virulence attributes expressed by *B. pertussis* (see below) and is known to be important in the early stages of respiratory tract colonization. Strictly it is the haemolysin part of the molecule which belongs to the RTX family and its main function appears to be in translocation of the AC moiety into the cell where cAMP levels are elevated with ensuing pathophysiological sequelae.

Staphylococcal α-toxin

Staphylococci produce a range of toxins some of which we have already met. The α-toxin is easily the most studied from a biophysical point of view and is considered the main cytolysin produced by *S. aureus*. Like streptolysin-O and staphylococcal δ-toxin, it is secreted as a water-soluble protein and undergoes self-induced oligomerisation on cell membranes to form heptameric pores. In systemic staphylococcal infec-

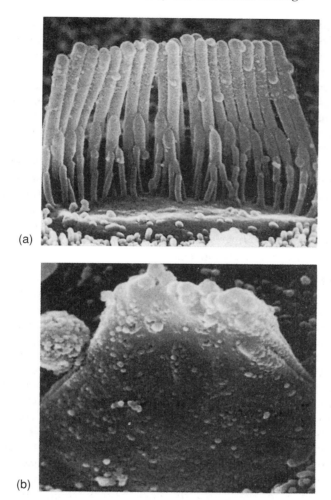

(a)

(b)

Fig. 8.3 The effect of pneumolysin on the hair cells of the inner ear of a guinea-pig. The effect of pneumolysin on the hair cells of the inner ear of a guinea-pig. (a) A scanning electron micrograph of normal hair cells. (b) Hair cells after exposure to pneumolysin; note disappearance of hairs. Hearing depends on the transmission to the hair cells of pressure waves generated in the fluid-filled chamber (scala tympani) of the cochlea. This causes lateral displacement of the hairs. Inelastic links between hairs in different rows results in membrane deformation, opening of ion channels and influx of ions. This generates an action potential in the underlying auditory nerves. (Kindly provided by Drs M. P. Osborne and S. D. Comis, Department of Physiology, The Medical School, University of Birmingham, UK.)

tions, death is most probably due to the potent α-toxin but in localised pyogenic infections – such as mastitis in cattle, goats, rabbits and mice – its role is most likely one of killing phagocytes or conferring surviv-ability on intracellular bacteria.

Detergent-like toxins

Staphylococcal δ-toxin acts in a manner similar to that of the choles-terol-binding cytolysins, with an important difference: the binding is nonspecific with no requirement for cholesterol. It initially forms small pores and then islands of membrane or large micelles; this gives rise to its perceived detergent-like properties. There is a family of closely related δ-toxins which inhibit the growth of gonococci. It is not often that one can ascribe a positive function to a toxin which is beneficial for the organism producing it. In this case δ-toxin(s) could have important ecological significance in the mixed culture that is characteristic of the real microbial world. Of great interest is the synergy that δ-toxin displays. Sublytic amounts of δ-toxin cause release of cell constituents without lysis. However, only 0.01 haemolytic units of staphylococcal β-toxin will cause lysis of cells in the presence of 0.004 lytic units of δ-toxin. This synergistic interaction could be the way in which staphy-lococcal toxins, which rarely exert their lethal effects in the majority of infections, exercise important cytolytic effects. Of less obvious signifi-cance is the fact that δ-toxin is a poor antigen; for a long time its anti-genicity was controversial. If δ-toxin were to prove of crucial importance as a cytolytic potentiator, then this could also partly explain why natural acquired immunity to staphylococcal infection is either non-existent or sufficiently low as to be easily overcome.

Binary toxins

These comprise two proteins, only one of which is toxic but the other is necessary at some stage for manifestation of toxicity. Two examples are given, *Serratia and Proteus* cytolysins and staphylococcal leukocidins.

Serratia marcescens is an opportunistic pathogen which causes a range of infections including respiratory and urinary tract infections; *Proteus mirabilis* causes acute pyelonephritis. Pathogenicity is multi-factorial but the accumulated evidence is that a membrane-active cytolysin plays an important role in disease causation by both patho-gens. *S. marcescens* haemolysin (ShlA) and *P. mirabilis* haemolysin (HpmA) represent a new type of cytolytic toxin: they are described as 'cell-associated' and have a specialised means of delivery from the bacterial cell to the target host cell. ShlA and HpmA each require a second protein, ShlB and HpmB, respectively. The B proteins form pores in the bacterial outer membrane facilitating the secretion of the corresponding A components, their concomitant activation and inser-tion into eukaryotic membranes. These A proteins are also cytotoxic, causing vacuolation in cells and release of a range of inflammatory mediators.

Formerly, the second example would have been limited to a discus-sion of staphylococcal 'Panton Valentine' (PV) leukocidin and staphylo-coccal γ-toxin; the latter is one of the much studied group (α, β, γ and δ) staphylococcal haemolysins. However, the problems generated by antibiotic-resistant staphylococci has stimulated a great deal of new research on this pathogen arising from which is the recognition that PV leukocidin and γ-toxin belong to a very large family of binary leukocidins. Each leukocidin consists of two proteins – S (so called because it elutes slowly) and F (it elutes fast) from an ion-exchange column. S binds first to as yet ill-defined cell receptors (important in defining target cell specificity), followed by F which acts synergistically with S to create functional pores in the target membrane. There are at least six class S proteins and five class F proteins which can give rise to *ca.* 30 biologically active combinations, a fact which could be highly significant in that some strains produce more than one binary leukocidin. Although various S-F combinations exhibit different target cell specificities, most are active against PMNs. It has been shown that PV, which is highly active against human PMNs, causes release of leukotriene B4, IL-8, histamine and tissue degradative enzymes, which would give rise to the chemotactic invasion of more PMNs and subsequent tissue damage.

Toxins with intracellular targets

Many toxins have intracellular targets. There is intense interest in seeking to understand the mechanism(s) of uptake of the active moieties of toxins whose targets are intracellular. This is driven by the desire to understand fundamental mechanisms in cell biology and to develop selective 'cytotoxic therapies' in clinical medicine as well as to unravel the molecular mechanisms of disease causation. To reach an intracellular target, a protein must first be translocated across the cytoplasmic membrane. There are at least three ways in which this can be achieved: self-translocation across cytoplasmic membrane; direct injection; and receptor-mediated endocytosis.

Self-translocation

There is only one example of self-translocation across cytoplasmic membrane known to date: the invasive adenylate cyclase of *B. pertussis* described above.

Direct injection

Recognition of a direct injection mechanism has resolved a problem in *Pseudomonas aeruginosa* pathogenicity. *P. aeruginosa* possesses many potential virulence determinants three of which have definitely been implicated as being important in pathogenesis: elastase (already referred to), and two ADP-ribosylating (ADPR; see below) proteins, exotoxin A (PEA, see below), and exoenzyme S (PES). PES is non-toxic when injected into animals or added to cells on its own. This problem has now been resolved. It is a single polypeptide with no receptor-binding component or translocation domain (see below): it is 'injected' directly across cell membranes by a mechanism functionally similar to that already described for the translocation of Tir by *E. coli* (see Ch. 2,

Fig. 2.7). PES is activated by a cytoplasmic protein FAS, and ADP-ribo-sylates the small G-protein Ras resulting in the collapse of the cytoskeleton (see Ch. 4; Fig. 4.1). Also included in this category are the 'phagocyte toxins' injected into phagocytes by *Yersinia enterocolitica* (Ch. 2). It is possible that other similar bacterial enzymes, e.g. *Clostridium botulinum* C3 which ADP-ribosylates G-protein Rho, will be shown to be internalised in a similar manner.

Receptor-mediated endocytosis

There are several variations on the receptor-mediated endocytosis theme reflecting the structure of the toxins; in some cases the process involves the subversion of normal processes used by the host cell to regulate movement and organisation of cellular membranes and substituent components. Toxins first bind to their respective receptors and become internalised via coated pits, vesicles or caveolae, into endosomes from which they still must escape into the cytoplasm.

Three types of toxin with intracellular targets have been recognised reflecting their genetic origin. Some toxins consist of a single peptide, the product of a single gene, which undergoes post-translational modification into A and B fragments which are covalently linked (Fig. 8.4).

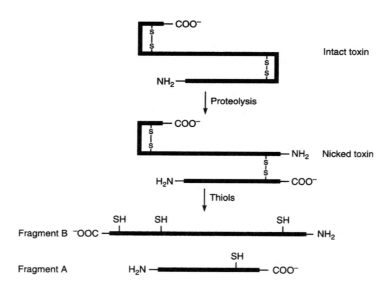

Fig. 8.4 Diphtheria toxin: post-translational modification of the single peptide into A and B fragments. The toxin is synthesised as a single polypeptide but is cleaved (nicked) by proteases into two fragments designated A and B, held together by an –S–S– bond. The latter is reduced during translocation of the A fragment into the cytosol. The B fragment consists of a receptor domain (R) which recognises the DT receptor – an EGF-like precursor which happens to be widely distributed throughout all organ systems – and a T domain which facilitates the translocation of A into the cytoplasm (see Fig. 8.5).

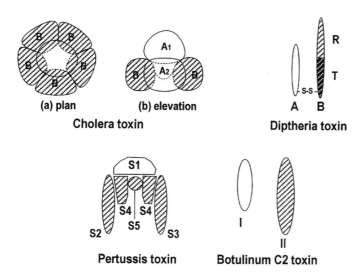

Fig. 8.5 Schematic structure of three types of A–B-type toxins. The hatched regions are the binding/translocation-facilitating parts ('B subunits'). Diphtheria toxin is synthesised as a single peptide (see Fig. 8.4). Cholera toxin (CT) is represented in plan and elevation views; *E. coli* LTs are structurally and functionally very similar to CT. CTA comprises CTA1–A2. A1 is the toxiphore which is held in association with B via A2. A2 has no known enzymic activity but plays some as yet undefined part in toxicity. Differences in CT A2 and LT A2 have been implicated as part of the reason for the lesser severity of disease caused by enterotoxigenic *E. coli*. Pertussis toxin B subunits are heterogeneous. Botulinum C2 toxin is a two-component binary toxin, in which two proteins do not form stable complexes prior to cell attachment. Not to scale. (Modified with permission of authors (I. H. Madshus and H. Stenmark) and publisher (Springer-Verlag GmbH & Co. KG, Heidelberg, Germany) from Fig. 1 in 'Entry of ADP-ribosylating toxins into cells', *Curr. Topics Microbiol. Immunol.*, 1992, **175**, 3, edited by K. Aktories.)

The A fragment is the 'active' toxiphore and the B fragment bears the receptor-binding domain and also mediates translocation of A into the cytoplasm. Examples include diphtheria toxin (DT), *Pseudomonas aeruginosa* exotoxin A (PEA) (Fig. 8.5), and the clostridial neurotoxins (BoNT and TeTx) (Fig. 8.6).

A second group of toxins are the products of separate genes giving rise to A and B subunits which noncovalently associate into stable complexes. They are also designated A-B type toxins, in which the number and nature of B subunits vary, but the connotations of A and B are as for DT. Examples include classical cholera toxin (CT), *E. coli* heat labile enterotoxins (LTs), pertussis toxin (PT), Shiga (ShT) and Shiga-like (ShLT) toxins (Fig. 8.5).

A third group of toxins are the products of separate genes giving rise to different proteins which are functionally equivalent to A and B

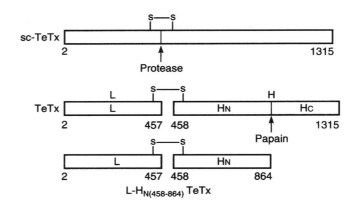

Fig. 8.6 Structure/nomenclature of tetanus and botulinum neurotoxins. sc-TeTx: single chain noncleaved tetanus toxin. TeTx: tetanus toxin after proteolytic activation. L and H: intact light and heavy chains, respectively. L-H$_{N(458-864)}$ TeTx: intact L chain linked to a fragment of H chain (from residue 458 at the N-terminal end to residue 864). H$_N$ and H$_C$: N-terminal and C-terminal parts, respectively, of the H chain. The corresponding nomenclature for botulinum toxin type A is BoNT/A, and for type B, BoNT/B, etc. (Reproduced with permission of the publisher (Gustav Fischer Verlag, Stuttgart, Germany) from 'Clostridial neurotoxins – proposal of a common nomenclature', *in* 'Bacterial Toxins, Fifth European Workshop', Veldhoven (B. Witholt *et al.*, eds, *Zentralbl. Bakteriol.*, 1992, Suppl. 23, p. 17.)

subunits. These proteins do not associate into stable complexes, but must act in concert to express toxicity and are known as binary (or bicomponent) toxins. Examples include anthrax toxins and *Clostridium botulinum* C2 toxin (Fig. 8.5), *Clostridium perfringens* iota toxin, and *Clostridium spiroforme* toxin.

Translocation of toxiphore into the cytoplasm

Direct escape from endosome

DT B fragment binds to its receptor (a precursor of heparin-binding epidermal growth factor (EGF)-like growth factor), undergoes conformational change in the acidified endosome and inserts into the endosomal membrane, pulling the C terminus of the A fragment across the membrane. The -S-S- bridge is exposed to the cytosol, reduced, thereby freeing A to enter the cytosol (Fig. 8.7). A similar mechanism operates with anthrax LF and EF toxins but in this case a third protein, protective antigen (PA) acts as the functional equivalent of DT B (Fig. 8.8).

Route to endoplasmic reticulum

For most if not all other toxins the route to an intracellular target is much more complex. There is direct, or persuasive indirect evidence that the following scenario (based on Olsnes *et al.*, 1999; Hirst, 1999) applies to CT, LTs, ShT, ShLTs, PEA, and probably to many others (see

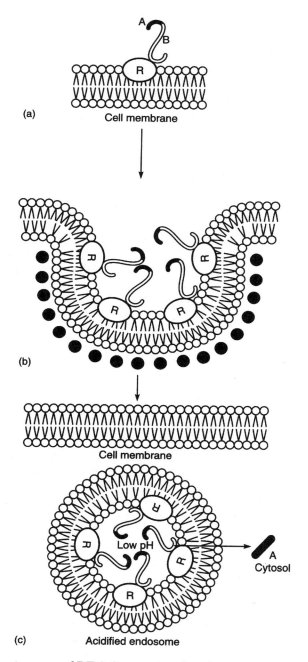

Fig. 8.7 Direct escape of DT A fragment and anthrax LF and EF toxins from acidified endosome. (a) DT binds via B fragment to its receptor (R) in the cell membrane. (b) These complexes migrate to clathrin-coated pits. (c) This gives rise to acidified endosomes which induces conformational changes in B, insertion of B into the membrane and escape of A fragments into the cytosol. The only other known examples of direct escape of toxin into the cytosol are anthrax EF and LF (see Fig. 8.8).

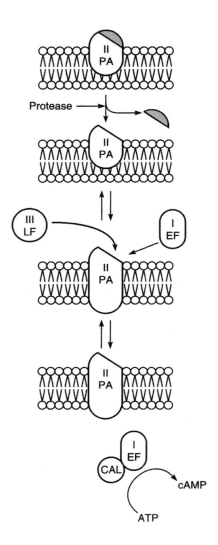

Fig. 8.8 Mode of entry and action of anthrax toxins. Protective antigen (PA) interacts with the cell membrane and forms heptameric oligomers. After proteolytic cleavage, PA sites are exposed which bind edema factor (EF), thereby facilitating translocation of EF into the cytosol directly from acidified endosomes as described for DTA (Fig. 8.7). EF must be rapidly inactivated since washing toxin-treated cells results in a rapid loss of adenylate cyclase activity. EF interacts with calmodulin (CAL) to become an active adenylate cyclase enzyme in nearly all cells. Interaction of PA with EF and subsequent internalisation of EF is blocked by prior binding of LF. In contrast to EF which is active in many cells, LF protease is only active in macrophages. This model explains the characteristic hypovolaemic shock syndrome (cAMP is a potent secretagogue), cytotoxicity to macrophages, and the immunogenicity of PA.

Fig. 8.9). The evidence is hardest for ricin, an A-B type toxic plant lectin (believed to have been located in the tip of an umbrella and used to poison a Bulgarian spy!). Newly formed toxin-containing endosomes enter those vesicular trafficking pathways which lead to the trans-Golgi network (TGN), through the Golgi and further into the endoplasmic reticulum (ER). This is the reverse of the normal secretory pathway and is therefore called 'retrograde transport'. This gives functional significance to the C-terminal sequence *lysine–aspartate–glutamate–leucine* (the KDEL motif) in the A subunit of cholera toxin and related sequences in LT and PEA. The KDEL motif is normally found in proteins which, having been processed in the Golgi, are returned to and are trapped by the ER which recognises the KDEL motif, thereby preventing such proteins being lost to the cell via exocytotic trafficking. The next part is speculative and described for CT. Either in the TGN or the ER, CTA is reduced freeing CTA$_1$, the CT toxiphore (Fig. 8.5). There is another ER pathway, the Sec61p secretion channel, through which aberrantly folded molecules are returned to the cytosol to proteasomes

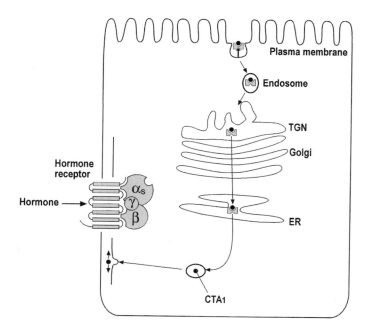

Fig. 8.9 Mode of entry of cholera toxin. This represents a much more complex route to an intracellular target. As described in the text it involves interaction of CTB subunits (shaded part of molecule) with ganglioside receptor GM1 (vertical arrow), receptor-mediated endocytosis, retrograde transport of the endosome to the trans-Golgi network (TGN), through the Golgi to the endoplasmic reticulum (ER), and anterograde transport of liberated CTA1 from ER in vesicles directed to the basolateral membrane, the intracellular location of adenyl cyclase. This mechanism is operative at least for several other toxins.

for degradation. Scrap proteins are normally ubiquitinylated at lysine residues which targets them to proteasomes. Since CTA_1 is lacking in lysine and is hydrophobic, it may well exploit this secretory pathway but remain attached to the cytosolic face of the ER. The next stage is marginally less speculative. It is known that *in vitro* CT is activated by ADP-ribosylating factors (ARFs) which are also known to be involved in vesicular trafficking. Thus, by anterograde transport (the term given to secretory pathways from ER to the plasma membrane) CTA_1 may well be ferried in vesicles to the basolateral membrane with which they fuse thereby depositing CTA_1 near to the target, adenylate cyclase (Fig. 8.9).

Intracellular targets

The targets for some of the intracellular toxins are listed in Table 8.3, and illustrated in Figs 8.10–8.14.

Fig. 8.10 ADP-ribosylation reaction. This enzymic reaction is common to a wide range of toxins with different target proteins (see Table 8.3).

Fig. 8.11 Inhibition of protein synthesis by diphtheria toxin (DT), *Pseudomonas aeruginosa* toxin A (PEA), Shiga toxin (ShT), Shiga-like toxins (ShLTs) and poliovirus. The schema shows a round of peptide elongation and illustrates the key role played by two enzymes, EF-1 and EF-2. EF-1-GTP interacts with aminoacyl-tRNA; this complex is docked into site A, EF-1-GTP becomes EF-1-GDP and is recycled as shown. After peptidyl transfer, EF-2-GTP catalyses transfer of the extended peptide to site P, and is itself autocatalytically converted to EF-2-GDP. DTA and PEA each ADP-ribosylates diphthamide (a modified histidine) in EF-2-GTP, which can no longer translocate the newly elongated peptide from the A site to the P site. The ShTA fragment is a specific *N*-glycosidase which cleaves an adenine residue from near the 3′ end of the 28S ribosomal RNA. This depurination results in failure of EF-1-dependent binding of aminoacyl-tRNA to site A and hence inhibits protein synthesis. Poliovirus achieves selective inhibition of host protein synthesis at an earlier stage than is depicted here. Host mRNA is first modified (capped) then bound to the small ribosomal subunit; poliovirus mRNA is not capped. The function of a cap-binding protein, which recognises and binds host mRNA to the ribosome, is inhibited by a poliovirus virion protein thereby allowing differential translation of virus messenger RNA. EF-1α, nucleotide-binding protein; EF-1αβγ, nucleotide exchange protein. (Modified with permission of authors and publisher (Elsevier Trends Journals, Cambridge, UK) from Fig. 1, Riis, B. *et al.*, 1990, *Trends Biochem. Sci.* **15**, 420–424.)

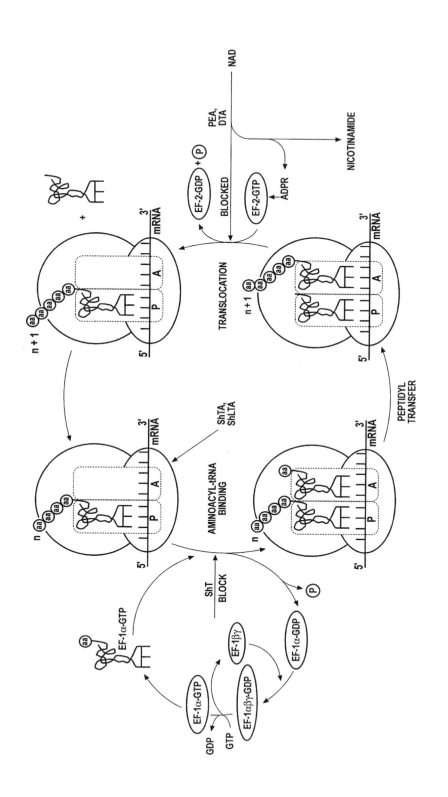

Table 8.3. Target proteins for intracellular toxins

Toxin group / Organism	Toxin	Preferred targets		Other targets
		GTP-binding proteins	ATP-binding proteins	
Ribosyltransferases (ADPRases)				
Corynebacterium diphtheriae	Diphtheria toxin (DT)	Elongation factor 2 (EF2); see Figs 8.10 and 8.11		
Pseudomonas aeruginosa	Exotoxin A (PEA)	Elongation factor 2 (EF2); see Figs 8.10 and 8.11		Vimentin[a]
Vibrio cholerae	Exotoxin S (PES)	Ras G-protein		
	Cholera toxin (CT)	α_s subunit of G_s ($\alpha_s\beta\gamma$) regulator of adenylyl cyclase;[b] see Fig. 8.12		
Escherichia coli	Heat-labile toxins LTI and LTII	α_s subunit of G_s ($\alpha_s\beta\gamma$) regulator of adenylyl cyclase; see Fig. 8.12		
	Cytotoxic necrotizing factor (CNF1)	Rho G-protein		
Bordetella pertussis	Pertussigen	α_i subunit of G_i ($\alpha_i\beta\gamma$) regulator of adenylyl cyclase;[c] see Fig. 8.12		
Clostridium botulinum	C2 toxin		Non-muscle actin, γ smooth muscle actin; see Fig. 8.14	
Iota group[d] Clostridium perfringens Clostridium spiroforme Clostridium difficile	Iota toxin C. spiroforme toxin ADPRase		All mammalian actin isoforms	
Clostridium botulinum Clostridium limosum	C3 ADPRase ADPRase (similar to C. botulinum C3)	Rho G-protein		

Glycosyltransferases (large clostridial toxins)			
Clostridium difficile	TcdA and TcdB	Rho, Rac G-proteins	
Clostridium sordelli	TcsL	Rac (and other) G-proteins	
Clostridium novyi	Tcnα	Rho, Rac G-proteins	
Shiga, Shiga-like toxins			
Shigella dysenteriae1	Shiga toxin (ShT)		Ribosomes; see Fig. 8.11
Escherichia coli	Shiga-like toxin (ShLT)		Ribosomes; see Fig. 8.11
'Invasive' adenylate cyclases			
Bacillus anthracis	Oedema factor (EF)		Activated by calmodulin; ATP
Bordetella pertussis	AC-Hly		Activated by calmodulin; ATP
Proteases			
Clostridium botulinum	Neurotoxin (BoNT)		Proteins involved in release of neurotransmitters; see Fig. 8.13
Clostridium tetani	Neurotoxin (TeTx)		
Bacillus anthracis	Lethal toxin (LF)		Protein kinase kinases 1 and 2

[a] Vimentin: intermediate filament protein.

[b] Cholera toxin A1–A2 will catalyse a range of reactions involving transfer of ADP-ribose to other substrates.

[c] Pertussigen will catalyse the ADP-ribosylation of G proteins involved in several transmembrane signalling events. This would account for many of its biological effects.

[d] *C. perfringens* and *C. spiroforme* toxins and *Clostridium difficile* ADP-ribosyltransferase form an iota subgroup, in that antibodies will cross-react within this group but not with *C. botulinum* C2 toxin. Only within the iota group are the binding components interchangeable. The iota group will also modify all mammalian actin isoforms.

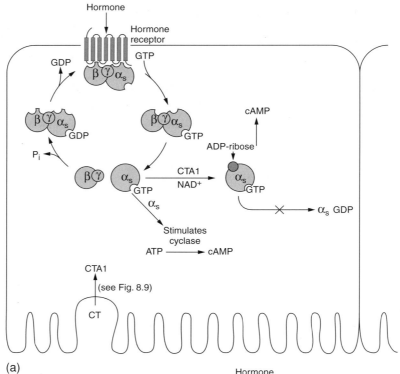

(a)

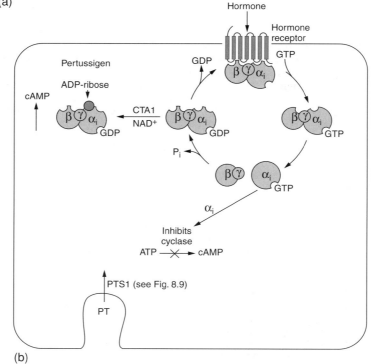

(b)

Another important example concerns *Helicobacter pylori*, which produces, in addition to urease, a vacuolating cytotoxin (VacA) and a protein encoded by the cytotoxin-associated gene A(*cagA*), which are encoded in a pathogenicity island. Both VacA and CagA are nearly always associated with strains isolated from severe forms of disease. The precise role of CagA is unclear but VacA belongs to the category of toxins with an intracellular site of action. However, the exact details have yet to be worked out.

Superantigens: toxins with multiple biological activities

The recognition of superantigenicity and its molecular basis has allowed us to classify into one major (still expanding) group the

---

Fig. 8.12 Mode of action of cholera toxin (CT), *E. coli* LT toxins, and pertussigen (PT). There are five main features in this diagram.

1. The production of cAMP by adenyl cyclase. Cyclic AMP is an important second messenger involved in the intracellular amplification of many cellular responses to external signals including hormones. The nature of the physiological response reflects the differentiation of the cell responding to the stimulus. For, example, in gut cells the response would be altered ion transport and hence fluid secretion; in muscle cells it would be glycogen breakdown in response to the call for more energy. The production of cAMP is controlled both positively (a) and negatively (b) at two levels. Interaction of hormone and receptor releases the heterotrimeric ($\alpha\beta\gamma$) G-protein regulator complex which, upon binding GTP, dissociates into α-GTP and $\beta\gamma$. The α-subunit may be stimulatory (α_s) and activate adenyl cyclase (as in (a)) or inhibitory (α_i) and inhibit adenyl cyclase (as in (b)); adenyl cyclase is not shown structurally in the diagram. In gut cells the receptor would be on the nonluminal basolateral side enabling enterocytes to respond to stimuli from the circulation.

2. The second level of control involves endogenous GTPase properties of both α_s and α_i subunits of the G-protein regulator: α_s-GDP and α_i-GDP are inactive.

3. The level of cAMP may be affected by physiological stimuli or by perturbation of the normal regulatory cycle as illustrated, by CT and LTs in enterocytes (a) or PT in a pancreatic B cell (b).

4. CTA1 ADP-ribosylates α_s-GTP which promotes continued dissociation of the heterotrimer and also inactivates the endogenous GTPase activity. Hence stimulation of the cyclase continues. LTs act in a similar manner.

5. PTS1 ADP-ribosylates the α_i-GDP$\beta\gamma$-heterotrimer which can no longer associate with the receptor or lose GDP to undergo another cycle of GTP activation; active cyclase can no longer be turned off. In pancreatic cells this results in loss of inhibition of insulin secretion.

Note: the α-GTP subunits are functionally analogous to the monomeric GPTases described in Fig.4.1. (Adapted with kind permission of the author (Gierschik, P.) and publisher (Springer-Verlag GmbH & Co. KG, Heidelberg, Germany) from Fig. 4 in 'ADP-ribosylation of signal-transducing guanine nucleotide-binding proteins by pertussis toxin', *Curr. Topics Microbiol. Immunol.*, 1992, **175**, 78, edited by K. Aktories.)

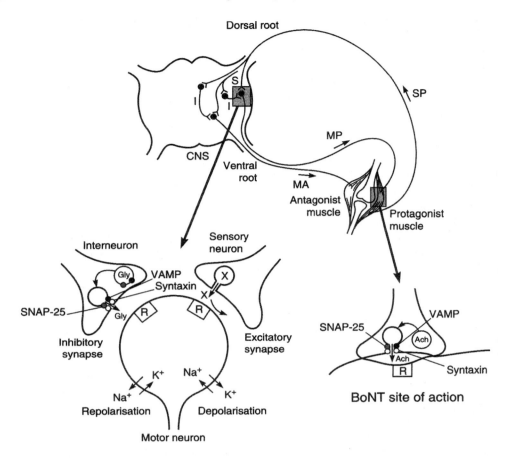

Fig. 8.13 Sites and mode of action of clostridial neurotoxins BoNT and TeTx. This figure has three main features.

1. Reflex arc (top). Mechanism for inhibiting the antagonists to a muscle contracting in response to stretch. Muscles are reciprocally innervated with sensory and motor neurons, although for clarity this is shown only for the protagonist muscle. On stretch, the stretch receptors generate an impulse which is transmitted along the afferent sensory (S) neuron of the protagonist (P) muscle. This SP neuron enters the spinal cord by the dorsal root and synapses with the motor neuron supplying the protagonist muscle (MP) and with an interneuron (I) which in turn synapses with the motor neuron supplying the antagonist muscle (MA); the efferent motor neurons leave the spinal cord by the ventral root. At the SP/MP synapse an excitatory transmitter is released which induces an impulse in MP which leads to contraction of protagonist muscle. However, excitation of I causes release of an inhibitory transmitter at the I/MA synapse which leads to relaxation of the antagonist muscle. Note that the basic reflex arc has been shown for simplicity but TeTx acts mainly on voluntary muscles.

2. A simplified version of the biochemical events occurring in synapses (lower left). Excitatory and inhibitory synapses, neurotransmitter release and

erythrogenic/pyrogenic toxins A and C* of *Streptococcus pyogenes* (SPEA, SPEC), staphylococcal enterotoxins (of which there are eight major serotypes, A to I, designated SEA, SEB, etc.) and staphylococcal toxic shock syndrome toxin (TSST-1 (human strains) and $TSST_{ovine}$). These proteins are superantigens by virtue of their ability to bind to major histocompatibility (MHC) class II molecules, outside the antigen-binding grove. They are presented as unprocessed proteins to certain T lymphocytes expressing specific T-cell receptor (TCR) motifs located in the variable domain of the β-chain (Vβ) of the TCR (see Chs 6 and 7). Nanogram to picogram quantities of superantigen will stimulate up to 20% of all T cells, compared with only 0.001 to 0.00001% T cells stimulated by conventional presentation of antigen to TCR. As a consequence of this huge proliferation of T cells and expression/release of aberrantly high levels of cytokines and other mediators, many biological systems are affected causing lethality/shock. This represents an important interference with a coordinated immune response, and the widespread polyclonal activation and cytokine release can be regarded as a microbial strategy, a 'diversion' of host immune defences. Ironically, the superantigen not only expands the circulating T-cell population, but also reacts with developing T cells in the thymus causing the same subpopulation to decline (see Ch. 7). It seems probable that these effects on immune cells represent a more important biological function of these toxins than the one responsible for the characteristics of disease; the latter may be no more than an 'accidental' phenomenon. It turns out that similar molecules are formed by mycoplasma and by certain retroviruses (e.g. the Mls antigen of mouse mammary tumour virus).

It has been shown experimentally (or proposed) that immune stimulation, cytokine release, induction of capillary leak, shock, and lethality are related to the superantigenicity of these proteins. However, there are other biological activities of these toxins which are not mediated by their superantigenicity. Some of these activities are common to all, and

---

* SPEB is a cysteine protease and is no longer classified as a superantigen.

---

action. Gly, glycine; R, receptors of neurotransmitters; X, hitherto uncharacterised (candidates include glutamate, dopamine, ATP, substance P, and somatostatin).

3. Sites of neurotoxin action (lower left and right). The predominant site of action of TeTx is the intermotor neuron synapse; the exocytotic machine is interfered with by the endopeptidase action of TeTx on VAMP. BoNT acts at the neuromuscular junction inhibiting the release of acetyl choline (Ach) by its proteolytic action on VAMP (types B, D and F), or SNAP (types A and E), or syntaxin (type C).

(Amplified from Figs 18 and 19, in *'Bacterial Toxins'*, 2nd edn, by J. Stephen and R. A. Pietrowski, 1986, pp. 60 and 62, Van Nostrand Reinhold (UK).)

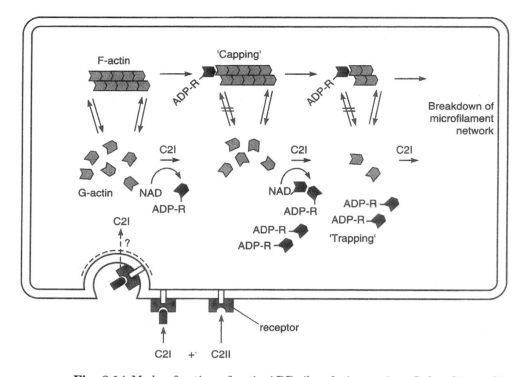

Fig. 8.14 Mode of action of actin-ADP-ribosylating toxins. *C. botulinum* C2 toxin component 2II binds to the cell membrane followed by C2I. The latter is internalised and upsets the equilibrium between polymerisation and depolymerisation of actin. ADP-ribosylation of actin inhibits its polymerisation and turns G-actin into a capping protein which binds to the fast-growing (concave) ends of actin filaments. Capping of the concave ends increases the critical concentration for actin polymerisation. Since the slow-growing (pointed) ends of actin filaments are free, depolymerisation of actin occurs at these ends. Released actin is substrate for the toxin and will be withdrawn from the treadmilling pool of actin by ADP-ribosylation, i.e. trapped. Both reactions will finally induce the breakdown of the microfilament network. (Reproduced with permission of the authors (K. Aktories *et al.*) and publisher (Springer-Verlag GmbH & Co. KG, Heidelberg, Germany) from Fig. 2 in 'Clostridial actin-ADP-ribosylating toxins', in *Curr. Topics Microbiol. Immunol.*, 1992 **175**, 107, edited by K. Aktories.)

others specific to certain members of this group. The red skin rash elicited by the streptococcal toxins (which gave rise to the original nomenclature 'erythrogenic' toxins) is regarded as a secondary hypersensitive effect, but pyrogenicity is the result of direct action on the hypothalamus as well as release of IL-1 and TNF-α from macrophages. Sensitivity to endotoxin shock can be increased up to 100 000 times in monkeys by SPEA/C and this may be due in part to an ability to impair the reticuloendothelial system which would result in inefficient clearing of circulating endotoxin.*

Staphylococci cause food poisoning on a world-wide scale; infection rates are under-reported probably because it is normally a self-limiting gastrointestinal disease. Onset of disease is rapid after consumption of enterotoxin-containing food. The main features of the disease are diarrhoea and severe vomiting, the latter being due to enterotoxin stimulation of the vagus nerve.

In addition, some strains of staphylococci cause toxic shock syndrome (TSS), a multisystem disease. Originally, TSS was seen characteristically in menstruating women whose tampons harboured multiplying staphylococci. It is due to a toxin called toxic shock syndrome toxin 1 (TSST-1; originally recognised as SEF, one of the so-called staphylococcal enterotoxins). Toxic shock syndrome is characterised by sudden onset of fever, vomiting, diarrhoea, an erythematous rash followed by peeling of the skin, hypotensive shock, impairment of renal and hepatic functions and occasionally death. The main symptoms of the disease have been reproduced in rabbits by implanting chamber-enclosed TSS strains in the rabbit uterus or peritoneum or by injection of TSST-1 into rabbits. Complex changes are observed including haemorrhage in kidney and liver, congestion and haematomas in the lungs, leakage of blood into the thymus, and fluid in the pericardial sac and in the gut lumen. These effects in rabbits are very similar to those seen in humans and would certainly explain the shock and diarrhoeal syndrome so characteristic of the disease. The lethal effect of TSST-1 is enhanced considerably by endotoxin.

We now know that TSS is not confined to menstruating women. Non-menstrual TSS presents with essentially the same signs as menstrual TSS and is caused by other staphylococcal enterotoxins (SEs). TSST-1 is isolated only from menstrual cases of TSS: this toxin has the ability to cross the vaginal mucosal barrier whereas the other SEs do not. Moreover, TSS is no longer thought to be caused exclusively by staphylococci. Streptococcal TSS (STSS), a life-threatening disease caused by streptococci, is now a well-recognised clinical entity

---

* There are numerous examples of synergistic reactions between toxins of the same or different species. *Bacillus cereus* makes a phosphatidyl choline-preferring phospholipase and a sphingomyelinase which are separately nontoxic; in concert they are haemolytic and termed cereolysin A-B. Another entirely different type of synergistic interaction is that of the increase in toxicities of staphylococcal α- and γ-toxins, diphtheria toxin and endotoxin for neonatal ferrets preinfected with influenza virus. Increases were 14-, 3-, 219- and 84-fold respectively. No increase in viral replication was observed. Neonates died suddenly without clinical symptoms as in human babies dying from sudden infant death syndrome (SIDS). Pathological examination showed inflammation of the upper respiratory tract, lung oedema and collapse, and early bronchopneumonia in animals receiving the dual challenge but not those receiving either toxin or virus on their own. Thus, some bacterial toxins in conjunction with influenza virus could be one of the several causes of SIDS.

probably corresponding to the severe cases of scarlet fever described in the older literature.

There is another class of potent immunogens not normally classified under 'superantigens'. Cholera toxin (CT) and *E. coli* LTs – and their respective B subunits – are extremely potent antigens eliciting extraordinary high levels of antitoxin without the need for conventional adjuvants. The phenomenology is highly complex and under active study. Understanding of the basic immunological mechanisms could lead to important practical applications, since the inherent adjuvanticity of these potent immunogens is demonstrably effective in some settings with unrelated antigens.

Significance of toxins in disease

It is important to point out that, while the outstanding advances made in our knowledge of toxin structure and mode of action at the cellular level can be exploited in a remarkable way (see below), it is important to remember that such knowledge by itself does not tell the whole story of the pathogenesis of infectious disease. To illustrate this some examples are given below.

Cholesterol-binding cytolysins

It is obvious from our consideration of these toxins that the elucidation of their lytic activities towards red cells in terms of a fundamentally similar mechanism, by itself tells us nothing about their respective roles in disease. Moreover, as already outlined above, pneumolysin is now known to be a multifunctional molecule whose relevance in disease varies with the infection setting!

Corynebacterium diptheriae

Corynebacterium diphtheriae produces diphtheria toxin which is of unquestionable importance in causing diphtheria. Sustained active immunisation with DT toxoid has made diphtheria a clinical rarity in advanced countries. Failure to continue this policy resulted in a huge diphtheria epidemic in the early 1990s in the states comprising the former USSR. *C. diphtheriae* organisms multiply on the epithelial surfaces of the body (nose, throat, skin) but do not penetrate deeply into underlying tissues. The infection on the body surface causes necrosis of mucosal cells with an inflammatory exudate and the formation of a thick 'membrane' (hence the name *C. diphtheriae*: Gr., *diphthera* = membrane) and if the infection spreads into the larynx there may be respiratory obstruction. The toxin probably assists colonisation of the throat or skin by killing epithelial cells and polymorphs. DT can also be disseminated from the infection site and has important actions, especially on the heart and nervous system. However, it is a fact that the strain used for commercial production of toxin for vaccine purposes is the avirulent PW8 strain. Five per cent of all the protein made by this strain, and 75% of all protein secreted by this strain is DT and yet it is avirulent! It is the production of DT under *in vivo* conditions that matters. DT is encoded by a lysogenic corynephage β whose transcription is controlled by an iron-dependent repressor, emphasising the importance of *C. diphtheriae* Fe metabolism *in vivo*.

Pseudomonas aeruginosa

Pseudomonas aeruginosa, is common in soil and water and can occasionally be isolated from the faeces of normal, healthy individuals. It is virtually harmless for healthy adults, but its ability to multiply in almost any moist environment and its resistance to many antibiotics have made the bacterium a major cause of hospital-acquired infection. This is particularly so among patients with impaired host defence mechanisms such as those with chronic illness, genetic immunodeficiencies, those under treatment with immunosuppressive drugs, or patients suffering from extensive burns. *P. aeruginosa* causes localised infection in the urinary tract, respiratory tract, burns and wound infections. In severely debilitated patients these localised infections may develop into general septicaemia, with mortality in such cases approaching 100%. However, unlike *C. diphtheriae*, the organism elaborates several potentially toxic extracellular products, including a phospholipase, several proteases, lipase, haemolysin, enterotoxin, lipopolysaccharide endotoxin, elastase, exoenzyme S (PES), and exotoxin A (PEA). While there is definite evidence to implicate the last three, the roles of the others are less well defined; mechanisms of pathogenicity of *P. aeruginosa* are complicated and still unclear.

Shigella dysenteriae 1

Shigella dysenteriae 1 is the cause of bacillary dysentery. For a long time it was thought that Shiga toxin (ShT) was the principal cause of this disease. However, in Ch. 2, the importance of gut invasiveness in *Shigella* infections was emphasised. While it is not at all clear how ShT can be involved in the watery diarrhoea phase of dysentery, it is perceived as exacerbating the bloody diarrhoea phase rather than initiating it. In contrast there is now evidence that Shiga-like toxins (ShLTs) are important in haemorrhagic colitis (HC) and haemolytic urea syndrome (HUS) caused by ShLT-producing strains of enterohaemorrhagic (EHEC) *E. coli*: EHEC has the capacity to progress disease beyond a watery diarrhea stage to HC and HUS; the latter is characterised by renal failure, thrombocytopenia and microangiopathic haemolytic anaemia. This virulence attribute is due to possession of one or more of a family of ShLTs. As in dysentery, the role of ShLT in the causation of watery diarrhoea is controversial and indeed it may not be absolutely necessary. However, there is no doubt that ShLT is either responsible for, or severely exacerbates the bloody diarrhoea in HC. Recently the first histochemical demonstration was made of ShLTs bound to renal tubules in the kidney of a child who died as a result of HUS associated with *E. coli* O157:H7 infection.

Vibrio cholerae

The classic paradigm for bacterial watery diarrhoea is cholera caused by *V. cholerae* in the small intestine. *V. cholerae* colonises the upper small intestine by adhering to epithelial cells. Water and electrolytes are lost through the intact epithelial cells into the small intestine. As the multiplying bacteria increase in numbers and more and more epithelial cells are affected, the absorptive capacity of the colon is over-whelmed and there is profuse watery diarrhoea, as much as 1 litre h^{-1} in severe cases. The massive loss of isotonic fluid with excess of sodium bicarbonate and potassium leads to hypovolaemic shock, acidosis and haemoconcentration. Anuria develops, and the collapsed, lethargic patient may die in 12–24 h. Lives are saved by replacing the lost water and salts; but the patient recovers as affected cells are shed and replaced in the normal fashion. The infection is particularly severe in children who easily develop low levels of plasma potassium. However, on a global scale this greatly feared disease, cholera, is only responsible for less than 1% of the total deaths due to diarrhoea.

Despite being arguably the most studied pathogen over the last three decades, the basis of *V. cholerae* pathogenicity and the detailed mechanisms underlying the dramatic diarrhoeal secretion induced by this organism are still not fully understood. Very recently, spectacular advances have been made in the molecular biology of *V. cholerae*. Chromosomal DNA of virulent *V. cholerae* contains two essential genetic elements which are important in *V. cholerae* virulence: CTXφ (the genome of a filamentous bacteriophage) which encodes the cholera toxin (CT), and a large pathogenicity island VPI (for *Vibrio cholerae* pathogenicity island). VPI is now known to be the integrated genome of another large filamentous bacteriophage (VPIφ) and encodes the toxin co-regulated type IV pilus (Tcp). Of the numerous colonization factors known to be produced by *V. cholerae* only Tcp has been proven to be important in human disease. Tcp is a remarkable entity; its subunit TcpA is a coat protein of VPIφ, but it also acts as a receptor for CTXφ and mediates interbacterial adherence. Thus, as a result of sequential infection by two 'pathophages', *V. cholerae* acquires the ability to colonise the human gut and secrete classical cholera toxin, which is a potent enterotoxin. The integration into the chromosome of these phage genomes brings their expression under the control of regulatory genes in the ancestral chromosome, whilst the replication of phages enables their interbacterial spread. CT is an 'AB' type toxin in which the pentameric B subunit recognises and binds to its cell receptor (GM1 ganglioside) thereby initiating the internalisation of the active A subunit (CTA$_1$) and elevation of cAMP.

Such elegant work implies that cholera diarrhoea is a purely patho-physiological disease and that CT is the only determinant of disease;

that is too simplistic. CT may be the major diarrhoeagenic toxin but there are at least eight other toxins which have been potentially implicated in cholera diarrhoea, of which we have already mentioned one, ZOT (Fig. 8.1). Moreover, studies on human jejunal biopsies show that cholera is not a purely pathophysiological disease but a pathological one, involving changes in the microvasculature and enteric nerve fibres, degranulation of argentaffin cells, mucosal mast cells and eosinophils; the extent of these changes correlated with clinical severity of disease. A *V. cholerae* vaccine strain produced by the deletion/mutation of all known toxin genes yielded a vaccine strain which, although less reactogenic than wild-type virulent strains, still produced a significant diarrhoea suggesting the involvement of an inflammatory component (as yet undefined) in the causation of cholera diarrhoea. There is also experimental evidence to implicate the enteric nervous system (ENS) in cholera diarrhoea. It has also been shown that CT administered to rat jejunum elicited a secretory response in both the jejunum *and* colon, which suggests neurological transmission of the locally induced secretory stimulus to distal colon.

Despite the undoubted importance of CT in the causation of the disease, and the potent antigenicity of CT, it is now recognised that protective immunity is very largely antibacterial. It is stopping effective colonisation which is important rather than neutralisation of the toxin. This has been partially achieved by using killed whole cell vaccines. Several attempts have been made in the laboratory to manipulate virulent strains genetically (in practice this means deleting or inactivating the known toxin genes) such that the attenuated strain will colonise the gut and stimulate local immune responses, and thereby prevent colonisation of the gut by virulent strains. To date, attenuated strains have been developed which fulfil these criteria, but these induce a mild transient diarrhoea, which has prevented their adoption into vaccination programmes, and incidentally, reinforce the argument that cholera is not wholly about CT.

Bordetella pertussis toxin (pertussigen; PTx)

Whooping cough (pertussis) is a severe respiratory tract infection characterised by prolonged paroxysmal coughing, attacks of which continue long after infection has cleared. The disease is capable of striking all ages but is particularly prevalent and severe in young children, where hospitalisation is required in about 10% of cases. The causative agent, *B. pertussis*, is transmitted aerially from the respiratory tract of an infected individual to that of a susceptible host. The organism attaches via several adhesins – filamentous haemagglutinin, fimbriae and the 69 kDa outer-membrane protein, pertactin – to the mucosal surface between cilia, and multiplies there during the incubation period of the disease, which is commonly around 7 days. The infection then manifests as a slight fever and catarrh which is often indistinguishable from a common cold. However, 1–2 weeks later bouts of uncontrollable coughing begin. It is this paroxysmal coughing, along with the notorious 'whoop' as the child attempts to draw breath, which characterises the disease. The paroxysmal coughing stage often lasts for several weeks and no treatment is fully effective in controlling the symptoms. The only proven means of controlling whooping cough is vaccination but, in the UK at least, sporadic reports of vaccine-induced brain damage in infants has diminished public acceptance of the vaccine. However, it should be noted that permanent encephalopathy (brain damage) is a recognised though rare consequence of whooping cough infection.

Without doubt, PTx, whose biochemical mode of action is described above, is an exceedingly important virulence determinant of *B. pertussis*: PTx toxic activities including histamine sensitisation, hyperinsulinaemia followed by hypoglycaemia, induction of leukocytosis, IgE induction are all observed after infection and administration of PTx; these toxic properties of PTx are abolished when the ADPR activity of PTx is inactivated. PTx non-toxic activities – mitogenicity, haemagglutination, platelet activation, mucosal adjuvanticity – are triggered by PTx B subunits. Much current work is being devoted to producing immunogenic, completely nontoxic preparations of pertussis toxin by genetic manipulation of the gene encoding the S1 subunit (Fig. 8.5); in clinical trials in Italy, such engineered vaccines have been shown to be both safe and effective as judged by antibody titres to pertussis toxin.

However, *Bordetella pertussis* also produces other potentially important toxins including AC–Hly involved in colonisation (see above), dermonecrotic toxin (DNT; formerly known as heat-labile toxin), and two non-protein toxins – tracheal cytotoxin (TCT) and endotoxin. DNT is lethal for mice and causes skin lesions in rabbits and guinea-pigs. It is of doubtful significance in humans but important in atrophic rhinitis in pigs caused by *Bordetella bronchiseptica*.* TCT is a small glycopeptide which destroys ciliated epithelial cells and is almost certainly responsible for some of the observed histopathological damage in *B. pertussis* infections.

Clostridial neurotoxins

Tetanus occurs in man and animals when *Clostridium tetani* spores germinate in an infected wound and produce their toxin; all strains of *C. tetani* produce the same toxin. Spores are ubiquitous in faeces and soil and require the reduced oxygen tension for germination provided locally in the wound by foreign bodies (splinters, fragments of earth or clothing) or by tissue necrosis as seen in most wounds, the uterus after septic abortion, or the umbilical stump of the newborn. The site of

---

* DNT is similar to the dermonecrotic toxin of *Pasteurella multicoda* also involved in porcine atrophic rhinitis. They inactivate the GTPase activities of Rho proteins resulting in cytoskeletal changes affecting osteoblasts.

infection may be a contaminated splinter just as well as an automobile or battle injury. It also reaches the central nervous system by travelling up other peripheral nerves following blood-borne dissemination of the toxin through the body. The motor nerves in the brain stem are short and therefore the cranial nerves are among the first to be affected, causing spasms of eye muscles and jaw (lockjaw). There is also an increase in tonus of muscles round the site of infection, followed by tonic spasms. In generalised tetanus there is interference with respiratory movements, and without skilled treatment the mortality rate is about 50%.

Botulism* is caused by *Clostridium botulinum*, a widespread saprophyte present in soil and vegetable materials. *C. botulinum* contaminates food, particularly inadequately preserved meat or vegetables, and produces a powerful neurotoxin. The toxin is destroyed at 80°C after 30 min – of great importance to the canning industry – and there are at least seven antigenically distinct serotypes (A–G) produced by different strains of bacteria but which have a pharmacologically similar mode of action. It is absorbed from the intestine and acts on the peripheral nervous system, interfering with the release of acetylcholine at cholinergic synapses of neuromuscular junctions. Somewhere between 12 and 36 hours after ingestion there are clinical signs suggesting an acute neurological disorder, with vertigo, cranial nerve palsies and finally death a few days later with respiratory failure. A less typical form of botulism occurs in small infants. The spores, present in honey applied to rubber teats, appear to colonise the gut, so that the toxin is produced *in vivo* after ingestion.

However, some puzzles remain to be resolved. Tetanus and botulinum toxins enter neuronal tissue preferentially at motoneuronal endplates, but the nature of the relevant toxin receptors is still not known. It remains unclear why botulinum toxin acts directly at the site of uptake and not, as observed with tetanus toxin, in the central nervous system, although a considerable amount of botulinum toxin (like tetanus toxin) is retrogradely transported. Botulinum toxin blocks the release of acetylcholine at neuromuscular junctions to cause flaccid paralysis. Likewise, one might ask why tetanus toxin fails to act at the motoneuronal junction at concentrations which would completely block release of neurotransmitter from GABAergic synapses. To reach inhibitory interneurons – its principal site of action – tetanus toxin must leave α-motoneurons after the primary uptake step, traverse the synaptic cleft to interneurons, leave those again in order to become finally internalised again from presynaptic membranes – a route identical to that of several neurotropic viruses. It

---

* *Botulus* (Latin) = sausage. In 1793 a large sausage was eaten by 13 people in Wildbad in Germany; all became ill, and six died. The disease was subsequently referred to as botulism.

acts by blocking the release of inhibitory transmitters (glycine or GABA) resulting in a failure to relax the affected muscle – patho-physiological 'tetanus'. Only in rare cases does it act peripherally like botulinum toxin.

Anthrax toxin

Anthrax is a disease of animals, particularly sheep and cattle, and to a lesser extent man, caused by infection with *Bacillus anthracis*. Infection takes place following the ingestion of spores, the inhalation of spores, or in most cases by the entry of spores through abraded skin. The spores germinate inside macrophages and then the bacteria form a toxin which kills macrophages, increases vascular permeability and gives rise to local oedema and haemorrhage. Infection of the skin in man leads to the formation of a lesion (malignant pustule; a *black* eschar, hence *B. anthracis*; Gr. *anthrakos* = coal) consisting of a necrotic centre surrounded by vesicles, blood-stained fluid and a zone of oedema and induration. In severe infections (nearly all cases of inhalation anthrax are fatal) there is septicaemia with toxic signs, loss of fluid into tissues, with widespread oedema and eventually death. Anthrax in man occurs mainly in those whose work brings them into contact with infected animals. It is not a common disease in the UK, and the usual source of infection is imported bones, hides, skins, bristles, wool and hair, or imported fertilisers made from the blood and bones of infected animals.

The anthrax toxin complex consists of three components, factor I (oedema factor; EF), factor II (protective antigen; PA) and factor III (lethal factor; LF), none of which are toxic by themselves, but in binary combinations exhibit two types of activity. PA and LF form a binary proteolytic cytotoxin which kills macrophages (see Ch. 4) but not any other cell type, whereas PA and EF form a binary toxin which will elevate cAMP levels (Fig. 8.8) in nearly all types of cell.

These combined binary toxic activities explain much, but not quite all of the events which occur in anthrax infections. Virulent *B. anthracis* contains two plasmids: plasmid pXO1, which encodes the toxin genes, and plasmid pXO2, which encodes capsule genes. Loss of either results in a dramatic loss of virulence. Avirulence due to loss of toxin genes is understandable in terms of the above outline. The capsule contains poly-D-glutamic acid which renders the organism resistant to phagocytosis; loss of capsule renders the organism avirulent even though it still possesses toxin genes! The most effective field vaccine is the Sterne strain used to protect animals. This is a pOX2⁻ strain which cannot establish an infection focus due to the loss of the antiphagocytic capsular poly-D-glutamic acid. However, this strain manages to express sublethal doses of toxin, the PA component of which is the protective immunogen, a fact readily understood in terms of the model for cell entry described in Fig. 8.8. If you block attachment of PA with anti-PA antibodies, then EF and LF will be nontoxic.

Streptococcal and staphylococcal superantigens

Streptococci cause a wide spectrum of diseases including simple skin pimples, uncomplicated tonsillitis, scarlet fever, rheumatic fever, severe invasive infections like necrotising fasciitis and STSS. We have touched on a few of the known streptococcal virulence determinants, and sought to explain much of the pathology caused by streptococci in terms of these determinants, despite the fact that the severity of streptococcal disease can range from simple pimples to STSS! In addition to the status of the host, full explanations must clearly reflect the particular strain of the organism, its initial site of lodgement, and the efficiency with which it produces its particular combination of virulence determinants *in vivo*. The virulence determinants must include surface antiphagocytic M protein and maybe some of the huge number of other proteins known to be secreted by streptococci.

Exactly the same kind of considerations apply to staphylococci and the diseases they cause. Staphylococci also produce a plethora of putative virulence determinants in addition to those discussed here.

Clostridium difficile

Clostridium difficile represents a classic example of the difficulty in interpreting disease mechanisms in terms of characterised enzyme activities ascribed to toxins relevant in disease. *C. difficile* is now established as the most common nosocomial enteric pathogen causing pseudomembranous colitis, antibiotic-associated colitis and antibiotic-associated diarrhoea. The most important defence against this opportunistic pathogen is the normal colonic microflora, although the microbial species responsible for, and the mechanisms whereby they suppress the growth of *C. difficile* are still not understood. Disruption of the normal ecosystem by antibiotics can result in colonisation by *C. difficile* which, if of the right pathotype, will cause diarrhoea or, more seriously, pseudomembranous colitis. Production of proteolytic and hydrolytic enzymes and capsule, expression of fimbriae and flagella, chemotaxis and adhesion to gut receptors, may all play a part in the pathogenesis of *C. difficile*-induced disease by facilitating colonization or by directly contributing to tissue damage. However, toxins A and B (TcdA, TcdB) are thought to be the primary virulence determinants of this pathogen in the context of antibiotic-associated gastrointestinal disease. The toxins have identical enzyme specificities – they glucosylate the same serine residue in target proteins – but yet they have very different biological properties. B is *ca.* 1000-fold more cytotoxic to cultured cells than A, but does not cause fluid secretion in the gut on its own, whereas A does. A powerful case can be marshalled to implicate toxin A as the major effector in *C. difficile* diarrhoeal disease and of the colitis so characteristic of *C. difficile* infections. Several attempts have been made to explain these superficially discrepant data but perhaps the most obvious explanation is that glucosyltransferase is not the primary mechanism of toxicity!

One alternative explanation is that the main effect of toxin A is to upregulate the secretion of IL-8 from colonocytes and downregulate the exocytosis of mucin. This would result in the recruitment of inflammatory and immune cells (seen in pseudomembranous colitis) with consequential indirect mucosal damage. The depression of stimulated mucin secretion (observed in toxin A treated cells *in vitro*) could well be explained as a secondary effect in terms of the enzymatic activity of the toxin which by disruption of the cytoskeleton would impair exocytosis.

Other clostridial toxins

There is a plethora of toxins produced by numerous clostridial pathogens important in both human and veterinary medicine. The clostridial genus comprises a large number of toxigenic species, some of which are known to produce several toxins, extracellular enzymes, and other factors which are as yet recognised only by a letter of the Greek alphabet. All the α-toxins are dermonecrotic and the others have haemolytic, enzymatic properties. Toxin production *in vitro* is used as the basis of typing clostridia, and the picture is complex. No attempt will be made to describe every disease in man or animals associated with clostridia; only those are selected which best serve to illustrate the involvement of some recognisable toxins. In the case of sheep diseases – lamb dysentery, struck, enterotoxaemia, black disease, braxy, black quarter – some of the best evidence for implicating relevant toxins comes from field studies using multivalent vaccines (based on toxoids of these toxins).

Clostridium perfringens type C causes pig bel in man, essentially due to the production of β-toxin. This is a rare disease in developed societies but a public health hazard in Papua New Guinea. Three factors are responsible: the ubiquity of *C. perfringens* type C in the soil and faeces of man and pigs; the high-carbohydrate, low-protein nature of the staple diet; and the sporadic consumption of large quantities of pork on occasions of celebration. The latter dietary change promotes a proliferation of clostridia in the intestine which may lead to intestinal gangrene and death. β-Toxin damages the mucosa, reduces mobility of villi and causes more bacteria to become attached to the villi. More toxin is absorbed and the mucosa and underlying intestinal wall become necrotic, leading to death in many cases. The influence of diet is additionally important in that low-protein diets cause decreased secretion of pancreatic proteolytic enzymes and sweet potato contains a trypsin inhibitor. These conditions promote the survival of β-toxin which is highly sensitive to proteolytic inactivation. Immunisation with β-toxoid preparations has dramatically lowered the incidence of this fatal disease in children.

Gas gangrene in man may be caused by several bacterial species separately or in concert. These include *Clostridium perfringens* type A, *C. novyi* types A and B, and *C. septicum*; *C. perfringens* and its α-toxin have already been discussed. Far less is known about *C. novyi* and *C. septicum* and their toxins in gas gangrene in man. Much more is known about the role these organisms play in diseases of animals, and multivalent vaccines confer a very high degree of immunity (particularly to sheep) against several clinically identifiable but separate diseases. A few of these diseases are described below.

Clostridium perfringens type B causes lamb dysentery. This is an acute, fatal disease of young lambs occurring during the first week of life and caused by absorption of toxin(s) generated by *C. perfringens* type B in the small intestine.

Clostridium perfringens type C causes struck in sheep, a disease occurring in the Romney marshes of Kent, but rare in other areas in the world. The pathological changes observed differ markedly from other enterotoxaemias and include enteritis.

Clostridium perfringens type D enterotoxaemia in sheep is another acute fatal disease. The most constant lesion is subendocardial haemorrhage around the mitral valve. *C. perfringens* type D ε-toxin or its protoxin are recoverable from intestinal contents.

Clostridium novyi type B causes black disease of sheep or infectious necrotic hepatitis. This is an acute infectious disease of sheep (occasionally cattle) caused by the absorption of the α-toxin elaborated by the organism in necrotic foci in the liver, and is nearly always associated with invasion of the liver by immature liver flukes. How *C. novyi* gets to the liver in the first place is not known but it is readily demonstrable in livers of normal sheep in areas where the disease is prevalent. Experimental reproduction in guinea-pigs of a similar disease is possible by the combined action of *C. novyi* spores and liver fluke infestation.

Clostridium novyi type D causes a rapidly fatal disease in cattle (redwater disease) similar to, and regarded by some as an atypical manifestation of, black disease. The characteristic lesions include jaundice, various haemorrhagic manifestations, and anaemic infarcts in the liver; active liver fluke infestation may or may not be present. In culture this organism produces β-toxin, which explains the haemoglobinuria, but no α-toxin.

Clostridium septicum causes braxy in sheep. The role of *C. septicum* in this acute, fatal disease is assumed because of its association with the characteristic haemorrhagic inflammatory lesion in the abomasum. The disease has not been reproduced experimentally with *C. septicum* but can be prevented by immunisation with sterile toxoids derived from this organism.

Clostridium chauvoei causes black quarter in sheep and cattle. This is a gas gangrene-type infection of muscles and associated connective tissues in cattle and sheep; *C. chauvoei* is also the causative agent of parturient gas gangrene in sheep. The initial stimulus which activates the infection in cattle is not known, since the disease is hardly ever associated with any overt wounding. Washed spores alone do not cause disease when injected, but do in conjunction with a tissue-necrotising agent. In sheep, wounding caused by parturition, castration, tailing, shearing, vaccination, as well as accidental damage, will create a focus within which *C. chauvoei* can multiply.

Exploitation of native toxins and toxin chimeras as therapeutic agents

Reference has already been made to the potential exploitation of non-toxic B subunits of CT and LT. However, the most toxic substances known to man are now being used as therapeutic agents to treat focal dystonias such as neck twists and eye squints, or eyelid closure and, more recently, some childhood palsies. The preparations are made from BoNT A and consist of toxin–haemagglutinin complex – the form in which the toxin is usually produced by the organism. This complex is less toxic than purified neurotoxin when administered parenterally but relatively more toxic when given orally; the haemagglutinin apparently protects the neurotoxin from proteolytic degradation in the gut. The effects of BoNT in relieving muscle spasms are not permanent but last for several months. Treatment has to be repeated. To date the successes significantly outweigh the failures and no long-term adverse effects have been reported. Antibody to the toxin has not so far been detected in the sera of the majority of patients.

Another exciting development concerns the development and use of modified toxins as therapeutic agents. For decades, attempts have been made to make immunotoxins by coupling native toxins to antibodies specific to some surface antigen on tumour cells, with little practical success, as yet. Several attempts have been made to modify diphtheria toxin (DT), adenyl cyclase of *Bordetella pertussis*, *Pseudomonas aeruginosa* exotoxin A, cholera toxin and the LF anthrax toxin for selective intracellular delivery of extraneous proteins. The most successful attempt to date has been the development by Murphy and colleagues in the USA of an anticancer agent developed from DT. Their DT chimera has now been approved by the Food and Drug Administration for treatment of refractory cutaneous T-cell lymphoma (CTCL). DT comprises three functional domains: C-domain which is the toxiphore (A fragment), T-domain (in B fragment) involved in the translocation of C to the intracellular target, and R-domain (in B fragment) which recognises and binds to the receptor on sensitive cells. For DT, the receptor is an EGF-like precursor which happens to be widely distributed throughout all organ systems. DT was genetically engineered by substituting that portion of the DT structural gene encoding R domain with cDNA encoding IL-2, such that the hybrid toxin would target IL-2-receptor positive cells. The resultant fusion toxin bears a new 'cellular address', but retains all of the other biological properties of the native DT molecule as well as a three-dimensional structure nearly identical with native DT. This chimeric toxin has been shown to be safe and well tolerated and is successful in treating CTCL which expresses the high-affinity receptor for IL-2. Such constructs are potent against human CTCL and have achieved partial to durable remission of this tumour. An era of new 'magic bullets'?

Fungal exotoxins

Many fungi contain substances that are harmful when taken by mouth, and there are two diseases that result from the ingestion of food containing preformed fungal toxins. As with *C. botulinum*, the disease is caused without the need for infection. *Aspergillus flavus* infects ground nuts (monkey nuts) and produces a very powerful toxin (aflatoxin). Contaminated (badly stored) ground nuts used to prepare animal feeds caused the death of thousands of turkeys and pigs in the UK in 1960 and the survivors of intoxication nearly all developed liver cancer. Human disease has not yet been associated with this toxin. *Claviceps purpurae* is a rust fungus affecting rye, and it produces toxins (ergotamine especially) that give rise to ergot poisoning when contaminated grain is eaten. Mushrooms and toadstools have long been recognised as sources of poisons and hallucinogens.

Cell-associated toxins

Unlike the toxins already discussed in this chapter, there is a group of toxins which are distinct structural components and are not released into the surrounding medium in any quantity except upon death and lysis of the bacteria. Several protein toxins (e.g. *Clostridium difficile* toxin A, tetanus toxin, pneumolysin) are released during the decline phase of batch culture – probably on autolysis – and these are classified as exotoxins. Here we deal with toxins which are known to comprise well-recognised structural entities which on a priori grounds must have key functions in the organism: they are found in the outer membranes of Gram-negative organisms. There are two chemically distinct types of toxin considered: lipopolysaccharide (endotoxin; LPS) and protein. The bulk of this section is taken up with endotoxin.

Many pathogenic organisms, however, are pathogenic by virtue of possessing various types of surface structure important in conferring virulence. These include, for example, adhesins which are important in colonising body surfaces or a variety of surface molecules (which may or may not be inside capsules) that render them resistant to phagocytosis. But the majority of adhesins and antiphagocytic determinants are themselves nontoxic.

The Gram-negative bacterial cell wall is subject to considerable variations in both the composition of LPS and in the number and nature of the proteins found in the outer-cell membrane. Apart from the examples given in Ch. 7 in relation to the gonococcus, such phenotypic variation in LPS has rarely been examined in the context of pathogenicity. However, the examination of cell-bound proteins of *Yersinia pestis* from organisms grown *in vivo* led to the discovery of a toxin lethal for mice and guinea-pigs.

Protein toxins of *Yersinia pestis*

Plague is one of the most deadly diseases of man and has, over several thousands of years, claimed millions of lives. In the fourteenth century, 'the black death' wiped out a quarter of the population of Europe before spreading through the Middle East and Asia. Fortunately, however, the last 60 years or so have seen a drastic decrease in outbreaks of plague, although the threat of another epidemic is still with us.

The causative organism of plague, *Yersinia pestis*, is primarily a parasite of rodents in which it is endemic in many areas of the world. Only when man comes into close proximity with infected rodents do outbreaks of human plague occur. The disease is spread from rat to rat and from rat to man by fleas as already described in Chs 3 and 4. The principal features of human plague can be reproduced in guinea-pigs and mice. Monkeys show shock-like signs only during the terminal period of 6–10 h, when they become quiet, progressively weak, prostrate and hypothermic; for the previous 2–4 days infected animals are lively and vigorous. In the terminal stages blood pressure drops rapidly but there is no evidence of oligaemia (low blood volume) caused by haemorrhage, or of oedema, suggesting that vascular collapse must be associated with a vasodilatory factor(s), resulting in pooling of blood. In this respect monkeys differ from humans and guinea-pigs.

The symptoms of plague – high fever and vascular damage – are characteristic of intoxication with endotoxin. However, it is extremely unlikely that endotoxin alone is the main toxin involved in plague. It is much more likely to act in conjunction with one or more other potentially toxic fractions from *Y. pestis*. Plague murine toxin is a protein which, although highly lethal for mice and rats, is relatively nontoxic for guinea-pigs, rabbits, dogs and monkeys. A completely separate guinea-pig toxin complex exists comprising at least two cell wall/membrane protein components, one of which will kill mice, although both are needed to kill guinea-pigs. However, the nature of the toxin or toxins of *Y. pestis* and their role in the human disease syndrome are still far from clear. As has been the case with anthrax, it would be highly valuable to rework the early experimental pathology in the light of the wonderful molecular biological insights we now have of *Yersinia* as outlined in Chs 2 and 3. The classical approach to such a problem is to prepare a specific toxoid and to determine whether injection of this confers immunity to the disease. Needless to say, the severity of human plague renders such experiments impossible. Until such questions can be answered, we are left to argue whether, in the context of plague, man resembles a mouse, a monkey or a guinea-pig.

Endotoxins

Endotoxins are part of the outer membrane of Gram-negative bacteria. It has been known for many years that the cells (alive or dead) or cell extracts of a wide variety of Gram-negative bacteria are toxic to man and animals. The literature on this subject is vast, sometimes confusing and often controversial; here we can give no more than a brief outline. Some of the diseases in which endotoxin may play an important role include typhoid fever, tularaemia, plague and brucellosis, and a variety of hospital-acquired infections caused by opportunistic Gram-negative pathogens, which include *Escherichia coli, Proteus, Pseudomonas aeruginosa, Enterobacter, Serratia* and *Klebsiella*. In addition, endotoxin has been intensively studied as a possible causative agent of shock arising from postoperative sepsis or other forms of traumatic injury in which the normal flora of the gut is often the source of endotoxin.

The toxins we have considered so far have been protein (or at least part protein) in nature but, in contrast, endotoxin is a complex lipopolysaccharide. It is also much more heat stable than protein toxins and much less easily toxoided. In addition to lethality, endotoxin displays a bewildering array of biological effects.

Location in cell envelope

The complex nature of the multilayered Gram-negative bacterial envelope is shown in Fig. 8.15 (see also Fig. 4.3). The outer membrane is composed of a bimolecular leaflet arrangement as are other membranes but has a different composition from the cytoplasmic membrane. The lipopolysaccharide (LPS) is unique in nature, only found in Gram-negative bacteria, and is, or contains within it, what we designate endotoxin. Immunoelectron microscopy indicates that LPS exists in the outer leaflet of the membrane and extends outward up to 300 nm; it is on, rather than in, the cell. Thus it is evident that the term endotoxin is a misnomer which derives from the era when toxins were considered to be either exotoxins, which were synthesised and secreted by the viable organism, or endotoxins, which were intracellular and released only upon lysis. Moreover, extraction with EDTA shows that approximately 50% of LPS is held noncovalently linked in the membrane. Extraction with a variety of different solvents yields material which is highly heterogenous and of apparent molecular weight $1-20 \times 10^6$. However, treatment with pyridine or addition of detergents reduces the polydispersity. The endotoxic glycolipid from the rough

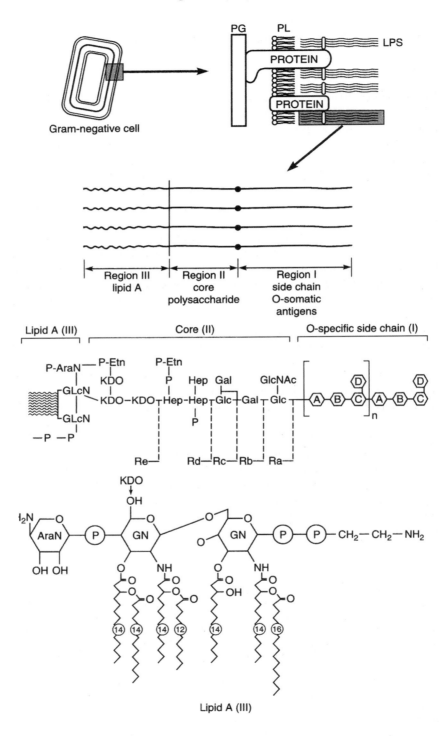

Lipid A (III)

mutant of *Salmonella minnesota*, R 595, has an M_r of 5900 for the basic unit, from which complex aggregate structures are derived.

Structure

Lipopolysaccharide consists of three regions: polysaccharide side chains, core polysaccharide, and lipid A which consists of a di-glucosamine backbone to which long-chain fatty acids are linked (Fig. 8.15). The relationship of this type of molecule to the outer membrane is also shown in Fig. 8.15. The long-chain fatty acids inter-digitate between the phospholipids in the outer leaflet and may also be linked (or interact) with lipoproteins, which in turn may or may not be covalently anchored to the rigid peptidoglycan (PG). The polysaccha-ride side chains project outwards.

This structure is not invariant. For example, many organisms when first isolated give rise to colonies with a smooth appearance on agar but on subculture produce colonies with a rough appearance. In general, 'smooth' strains of pathogenic species are more virulent than rough strains. This S → R conversion is accompanied by a loss of region I side chains, which contain the deoxy and dideoxy sugars found in these LPS complexes. In addition to these somewhat drastic changes involving loss of side chains, it is possible to induce major composi-tional changes by manipulating the growth rate of these organisms in a chemostat. Thus the LPS of *Salmonella enteriditis*, when grown with a mean generation time of 20 min is nearly totally deficient in tyvelose (a dideoxy sugar), possesses 85% of the galactose and 150% of the glucose contents of LPS obtained when the generation time is 50 min. These genotypic S organisms exhibit an R-phenotype in terms of their vastly reduced O-agglutinability (see below); such observations are potentially very important in the context of the *in vivo* phenotype and pathogenicity, since it is well known that the growth rate of *Salmonella typhimurium* in mice is 10–20 times lower than *in vitro*. Examples have already been given in Ch. 7 of changes in LPS structures *in vivo* in relation to *Neisseria gonorrheae* and *Neisseria meningitidis*.

Fig. 8.15 General structure of *Salmonella* lipopolysaccharide. See text for fuller explanation. Abbreviations: PG, peptidoglycan; PL, phospholipid; A–D, sugar residues; Glc, D-glucose; Gal, D-galactose; GlcN, D-glucosamine; GlcNAc, N-acetyl-D-glucosamine; Hep, L-glycerol-D-*manno*-heptose; KDO, 2-keto-3-deoxy-D-*manno*-octonate; AraN, 4-amino-L-arabinose; P, phosphate; EtN, ethanolamine; 〜〜〜 hydroxy and nonhydroxy fatty acids; Ra–e, incomplete forms of lipopolysaccharides. The structures indicated are typical of the Enterobacteriaceae and the Pseudomonadaceae. *Haemophilus influenzae, Neisseria meningitidis, Bordetella pertussis, Acinetobacter calcoaceticus* and *Bacteroides fragilis* have less complicated LPS structures in that they do not possess the equivalent of the O-somatic side chains. *Chlamydiae* possess only lipid A and the inner core region comprising lipid A and KDO.

Immunochemistry and seroclassification

The extent to which lipid A is common between different genera is uncertain, but it is not likely to vary tremendously. The core polysaccharide structure is the same or very similar within groups of the Enterobacteriaciae. Thus polysaccharides from salmonellae are similar to each other, but differ from those of *E. coli* strains. However, within a group such as the salmonellae, there is a wide variation in the composition and detailed structures of the side chains, a fact which is exploited in the Kauffman–White scheme for classifying salmonellae, giving rise to several thousand serotypes.

The side chains carry the O-somatic antigen specificities of which there are far more than can readily be accounted for on the basis of the known number of sugars involved in the basic repeating units. In the side chains are found a range of deoxy and dideoxy sugars. The general principles governing the relationship between the various chemotypes and serotypes are now well understood; the multiplicity of antibody specificities evoked may be explained in terms of antibodies which can recognise different aspects of one three-dimensional structure.

Biological properties

There is now direct evidence that lipid A is the primary toxiphore, but the polysaccharide plays an important part in conferring solubility upon, and optimising the size of micellar aggregates of LPS, hence affecting biological activity. However, the immune status of the test animal may affect toxicity: as normal animals produce antibodies to the antigenic determinants on the surface of normal gut organisms (including O-somatic antigens), some of the biological effects of endotoxin may be mediated by hypersensitivity mechanisms.

The range of biological properties of endotoxin is quite bewildering and the mode(s) of action very complicated. Included among those effects which might play a role in Gram-negative bacterial infections are abortion, pyrogenicity, tolerance (not immune tolerance), the Schwartzmann phenomenon, hypotension and shock, and lethality, but the precise part played by LPS in these phenomena in Gram-negative infections is far from clear. LPS causes the release of vasoactive substances, activates the alternative pathway of the complement cascade, and also activates factor XII (Hageman factor), the first step of the coagulation cascade, which sometimes results in disseminated intravascular coagulation (p. 287). Many, perhaps nearly all, the actions of LPS are due to the stimulation of cytokine release from macrophages and other cells. There is an effect on the circulation, leading ultimately to vascular collapse. The vascular regions most affected differ from species to species; in man and sheep the main changes are found in the lungs. LPS has powerful immunological actions, which is surely no accident; as well as activating the complement system, it induces IL-1 production and is a potent B-cell mitogen. Man is one of the most sensitive of all species to the pyrogenic action of endotoxin. A dose of 2 ng kg^{-1} of body weight injected intravenously into man causes the release of the endogenous pyrogen IL-1 and TNF from macrophages, which act on the hypothalamus to give an elevation of body temperature within an hour. It is possible that the pyrogenic action of LPS helps to generate fever in Gram-negative bacterial infections, but LPS is not the only bacterial factor capable of inducing a febrile response. For example, recall the streptococcal and staphylococcal superantigens discussed earlier in this chapter.

In spite of all these toxic actions, there have been suggestions that some of the responses to LPS (by macrophages, polymorphs) could be advantageous to the host, possibly assisting in the recognition and destruction of bacteria. Could it be that host responses to LPS are, like the complement or the clotting systems, useful in moderation but harmful in excess? There are reports that, when animals with less vigorous responses to LPS are infected, they suffer fewer symptoms, but permit greater growth of bacteria.

Very large numbers of Gram-negative bacteria are normally present in the intestines (see Ch. 2), their continued death and exit in the faeces being balanced by multiplication in the lumen. There is a continuous, inevitable low-grade absorption of endotoxin from the intestine.* Absorbed (endogenous) endotoxin enters the portal circulation and is taken up and degraded by reticuloendothelial cells, mainly Kupffer cells in the liver. Continuous exposure to endotoxin probably has profound effects on the immune system and on the histology of the intestinal mucosa, stimulating development of the immune system in the immature individual, but there are no obvious pathogenic consequences. Normal people have low levels of antibody to endotoxin as a result of this continuous exposure. The sick individual may be much more susceptible to endogenous endotoxin, perhaps because of defects in removal by Kupffer cells.

After trauma or after genito-urinary instrumentation, endotoxin is detectable in peripheral blood but this leads to no particular signs or symptoms. When large amounts of endotoxin enter the blood there are profound effects on blood vessels with peripheral vascular pooling, a drastic fall in blood pressure, collapse and sometimes death. Thus, if enough endotoxin enters the blood during massive Gram-negative bacterial sepsis, the vasomotor action of endotoxin becomes important

---

* In addition, various antigens are absorbed in small quantities from the intestine, and in normal individuals antibodies are formed against various food proteins and to some extent against resident intestinal bacteria (see Ch. 2). Kupffer cells remove any antigen–antibody complexes formed locally in the intestine and prevent them from entering the systemic circulation.

and shock intervenes.* In experimental animals endotoxin also causes vasodilation and haemorrhage into the intestinal mucosa, and sometimes haemorrhage into the placenta with abortion, but these actions do not appear to be important in all Gram-negative bacterial infections.

To summarise, endotoxin, although studied so carefully and for so long, has not yet been shown to play a definitive role as a toxin in the pathogenesis of any infectious disease. But, in spite of its effects on various host defence systems including polymorphs, lymphocytes, macrophages, complement, and on endothelial cells and platelets, its overall role in *infection* is still not clear. It can, however, cause shock when Gram-negative bacteria invade the blood. It is for this reason that considerable effort in recent years has gone into the development of antilipid A antibodies for use as therapeutic agents to combat shock in such situations; the success rate is only partial and the expense enormous. For that reason several groups are seeking to exploit the wealth of chemical and biophysical information available on LPS in attempts to develop synthetic derivatives which would neutralise the biological activity of lipid A. We await the outcome of such research. However, the characteristics of the O-antigen polysaccharide are sometimes important in determining virulence: certain chemotypes are important in resisting phagocytosis.

General observations on toxins

Considerable space has been given to toxins because they are being intensively investigated as possible virulence determinants. The account illustrates the complexity of host–microbe interactions when analysed at the molecular level. Most toxins are liberated from the microbial cell and can be studied with greater facility than many of the more elusive determinants of pathogenicity. But remember that microbes that replicate inside host cells are less likely to form powerful toxins because they cannot afford to damage at too early a stage the cell in which they are multiplying. Thus, toxins are not prominent products in intracellular infections due to mycobacteria, *Brucella, Rickettsiae, Mycoplasma* or *Chlamydia*, and viruses do not form toxins.

Although a single molecule of a toxin-like diphtheria toxin is enough to kill a cell, other toxins may do no more than impair cell function when present in sublethal concentrations. This can lead, for instance, to defective function in immune or phagocytic cells. Low concentrations of the streptococcal streptolysins will inhibit leucocyte chemotaxis. At even lower concentrations the toxins can be potent inducers of

---

* It must be remembered that endotoxin is only one of the pathways to shock in infectious diseases. Shock is also seen for instance in leptospiral and rickettsial infections, in gas gangrene, and in sepsis due to Gram-positive bacteria (see above).

cytokines. The pneumolysin of *S. pneumoniae* and anthrax lethal toxin make monocytes release IL-1 and TNF-α at 10^{-15} and 10^{-18} molar concentrations, respectively.

The ability to form toxins, whether encoded by plasmids or the microbial genome, is subject to selective forces. If toxin production puts a microorganism at a serious disadvantage, it will tend to disappear. If it is advantageous it will be maintained, and will spread through the microbial population, just as the genetic changes that confer resistance to antimicrobial drugs are selected for when these drugs are widely used. It is therefore not unreasonable to ask how many of the well-known toxins are actually useful to the microbe as well as being important in causing disease in the host (Table 8.4). However, microbes that multiply extracellularly must produce a variety of enzymes and other molecules involved in nutrition, adherence to substratum, and so on. In the case of free-living microbes, these substances, as well as substances that damage or interfere with competing organisms, are of major importance. Probably they cannot all be discarded when the parasitic mode of life is adopted. Many will have a toxic action. Yet, for the infecting microbe, these substances remain as unfortunate necessities, of no particular advantage and perhaps a disadvantage, in the parasitic way of life.

Indirect Damage via Inflammation

In infectious diseases there is nearly always a certain amount of direct microbial damage to host tissues, as discussed above. Host cells are destroyed or blood vessels injured as a direct result of the action of microbes or their toxins. Blood vessel injuries account for much of the disease picture in rickettsial infections (see above). Inflammatory materials are liberated from necrotic cells, whatever the cause of the necrosis. Also many bacteria themselves liberate inflammatory products and certain viruses cause living infected cells to release inflammatory mediators. Therefore it is not always clear how much of the inflammation is directly microbial rather than host in origin.* But inevitably the host (see Ch. 3) generates inflammatory and other tissue responses, and these responses sometimes account for the greater part of the tissue changes. Pathological changes can then be regarded as occurring indirectly as a result of these responses to the infection. Inflammation causes redness, swelling, pain and sometimes loss of

---

* For instance, peptidoglycan of *H. influenzae* type b causes acute inflammation when introduced into the cerebrospinal fluid of adult rats. Probably much of this is caused by inflammatory mediators from the host; TNF-α is detectable in cerebrospinal fluid of most cases of purulent bacterial meningitis in humans and there are raised concentrations in the serum of patients with *P. falciparum* malaria.

Table 8.4.   Examples of possible usefulness of toxins to microorganisms

Microorganism	Toxin	Disease production by toxin	Value of toxin to microorganism
C. diphtheriae	Diphtheria toxin	Epithelial necrosis Heart damage Nerve paralysis	Epithelial cell and polymorph destruction assists colonisation
Clostridium tetani Clostridium botulinum	Tetanus toxin Neurotoxin	Muscle spasm, lockjaw Paralysis	Could killing the host be worthwhile? A dead, putrefying corpse is a fine growth medium for these anaerobic, basically saprophytic bacteria
Shigella spp.	Shiga toxin	Exacerbates diarrhoea, dysentery Neurological effects	Diarrhoea aids transmission Nil?
V. cholerae	Cholera toxin	Diarrhoea	Diarrhoea aids transmission
B. anthracis	Anthrax toxin(s)	Oedema, haemorrhage Circulatory collapse	Kills phagocytes. Also a dead host, teeming with spores, can be a good source of infection
Legionella pneumophila	Proteases etc.	Contribute to lung pathology	Possible role in resisting phagocytic destruction by free-living amoebae
Staphylococcus pyogenes	TSSS-1 Enterotoxins	Toxic shock Diarrhoea, vomiting	All are powerful T-cell mitogens (superantigens, see Chs 7 and 8). Possible role in diverting T cells from antibacterial activity
Streptococcus pyogenes	'Erythrogenic toxin' (SPEA)	Scarlet fever	
Pseudomonas aeruginosa	Exotoxin A Proteases, elastase, etc.	Various clinical diseases	Possible role in free-living existence
Bordetella pertussis	Pertussigen	Whooping cough	Cough aids transmission; interferes with T-cell migrations?
Streptococcus pneumoniae	Pneumolysin	Promotes bacteraemia Sensorineural deafness	Weakens host defences (polymorphs, complement)
Yersinia pestis	Endotoxin Other toxins	Severe systemic disease	Kills phagocytes
Various Gram-negative bacteria	Endotoxin	Contributes to disease, septic shock	LPS acts as B-cell mitogen. Possible role in diverting B cells

function of the affected part (see Ch. 6), and is generally a major cause of the signs and symptoms of disease. Indirect damage attributable to the host immune response is discussed separately below. In most diseases direct and indirect types of damage both make a contribution to pathological changes, but in a given disease one or the other may be the most important.

In a staphylococcal abscess the bacteria produce inflammatory materials, but they also kill infiltrating polymorphs whose lysosomal enzymes are thereby liberated and induce further inflammation. This type of indirect nonimmunological damage is sometimes important in streptococcal infections. Virulent streptococci produce various toxins that damage phagocytes, and also bear on their surfaces substances that impede phagocytosis (see Ch. 4). Nevertheless, with the help of antibody, all streptococci are eventually phagocytosed and killed and the infection terminated. Unlike the staphylococci, however, killed group A streptococci pose a digestive problem for phagocytic cells. The peptidoglycan component of the streptococcal cell wall is very resistant to digestion by lysosomal enzymes. When streptococci are injected into the skin of a rabbit, for instance, streptococcal peptidoglycans persist in macrophages for as long as 146 days. Hence macrophages laden with indigestible streptococcal cell walls tend to accumulate in sites of infection. Lysosomal enzymes, including collagenase, leak from these macrophages, causing local destruction of collagen fibres and the connective tissue matrix. Macrophages secrete many other substances some of which may contribute to cell and tissue damage (see also pp. 95–96). Many macrophages eventually die or form giant cells, sometimes giving rise to granulomatous lesions (see p. 322). In this way persistent streptococcal materials sometimes cause chronic inflammatory lesions in the infected host. An additional immunopathological contribution to the lesions is to be expected if the host is sensitised to peptidoglycan components. Other pathogenic microorganisms that are digested with difficulty by phagocytes include *Listeria, Shigella, Candida albicans* and, of course, *Mycobacteria*, but the importance of this in the pathogenesis of disease is not generally clear.

Indirect Damage via the Immune Response (Immunopathology)

The expression of the immune response necessarily involves a certain amount of inflammation, cell infiltration, lymph node swelling, even tissue destruction, as described in Ch. 6. Such changes caused by the immune response are classed as immunopathological. Sometimes they are very severe, leading to serious disease or death, but at other times they play a minimal part in the pathogenesis of disease. With the possible exception of certain vertically transmitted virus infections and the transmissible 'prion' dementias (see Ch. 10), there are signs of an immune response in all infections. Therefore it is to be expected that there will nearly always be some contribution of the immune response to pathological changes.* Often the immunological contribution is small, but sometimes it forms a major part of the disease. For instance, in tuberculosis the pathological picture is dominated by the operation of a strong and persistent cell-mediated immunity (CMI) response to the invading bacillus. In the classical tubercle a central zone of bacilli with large mononuclear and giant cells, often with some necrosis, is surrounded by fibroblasts and lymphocytes. Mononuclear infiltrations, giant cells and granulomatous lesions (see p. 322) are characteristic pathological features of tuberculosis. When macrophages are killed by intracellular mycobacteria the lysosomal enzymes and other materials released from the degenerating cell contribute to chronic inflammation as in the case of the streptococcal lesions referred to above. There are no recognised toxins formed by tubercle bacilli, and there seems to be no single antigen or other component that accounts for virulence. Bacterial glycolipids (e.g. 'cord factor'), resistance to H_2O_2 (see pp. 91–92) and ability to utilise host Fe (see p. 387) have been correlated with pathogenicity, and inhibition of phagosome–lysosome fusion in macrophages (see p. 106) by release of unidentified bacterial components would also contribute to pathogenicity. However, none of these factors is by itself absolutely necessary for virulence, which in such a complex, ancient parasite is likely to be multifactorial. Now that the genome of *M. tuberculosis* has been completely sequenced, there will be opportunities for clearer definition of virulence determinants.

The mere enlargement of lymphoid organs during infectious diseases is a morphological change that can often be regarded as pathological. The lymph node swelling seen in glandular fever, for instance, is an immunopathological feature of the disease, and the same can be said of the striking enlargement of the spleen caused by chronic malaria and other infections in the condition known as tropical splenomegaly.

As often as not the relative importance of direct microbial damage as opposed to immune and nonimmune inflammatory reactions have not yet been determined, but the picture is clearer in most of the examples given below.

In one important human disease, pathological changes are certainly immunopathological in nature, but not enough is known about it to classify the type of reaction (see Table 8.5). This disease is rheumatic fever, which follows group A streptococcal infections of the throat. It is the commonest form of heart disease in many developing countries,

---

* A number of different microbial antigens are produced during most infections (see Ch. 6) and the possible immunological reactions are therefore numerous. For instance, at least 18 types of circulating malarial antigen are found in heavily infected individuals.

Table 8.5. Immunopathological reactions and infectious diseases

Reaction	Mechanism	Result	Example from infectious disease
Type 1 Anaphylactic	Antigen + IgE antibody attached to mast cells → histamine, etc. release	Anaphylactic shock Bronchospasm Local inflammation	Contribution to certain rashes? Helminth infections
Type 2 Cytotoxic	Antibody + antigen on cell surface → complement activation or ADCC	Lysis of cell bearing microbial antigens	Liver cell necrosis in hepatitis B?
Type 3 Immune complex	Antibody + extracellular antigen → complex	*Extravascular complex* Inflammation ± tissue damage *Intravascular complex* Complex deposition in glomeruli, joints, small skin vessels, choroid plexus → glomerulonephritis, vasculitis, etc.	Allergic alveolitis Glomerulonephritis in LCM virus infection (mice) or malaria (man) Prodromal rashes Fever
Type 4 Cell-mediated (delayed)	Sensitised T lymphocyte reacts with antigen; lymphokines liberated; cytotoxicity triggered	*Extracellular antigen* Inflammation, mononuclear accumulation, macrophage activation Tissue damage *Antigen on tissue cell* T lymphocyte lyses cell	Acute LCM virus disease in mice Certain virus rashes Tuberculosis, leprosy (granulomas) An *in vitro* classic, but difficult to demonstrate *in vivo*

where it currently affects 30 million children. Antibodies formed against a streptococcal cell wall or membrane component also react with the patient's heart muscle or valves, and myocarditis develops a few weeks later. Many strains of streptococci have antigens that cross-react with the heart, and repeated infections with different streptococci cause recurrent attacks of rheumatic fever. There is genetic predisposition to the disease, based either on a particular antigen present in the heart of the patient or on a particular type of antibody response. Chorea, a disease of the central nervous system, is a rare complication of streptococcal infection and antistreptococcal antibodies have been shown to react with neurons in the caudate and subthalamic nuclei of the brain.

A number of microorganisms have antigens similar to host tissue components (p. 187) so that in the course of responding immunologically to such infections the host is vulnerable to autoimmune damage (see ankylosing spondylitis, p. 220). The antibodies to host components such as DNA, IgG, myofibrils, erythrocytes, etc. that are seen in trypanosomiasis, *Mycoplasma pneumoniae*, and Epstein–Barr virus infections appear to result from polyclonal activation of B cells (see p. 199). It is not clear how important these autoimmune responses are in pathogenesis, but they reflect fundamental disturbances in immunoregulation.

Four types of immunopathology can be distinguished according to the classification of allergic reactions by Coombs and Gell, and microbial immunopathology will be described under these headings (see Table 8.5).

Type 1: anaphylactic reactions

These depend on the reactions of antigens with reaginic (IgE) antibodies attached to mast cells via the latter's Fc receptors. The reaction takes place mostly at the body surfaces, resulting in the release of histamine, eosinophil and neutrophil chemotactic factors, leukotrienes (see p. 76) and heparin from mast cells, and the activation of serotonin and plasma kinins. If the antigen–antibody interaction takes place on a large enough scale in the tissue, the histamine that is released can give rise to anaphylactic shock, the exact features depending on the sensitivity and particular reaction of the species of animal to histamine. Guinea-pigs suffer from bronchospasm and asphyxia, and in man there are similar symptoms, sometimes with a fall in blood pressure and shock. This type of immunopathology, although accounting for anaphylactic reactions to horse serum or to penicillin, is not important in infectious diseases. When the antigen–IgE antibody interaction takes place at the body surface there are local inflammatory events, giving rise to urticaria in the skin, and hay fever or asthma in the respiratory tract. This local type of anaphylaxis may play a part in the pathogenesis of virus infections of the upper respiratory tract (e.g. common cold, respiratory syncytial virus infections of infants), or in skin rashes in infectious diseases.

Type 1 reactions are common in helminth infections perhaps because IgE antibodies have an important role in protection against these parasites. The IgE-antigen reaction, by causing inflammation, summons up from the blood antimicrobial forces such as polymorphs, antibodies, and complement components. A dramatic Type 1 reaction can follow rupture of a hydatid cyst of *Echinococcus granulosus* (the dog tapeworm). Slow leakage of worm antigens means that mast cells are sensitised with specific IgE antibody, and the sudden release of antigen can cause life-threatening anaphylaxis. When the larvae of *Ascaris lumbri-*

coides pass through the lung on their journey from blood to intestine, they can give rise to IgE-mediated respiratory symptoms, with infiltration of eosinophils.

Type 2: cytolytic or cytotoxic reactions

Reactions of this type occur when antibody combines with antigen on the surface of a tissue cell, and either activates the complement sequence whose membrane attack complex (see p. 177) kills the cell, or triggers cytotoxicity by K cells (NK cells or phagocytes with Fc receptors). K (killer) cell cytolysis is referred to as antibody-dependent cellular cytotoxicity (ADCC). The antibody-coated cell is destroyed. As discussed in Ch. 6, the same reaction on the surface of a microorganism (e.g. enveloped virus) constitutes an important part of antimicrobial defences, often leading to the destruction of the microorganism. Cells infected with viruses and bearing viral antigens on their surface are destroyed in a similar way.

Clearly the antibody-mediated destruction of infected cells means tissue damage, and it perhaps accounts for some of the liver necrosis in hepatitis B, for instance, and probably in yellow fever. Infected cells can also be destroyed by sensitised lymphocytes or NK cells independently of antibody (see below).

In certain infections antibodies are formed against host erythrocytes and these cells are particularly sensitive to lysis. The haemolysis in malaria is caused by antibodies to parasite-derived antigens that have attached to red cells, rather than by autoantibodies to red cells themselves. In pneumonia due to *Mycoplasma pneumoniae* (atypical pneumonia), antibodies (cold agglutinins) are formed against normal human group O erythrocytes. Haemolytic anaemia is occasionally seen, and there is reticulocytosis (see Glossary) in 64% of patients. The lesions in the lungs are perhaps based on cell-mediated immunopathological reactions.

Type 3: immune complex reactions

The combination of antibody with antigen is an important event, initiating inflammatory phenomena that are inevitably involved in the expression of the immune response. In the infected host, these inflammatory phenomena are most of the time of great antimicrobial value (see Ch. 6). But there are nevertheless immunopathological features of the infection, and immune complex reactions sometimes do a great deal of damage in the infected individual. The mechanisms by which antigen–antibody reactions cause inflammation and tissue damage are outlined in Fig. 8.16. IgA immune complexes are generally less harmful. Antigens absorbed from the intestine can combine locally with IgA

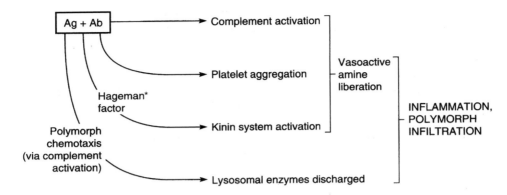

Fig. 8.16 Mechanisms of inflammation and tissue damage induced by antigen–antibody reactions. *Activation of Hageman factor can lead to blood coagulation.

antibody and the complex then enters the blood, to be filtered out in the liver and excreted harmlessly in bile (see p. 160).

When the antigen–antibody reaction takes place in extravascular tissues, there is inflammation and oedema with infiltration of polymorphs. If soluble antigen is injected intradermally into an individual with large amounts of circulating IgG antibody, the antigen–antibody reaction takes place in the walls of skin blood vessels, and causes an inflammatory response. The extravasating polymorphs degenerate and their lysosomal enzymes cause extensive vascular damage. This is the classical Arthus response. Antigen–antibody reactions in tissues are not usually as serious as this, and milder inflammatory sequelae are more common as in the case of allergic alveolitis (see below).

Glomerulonephritis and vasculitis

When the antigen–antibody reaction takes place in the blood to give circulating immune complexes, the sequelae depend to a large extent on size and on the relative proportions of antigen and antibody. If there is a large excess of antibody, each antigen molecule is covered with antibody and is removed rapidly by reticuloendothelial cells, which have receptors for the Fc portion of the antibody molecule (see Ch. 4). When equal amounts of antigen and antibody combine, lattice structures are produced, and these form large aggregates whose size ensures that they are also rapidly removed by reticuloendothelial cells. If, however, complexes are formed in antigen excess, the poorly coated antigen molecules are not removed by reticuloendothelial cells. They continue to circulate in the blood and have the opportunity to localise in small blood vessels elsewhere in the body. The mechanism is not clear, but complexes are deposited in the glomeruli of the kidneys, the choroid plexuses, joints and ciliary body of the eye. Factors may include local high blood pressure and turbulent flow (glomeruli), or the filtering function of the vessels involved (choroid plexus, ciliary body). In the glomeruli the complexes pass through the endothelial windows (Fig. 8.17) and come to lie beneath the basement membrane. The smallest-sized complexes pass through the basement membrane and seem to enter the urine. This is probably the normal mechanism of disposal of such complexes from the body.

Immune complexes are formed in many, perhaps most, acute infectious diseases. Microbial antigens commonly circulate in the blood in viral, bacterial, fungal, protozoal, rickettsial, etc. infections. When the immune response has been generated and the first trickle of specific antibody enters the blood, immune complexes are formed in antigen excess. This is generally a transitional stage soon giving rise to antibody excess, as more and more antibody enters the blood and the

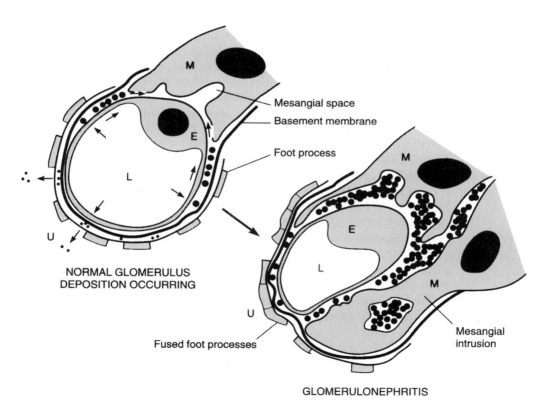

Fig. 8.17 Immune complex glomerulonephritis. Arrows indicate the movement of immune complex deposits, some moving through to the urine and others (larger deposits) being retained. M, mesangial cell; U, urinary space; L, lumen of glomerular capillary; E, endothelial cell (contains 100 nm pores or windows; see Fig. 3.2b).

infection is terminated. Sometimes the localisation of immune complexes and complement in kidney glomeruli* is associated with a local inflammatory response after complement activation. There is an infiltration of polymorphs, swelling of the glomerular basement membrane, loss of albumin, even red blood cells, in the urine and the patient has acute glomerulonephritis. This is seen following strepto-coccal infections, mainly in children (see below). As complexes cease to be formed the changes are reversed, and complete recovery is the rule. Repeated attacks or persistent deposition of complexes leads to irre-versible damage, often with proliferation of epithelial cells following the seepage of fibrin into the urinary space.

Under certain circumstances complexes continue to be formed in the blood and deposited subendothelially for long periods. This happens in certain persistent microbial infections in which microbial antigens are continuously released into the blood but antibody responses are only minimal or of poor quality (see below). Complexes are deposited in glomeruli over the course of weeks, months or even years. The normal mechanisms for removal are inadequate. The deposits, particularly larger complexes containing high molecular weight antigens or anti-bodies (IgM) are held up at the basement membrane and accumulate in the subendothelial space together with the complement components. As deposition continues, they gradually move through to the mesangial space (Fig. 8.17) where they form larger aggregates. Mesangial cells, one of whose functions is to deal with such materials, enlarge, multiply and extend into the subepithelial space. If these changes are gradual there are no inflammatory changes, but the structure of the basement membrane alters, allowing proteins to leak through into the urine. Later the filtering function of the glomerulus becomes progressively impaired. In the first place the glomerular capillary is narrowed by the mesangial cell intrusion. Also, the filtering area is itself blocked by the mesangial cell intrusion, by the accumulation of complexes (Fig. 8.17), and by alterations in the structure of the basement membrane. The foot processes of epithelial cells tend to fuse and further interfere with filtration. The pathological processes continue, some glomeruli ceasing to produce urine, and the individual has chronic glomerulonephritis.

Circulating immune complex deposition in joints leads to joint swelling and inflammation but in choroid plexuses there are no apparent pathological sequelae. Circulating immune complexes are also deposited in the walls of small blood vessels in the skin and else-where, where they may induce inflammatory changes. The prodromal rashes seen in exanthematous virus infections and in hepatitis B are probably caused in this way. If the vascular changes are more marked, they give rise to the condition called erythema nodosum, in which there

---

* Cells in kidney glomeruli, in joint synovium and in choroid plexuses bear Fc or C3b receptors. This would favour localisation in these tissues.

are tender red nodules in the skin, with deposits of antigen, antibody and complement in vessel walls. Erythema nodosum is seen following streptococcal infections and during the treatment of patients with leprosy. When small arteries are severely affected, for instance in some patients with hepatitis B, this gives rise to periarteritis nodosa.

Immune complex glomerulonephritis occurs as an indirect immuno-pathological sequel to a variety of infections. First there are certain virus infections of animals. The antibodies formed in virus infections generally neutralise any free virus particles, thus terminating the infection (see Ch. 6), but the infection must persist if antigen is to continue to be released into the blood and immune complexes formed over long periods. Non-neutralising antibodies help promote virus persistence because they combine specifically with virus particles, fail to render them noninfectious, and at the same time block the action of any good neutralising antibodies that may be present. Immune complexes in antigen excess are formed in the blood when the persis-tent virus or its antigens circulates in the plasma and reacts with anti-body which is present in relatively small amounts. Virus infections with these characteristics are included in Table 8.6. In each instance complexes are deposited in kidney glomeruli and sometimes in other blood vessels as described above. In some there are few if any patho-logical changes (LDV and leukaemia viruses in mice) probably because there is a slow rate of immune complex deposition, whereas in others glomerulonephritis (LCM virus in mice, ADV in mink) or vasculitis (ADV in mink) is severe.

A persistent virus infection that induces a feeble immune response forms an ideal background for the development of immune complex glomerulonephritis, but there are no known viral examples in man.

Table 8.6. The deposition of circulating immune complexes in infectious diseases

Microbe	Host	Kidney deposits	Glomerulo-nephritis	Vascular deposits
Leukaemia virus	Mouse, cat	+	±	−
Lactate dehydrogenase virus (LDV)	Mouse	+	±	−
Lymphocytic choriomeningitis virus (LCM)	Mouse	++	+	±
Aleutian disease virus (ADV)	Mink	+	+	++
Equine infectious anaemia virus	Horse	+	+	+
Hepatitis B virus	Man	+	−	+
Streptococcus pyogenes	Man	+	+	−
Malaria (nephritic syndrome)	Man	+	+	−
Treponema pallidum (nephritic syndrome in secondary syphilis)	Man	+	+	?
Infectious causes of chronic glomerulonephritis[a]	Man	++	++	−

[a] Nephrologists and pathologists distinguish ten different types of glomerulonephritis, some of them infectious in origin, the immune complexes being deposited directly from blood or formed locally in glomeruli.

There are one or two other microorganisms that occasionally cause this type of glomerulonephritis, and it is seen, for instance, in chronic quartan malaria and sometimes in infective endocarditis. In both these examples microbial antigens circulate in the blood for long periods. However, immune complex deposition does not necessarily lead to the development of glomerulonephritis, and immune complexes are detectable in the glomeruli of most normal mice and monkeys. Even in persistent virus infections the rate of deposition may be too slow to cause pathological changes as with LDV and leukaemia virus infections of mice (see Table 8.5). During the acute stage of hepatitis B in man, when antibodies are first formed against excess circulating viral antigen (hepatitis B surface antigen), immune complexes are formed and deposited in glomeruli. However, the deposition is short-lived and there is no glomerulonephritis. Persistent carriers of the antigen do not generally develop glomerulonephritis, because their antibody is usually directed against the 'core' antigen of the virus particle, rather than against the large amounts of circulating hepatitis B surface antigen.

Immune complex glomerulonephritis occurs in man as an important complication of streptococcal infection, but this is usually acute in nature with complement activation and inflammation of glomeruli, as referred to above. Antibodies formed against an unknown component of the streptococcus react with circulating streptococcal antigen, perhaps also with a circulating host antigen, and immune complexes are deposited in glomeruli. Streptococcal antibodies cross-reacting with the glomerular basement membrane or with streptococcal antigen trapped in the basement membrane may contribute to the picture. Deposition of complexes continues after the infection is terminated, and glomerulonephritis develops a week or two later. The streptococcal infection may be of the throat or skin, and *Streptococcus pyogenes* types 12 and 49 are frequently involved.

Kidney failure in man is commonly due to chronic glomerulonephritis, and this is often of the immune complex type, but the antigens, if they are microbial, have not yet been identified. It is possible that the process begins when antigen on its own localises in glomeruli, circulating antibody combining with it at a later stage. The antibody is often IgA ('IgA nephropathy') which could be explained as follows. Antigen in intestinal or respiratory tract combines locally with IgA, and the complex enters the blood. Here, for unknown reasons, it is not removed in the normal way by the liver, and thus has the opportunity to localise in glomeruli.

Allergic alveolitis

When certain antigens are inhaled by sensitised individuals and the antigen reaches the terminal divisions of the lung, there is a local antigen–antibody reaction with formation of immune complexes. The resulting inflammation and cell infiltration causes wheezing and respiratory distress, and the condition is called allergic alveolitis. Persistent inhalation of the specific antigen leads to chronic pathological changes with fibrosis and respiratory disease. Exposure to the antigen must be by inhalation; when the same antigen is injected intradermally, there is an Arthus type reaction (see p. 282), and IgG rather than IgE antibodies are involved.

There are a number of microorganisms that cause allergic alveolitis. Most of these are fungi. A disease called farmer's lung occurs in farm workers repeatedly exposed to mouldy hay containing the actinomycete *Micromonospora faeni*. Cows suffer from the same condition. A fungus contaminating the bark of the maple tree causes a similar disease (maple bark stripper's disease) in workers in the USA employed in the extraction of maple syrup. The mild respiratory symptoms occasionally reported after respiratory exposure of sensitised individuals to tuberculosis doubtless have the same immunopathological basis.

Other immune complex effects

In addition to their local effects, antigen–antibody complexes generate systemic reactions. For instance, the fever that occurs at the end of the incubation period of many virus infections is probably attributable to a large-scale interaction of antibodies with viral antigen, although extensive CMI reactions can also cause fever. The febrile response is mediated by endogenous pyrogen IL-1 and TNF liberated from polymorphs and macrophages, as described on p. 329. Probably the characteristic subjective sensations of illness and some of the 'toxic' features of virus diseases are also caused by immune reactions and liberation of cytokines.

Systemic immune complex reactions taking place during infectious diseases very occasionally give rise to a serious condition known as disseminated intravascular coagulation. This is seen sometimes in severe generalised infections such as Gram-negative septicaemia, meningococcal septicaemia, plague, yellow fever and fevers due to hantaviruses (see Table A.5). Immune complex reactions activate the enzymes of the coagulation cascade (Fig. 8.16), leading to histamine release and increased vascular permeability. Fibrin is formed and is deposited in blood vessels in the kidneys, lungs, adrenals and pituitary. This causes multiple thromboses with infarcts, and there are also scattered haemorrhages because of the depletion of platelets, prothrombin, fibrinogen, etc. Systemic immune complex reactions were once thought to form the basis for dengue haemorrhagic fever. This disease is seen in parts of the world where dengue is endemic, individuals immune to one type of dengue becoming infected with a related strain of virus. They are not protected against the second virus, although it shows immuno-logical cross-reactions with the first one. Indeed the dengue-specific antibodies enhance infection of susceptible mononuclear cells, so that larger amounts of viral antigen are produced (see p. 173). It was thought that after virus replication, viral antigens in the blood reacted massively with antibody to cause an often lethal disease with haemor-rhages, shock and vascular collapse. However, it has proved difficult to demonstrate this pathophysiological sequence, and the role of circu-lating immune complexes and platelet depletion remains unclear. Perhaps in this and in some of the other viral haemorrhagic fevers the virus multiplies in capillary endothelial cells. Disease seems due to cytokines liberated from infected mononuclear cells.

Immune complex immunopathology is probable in various other infectious diseases. For instance, the occurrence of fever, polyarthritis, skin rashes and kidney damage (proteinuria) in meningococcal menin-gitis and gonococcal septicaemia indicates immune complex deposi-tion. Circulating immune complexes are present in these conditions. Certain African arthropod-borne viruses with exotic names (Chikungunya, O'nyong-nyong) cause illnesses characterised by fever, arthralgia and itchy rashes, and this too sounds as if it is immune complex in origin. Immune complexes perhaps play a part in the oedema and vasculitis of trypanosomiasis and in the rashes of secondary syphilis.

Sensitive immunological techniques are available for the detection of circulating complexes and for the identification of the antigens and antibodies in deposited complexes. The full application of these tech-niques will perhaps solve the problem of the aetiology of chronic glomerulonephritis in man.

Type 4: cell-mediated reactions

Although antibodies often protect without causing damage the mere expression of a CMI response involves inflammation, lymphocyte infil-tration, macrophage accumulation and macrophage activation as described in Ch. 6. The CMI response by itself causes pathological changes, and cytokines such as TNF play an important part. This can be demonstrated, as a delayed hypersensitivity reaction by injecting tuberculin into the skin of a sensitised individual. The CMI response to infection dominates the pathological picture in tuberculosis, with mononuclear infiltration, degeneration of parasitised macrophages, and the formation of giant cells as central features. These features of the tissue response result in the formation of granulomas (see Glossary) which reflect chronic infection and accompanying inflamma-tion. There is a ding-dong battle as the host attempts to contain and control infection with a microorganism that is hard to eliminate. The granulomas represent chronic CMI responses to antigens released locally. Various other chronic microbial and parasitic diseases have granulomas as characteristic pathological features. These include chlamydial (lymphogranuloma inguinale), bacterial (syphilis, leprosy, actinomycosis), and fungal infections (coccidiomycosis). Antigens that are disposed of with difficulty in the body are more likely to be important inducers of granulomas. Thus, although mannan is the dominant antigen of *Candida albicans*, glucan is more resistant to breakdown in macrophages and is responsible for chronic inflammatory responses.

The lymphocytes and macrophages that accumulate in CMI responses also cause pathological changes by destroying host cells. Cells infected with viruses and bearing viral antigens on their surface are targets for CMI responses as described in Chs 6 and 9. Infected cells, even if they are perfectly healthy, are destroyed by the direct action of sensitised T lymphocytes, which are demonstrable in many viral infections. In spite of the fact that the *in vitro* test system so clearly displays the immunopathological potential of cytotoxic T cells, this is not easy to evaluate in the infected host. It may contribute to the tissue damage seen, for instance, in hepatitis B infection and in many herpes and poxvirus infections. In glandular fever, cytotoxic T cells react against Epstein–Barr virus-infected B cells to unleash an immunological civil war that is especially severe in adolescents and young adults. Antigens from *Trypanosoma cruzi* are known to be adsorbed to uninfected host cells, raising the possibility of autoimmune damage in Chagas' disease, caused by this parasite.* It is also becoming clear that cells infected with certain protozoa (e.g. *Theileria parva* in bovine lymphocytes) have parasite antigens on their surface and are susceptible to this type of destruction. Little is known about intracellular bacteria.

The most clearly worked out example of type 4 (CMI) immunopathology is seen in LCM virus infection of adult mice. When virus is injected intracerebrally into adult mice, it grows in the meninges, ependyma and choroid plexus epithelium, but the infected cells do not show the slightest sign of damage or dysfunction. After 7–10 days, however, the mouse develops severe meningitis with submeningeal and subependymal oedema, and dies. The illness can be completely prevented by adequate immunosuppression, and the lesions are attributable to the mouse's own vigorous CD8$^+$ T-cell response to infected cells.

---

* Chagas' disease, common in Brazil, affects 12 million people, and is transmitted by blood-sucking bugs. After spreading through the body during the acute infection, the parasitaemia falls to a low level and there is no clinical disease. Years later a poorly understood chronic disease appears, involving heart and intestinal tract, which contain only small numbers of the parasite but show a loss of autonomic ganglion cells. An autoimmune mechanism is possible (see p. 188), because a monoclonal antibody to *T. cruzi* has been obtained that cross-reacts with mammalian neurons.

These cells present processed LCM viral peptides on their surface in conjunction with MHC I proteins, and sensitised CD8$^+$ T cells, after entering the cerebrospinal fluid and encountering the infected cells, generate the inflammatory response and interference with normal neural function that cause the disease. The same cells destroy infected tissue cells *in vitro*, but tissue destruction is not a feature of the neuro-logical disease. In this disease the CD8$^+$ T cells probably act by liber-ating inflammatory cytokines. It may be noted that the brain is uniquely vulnerable to inflammation and oedema, as pointed out earlier in this chapter. The infected mouse shows the same type of lesions in scattered foci of infection in the liver and elsewhere, but they are not a cause of sickness or death. LCM infection of mice is a classical example of immunopathology in which death itself is entirely due to the cell-mediated immune response of the infected individual. This response, although apparently irrelevant and harmful, is nevertheless an 'attempt' to do the right thing. It has been shown that immune T cells effectively inhibit LCM viral growth in infected organs. However, a response that in most extraneural sites would be useful and appro-priate turns out to be self-destructive when it takes place in the central nervous system.

Another type of T cell-mediated immune pathology is illustrated by influenza virus infection of the mouse. When inoculated intranasally, the virus infects the lungs and causes a fatal pneumonia in which the airspaces fill up with fluid and cells. The reaction is massive and the lungs almost double in weight. Effectively the animal drowns. The cause is an influx of virus-specific CD8$^+$ T cells. Normally when an appropriate number of T cells had entered the lungs, the T cells would issue a feedback response to prevent such overaccumulation, but it is thought that influenza virus infects the T cells and inhibits this control process, so that the lungs are eventually overwhelmed. The virus does not multiply in or kill the infected T cells, and it is presumed that it undergoes limited gene expression.

One human virus infection in which a strong CMI contribution to pathology seems probable is measles. Children with thymic aplasia show a general failure to develop T lymphocytes and cell-mediated immunity, but have normal antibody responses to most antigens. They suffer a fatal disease if they are infected with measles virus. Instead of the limited extent of virus growth and disease seen in the respiratory tract in normal children, there is inexorable multiplication of virus in the lung, in spite of antibody formation, giving rise to giant cell pneu-monia. This indicates that the CMI response is essential for the control of virus growth. In addition there is a total absence of the typical measles rash, and this further indicates that the CMI response is also essential for the production of the skin lesions. Cell-mediated immune responses also make a contribution to the rashes in poxvirus infections.

Other Indirect Mechanisms of Damage

Stress, haemorrhage, placental infection and tumours

Sometimes in infectious diseases there are prominent pathological changes which are not attributable to the direct action of microbes or their toxins, nor to inflammation or immunopathology. The stress changes mediated by adrenal cortical hormones come into this category. Stress is a general term used to describe various noxious influences, and includes cold, heat, starvation, injury, psychological stress and infection. An infectious disease is an important stress, and corticosteroids are secreted in large amounts in severe infections (see also Ch. 11). They generally tend to inhibit the development of pathological changes, but also have pronounced effects on lymphoid tissues, causing thymic involution and lymphocyte destruction. These can be regarded as pathological changes caused by stress. It was the very small size of the thymus gland as seen in children dying with various diseases, especially infectious diseases, that for many years contributed to the neglect of this important organ, and delayed appreciation of its vital role in the development of the immune system.

Appreciation of the effects of stress on infectious diseases and the immune response in particular has led to the establishment of the science of neuroimmunology. Properly controlled experiments are difficult to mount but it is clear that the nervous system affects the functioning of the immune system. The pathways of this communication are still poorly understood, but there is a shared language for immune and neural cells. For example, neural cells as well as immune cells have receptors for interleukins, and lymphocytes and macrophages secrete pituitary growth hormone. Work on *Mycobacterium bovis* grew out of observations from the turn of the century that stress appears to increase the death rate in children with tuberculosis (TB). In one type of experiment mice were stressed by being kept in a restraining device where movement was virtually impossible. This resulted in the reduction of expression of MHC class II antigens on macrophages, which correlated with increased susceptibility to infection. Similarly stressing mice infected with influenza virus caused several immunosuppressive events including reduction of inflammatory cells in the lung, and decreased production of IL-2. Suppression of antibody responses is found in people suffering a type of stress familiar to students – examinations! The best responses to hepatitis B vaccine in students immunised on the third day of their examinations were found in those who reported the least stress. Finally, in a double-blind trial at the Common Cold Research Unit in England with five different respiratory viruses, it was ascertained in human volunteers that stress gave a small but statistically significant increased likelihood of an individual developing clinical disease.

Pathological changes are sometimes caused in an even more indirect way as in the following example. Yellow fever is a virus infection transmitted by mosquitoes and in its severest form is characterised by devastating liver lesions. There is massive mid-zonal liver necrosis following the extensive growth of virus in liver cells, resulting in the jaundice that gives the disease its name. Destruction of the liver also leads to a decrease in the rate of formation of the blood coagulation factor, prothrombin, and infected human beings or monkeys show prolonged coagulation and bleeding times. Haemorrhagic phenomena are therefore characteristic of severe yellow fever, including haemorrhage into the stomach and intestine. In the stomach the appearance of blood is altered by acid, and the vomiting of altered blood gave yellow fever another of its names, 'black vomit disease'. Haemorrhagic phenomena in infectious diseases can be due to direct microbial damage to blood vessels, as in certain rickettsial infections (see p. 140) or in the virus infection responsible for haemorrhagic disease of deer. They may also be due to immunological damage to vessels as in the Arthus response or immune complex vasculitis, to any type of severe inflammation, and to the indirect mechanism illustrated above. Finally there are a few infectious diseases in which platelets are depleted, sometimes as a result of their combination with immune complexes plus complement, giving thrombocytopenia and a haemorrhagic tendency (see also disseminated intravascular coagulation, p. 287). Thrombocytopenic purpura is occasionally seen in congenital rubella and in certain other severe generalised infections.

Infection during pregnancy can lead to foetal damage or death not just because the foetus is infected (p. 333), but also because of infection and damage to the placenta. This is another type of indirect pathological action. Placental damage may contribute to foetal death during rubella and cytomegalovirus infections in pregnant women.

Certain viruses undoubtedly cause tumours (leukaemia viruses, human papillomaviruses, several herpes viruses in animals – see Table 8.1) and this is to be regarded as a late pathological consequence of infection. As was discussed in Ch. 7 the tumour virus genome can be integrated into the host cell genome whether a tumour is produced or not, so that the virus becomes a part of the genetic constitution of the host. Sometimes the host cell is transformed by the virus and converted into a tumour cell, the virus either introducing a transforming gene into the cell, activating expression of a pre-existing cellular gene, or inactivating the cell's own fail-safe tumour suppressor gene. The transforming genes of DNA tumour viruses generally code for T antigens which are necessary for transformation, and the transforming genes of RNA tumour viruses are known as *onc* genes.*

---

* *Onc* genes (oncogenes) are also present in host cells, where they play a role in normal growth and differentiation, often coding for recognised growth factors (e.g. human platelet-derived growth factor). They can be activated and the cell transformed when tumour viruses with the necessary 'promoters' are brought into the cell. The *onc* genes of the RNA tumour viruses themselves originate from cellular oncogenes which were taken up into the genome of infecting viruses during their evolutionary history.

Transformation has been extensively studied *in vitro*, and the features of the transformed cell described (changed surface and social activity, freedom from the usual growth restraints).

Dual infections

Simultaneous infection with two different microorganisms would be expected to occur at times, merely by chance, especially in children. On the other hand, a given infection generates antimicrobial responses such as interferon production and macrophage activation which would make a second infection less likely. Dual infections are commonest when local defences have been damaged by the first invader. The pathological results are made much more severe because there is a second infectious agent present. This can be considered as another mechanism of pathogenicity. Classical instances involve the respiratory tract. The destruction of ciliated epithelium in the lung by viruses such as influenza or measles allows normally nonpathogenic resident bacteria of the nose and throat, such as the pneumococcus or *Haemophilus influenzae*, to invade the lung and cause secondary pneumonia. If these bacteria enter the lung under normal circumstances, they are destroyed by alveolar macrophages or removed by the mucociliary escalator. In at least one instance the initial virus infection appears to act by interfering with the function of alveolar macrophages. Mice infected with parainfluenza 1 (Sendai) virus show greatly increased susceptibility to infection with *Haemophilus influenzae*, and this is largely due to the fact that alveolar macrophages infected with virus show a poor ability to phagocytose and kill the bacteria. Specialised respiratory pathogens such as influenza, measles, parainfluenza or rhinoviruses damage the nasopharyngeal mucosa and can lead in the same way to secondary bacterial infection, with nasal catarrh, sinusitis, otitis media or mastoiditis. The normal microbial flora of the mouth, nasopharynx or intestine are always ready to cause trouble if host resistance is lowered, but under normal circumstances they hinder rather than help other infecting microorganisms (see Ch. 2).

One interesting example of exacerbation of infection occurs in mice dually infected with influenza virus and microorganisms such as *Streptococcus aureus* or *Serratia marcescens*. Under these conditions animals suffer a more severe viral infection. This results from the need to proteolytically cleave the viral haemagglutinin protein which is done by a cellular enzyme. If the appropriate protease is in short supply or lacking completely, virions are formed but they are not infectious. Under these circumstances the haemagglutinin can be cleaved extracellularly by microbial proteases with resulting increased amounts of infectious virus and disease.

As a final example of dual infections, microorganisms that cause immunosuppression can activate certain pre-existing chronic infections. In measles, for instance, there is a temporary general depression of CMI; tuberculin-positive individuals become tuberculin negative, and in patients with tuberculosis the disease is exacerbated. In the acquired immunodeficiency syndrome (AIDS; see p. 191) immunosuppression by HIV activates a variety of pre-existing persistent infections.

Diarrhoea

Diarrhoea deserves a separate section, since it is one of the commonest types of illness in developing countries and a major cause of death in childhood. Particularly in infants, who have a very high turnover of water relative to their size, the loss of fluid and salt soon leads to life-threatening illness. In 1998, diarrhoea was responsible for 2.2 million deaths world-wide in children under 5 years old. In villages in West Africa and Guatemala, the average 2–3-year-old child has diarrhoea for about 2 months in each year.* Diarrhoea also interacts with malnutrition and can cause stunted growth, defective immune responses and susceptibility to other infections (pp. 377–379). Fluid and electrolyte replacement is a simple, highly effective, life-saving treatment that can be used without determining the cause of the diarrhoea. Oral rehydration therapy (ORF) means giving a suitable amount of salt and sugar in clean water, and this is something that can be done by the mother. Diarrhoea is also a common affliction of travellers from developed countries, and business deals, athletic successes and holiday pleasures can be forfeited on the toilet seats of foreign lands. The most reliable prophylaxis is to 'cook it, peel it, or forget it'. Most attacks of diarrhoea are self-limiting. Diarrhoea means the passage of liquid faeces,† or faeces that take the shape of the receptacle rather than have their own shape. This could arise because of increased rate of propulsion by intestinal muscles, giving less time for reabsorption of water in the large bowel, or because there was an increase in the amount of fluid held or produced in the intestine. In many types of infectious diarrhoea the exact mechanism is not known. Diarrhoea, on the one hand, can be

* Diarrhoea on a massive scale is not always confined to developing countries. There was a major outbreak of *Cryptosporidium* infection in Milwaukee, USA, in 1993 with more than 400 000 cases; 285 of these were diagnosed in the laboratory and they suffered watery diarrhoea (mean 12 stools a day) for a mean of 9 days. The small (4–5 mm) oocysts, probably from cattle, had entered Lake Michigan, and then reached the community water supply because of inadequate filtration and coagulation treatment.

† Liquid faeces are not abnormal in all species. The domestic cow experiences life-long diarrhoea, but presumably does not suffer from it.

regarded as a microbial device for promoting the shedding and spreading of the infection in the community, or, on the other hand, as a host device to hasten expulsion of the infectious agent. Diarrhoea is a superb mechanism for the dissemination of infected faeces (see p. 58) and there is no doubt that strains of microbes are selected for their diarrhoea-producing powers. The advantages to the host of prompt expulsion of the infectious agent was illustrated when volunteers infected with *Shigella flexneri* were given Lomotil, a drug that inhibits peristalsis. They were more likely to develop fever and had more difficulty in eliminating the pathogen.

Before attempting to explain the pathophysiology of diarrhoeal disease, the normal structure and function of gut will be considered. The main function of the gut is the active inward transport of ions and nutrient solutes which is followed by the passive movement of water (Fig. 8.18). The driving force is the Na^+/K^+ ATPase situated in the basolateral membrane of enterocytes on the villus (Fig. 8.18), which maintains a low intracellular $[Na^+]$, thus creating the electrochemical gradient favourable for Na^+ entry and a high regional $[Na^+]$ in the intercellular spaces; Cl^- follows Na^+. A similar situation exists in crypt cells: Na^+/K^+ ATPase drives secretion. The key difference is the location of the carrier systems responsible for the facilitated entry of the actively transported species. In villus cells the carriers are present in the brush border, whereas in crypt cells they are located in the basal membrane: this is responsible for the vectorial aspects of ion/fluid traffic in villus/crypt assemblies. However, it is clear that several factors in addition to enterocytes are involved in regulating fluid transport in the gut; these include the enteric nervous system and the anatomy of the microcirculation. The latter plays a profoundly important role in the uptake of fluid. This is illustrated in Fig. 8.19, which shows the existence of zones of graded osmotic potential. At the tips of villi in adult human gut, osmolalities range from 700 to 800 mOsm kg^{-1} H_2O, which would generate huge osmotic forces. Thus, current perceptions are that enterocytes are responsible for generating this gradient and the blood supply acts as a countercurrent multiplier which amplifies the gradient in a manner analogous to the loops of Henle in the kidney. The hypertonic zone has been demonstrated directly in whole villi of infant mice in terms of the changing morphology of erythrocytes: in the lower regions of villi they show characteristic discoid morphology, whereas in the upper region they are crenated, indicating a hyperosmotic environment. The hypertonicity is dissipated if the blood flow is too slow and washed out if too fast. It is the villus unit rather than enterocytes by themselves that is responsible for fluid uptake. Another consequence of the microcirculatory anatomy is that villus tip regions are relatively hypoxic. In addition, neonatal brush borders contain disaccharidases (principally lactase) which break down nonabsorbable disaccharides (e.g. lactose) into constituent absorbable monosaccharides.

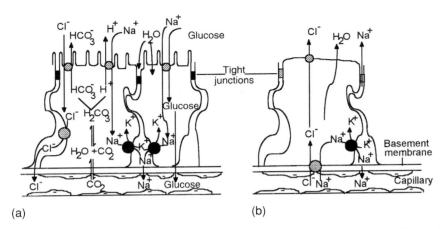

(a)   (b)

Fig. 8.18 Simplified schematic representation of electrolyte transport by ileal mucosal tissue and its consequence for (a) absorption and (b) secretion. Active processes involve the movement of ions and nutrient solutes; water follows passively.

(a) Two methods of Na^+ co-transport are shown involving a glucose-linked symport and two coupled antiports; the latter results in the co-transport of Cl^-. The coupled antiports are functionally linked via H^+ and HCO_3^-, the relative concentrations of which are a reflection of metabolic activity. These processes occur within the same cells but are shown separately for clarity. The driving force for Na^+ uptake is the low Na^+ concentration maintained by the Na^+/K^+ pump (ATPase) which creates the electrochemical gradient that promotes the inward movement of Na^+; Cl^- follows Na^+ by diffusion. Water is drawn osmotically across the epithelium paracellularly (i.e. across tight junctions) and/or transcellularly, the former pathway accounting for approximately 80% of fluid movement.

(b) Secretion is the result of the coupled entry of Na^+ and Cl^- across the basolateral membrane. Na^+ is recycled by the Na^+/K^+ pump and Cl^- exits by diffusing down an electrochemical gradient and across the undifferentiated crypt cell apical membrane; Na^+ follows Cl^- and water follows passively.

Note: (i) The driving force results from the same mechanism that powers absorption, i.e. the Na^+/K^+ pump located in the basolateral membrane; it is the location of the 'port' 'diffusion' systems that determines the vectorial aspects of ion movement. (ii) The tight junctions are less tight in the crypts than villi. (iii) The apical membrane of the crypt cell is undifferentiated and only acquires microvilli during ascent into villous regions. ●, Na^+/K^+ pump; ◐, symport, antiport or diffusion channel.

   Villus tips and crypts are regarded as the anatomical sites of physiological absorption and secretion respectively. Fluid transport is a bidirectional process in the healthy animal with net absorption in health and net secretion in disease. The balance between absorption and secretion is poised at different points throughout the intestinal tract reflecting differences in both structure and function. Proximal small intestine is relatively leaky; in contrast the colon is a powerfully absorptive organ.

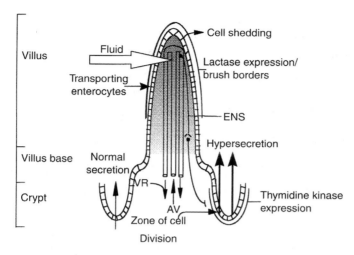

Fig. 8.19 Small intestinal villus: simplified schema of integrated structure and function. Note the central arterial vessel (AV) which arborizes at the tip into a capillary bed drained by a subepithelial venous return (VR). Movement of sodium into VR creates a concentration gradient between VR and AV, causing absorption of water from AV and surrounding tissue. This results in a progressive increase in the osmolarity of incoming blood moving into the tip region through to VR. Tip osmolarity is about three times higher than normal. Hyperosmolarity has been demonstrated in man and can be inferred in mice from the morphology of erythrocytes which changes during ascent of the same vessel from base to tip regions of villi. The intensity of shading indicates a vertical increase in osmolarity. The left crypt represents normal physiological secretion and the right crypt hypersecretion. ENS, the enteric nervous system, is depicted schematically and not anatomically.

Finally, crypts are the principal sites of cell regeneration, replacing cells which migrate up the epithelial escalator. The epithelium is renewed in approximately 3–5 days. At villus tips senescent cells are shed.

Diarrhoeal disease can result from interference with almost any one, or combination of these systems. The range of intestinal pathogens and the types of disease they cause is illustrated in Tables 8.7 and 8.8. The pathological/pathophysiological nature of some pathogen/host interactions is illustrated in Fig. 8.20. Noninvasive pathogens like *V. cholerae* and enterotoxigenic *E. coli* (ETEC) secrete toxins which perturb the ion transport systems. Invasive nonhistotoxic pathogens, such as some *Salmonella* strains (see Ch. 2) and rotavirus, invade villus tip cells which are then shed into the intestinal lumen. Invasive histotoxic pathogens, such as some strains of *Salmonella* (see Ch. 2), cause rapid toxin-mediated detachment of epithelial cells. Experimental rotavirus infections have been studied in great detail allowing us to delineate

Table 8.7. Production of diarrhoea by microorganisms shed in faeces

Infectious agent	Diarrhoea	Site of replication
Rotaviruses	+	Intestinal epithelium
Parvoviruses (dogs)	+	Intestinal epithelium (crypt cells)
Intestinal adenoviruses (types 40, 41)	+	Intestinal epithelium
Intestinal coronaviruses[a]	+	Intestinal epithelium
Norwalk virus group (caliciviruses)	+	Intestinal epithelium
Toroviruses (calves, horses, humans)	+	Intestinal epithelium and M cells (see Table A.5)
Vibrio cholerae	+	Intestinal lumen
Clostridium difficile	+	Intestinal lumen
Campylobacter jejuni	+	Intestinal epithelium
E. coli	+	Varies[b]
Shigella	+	Intestinal epithelium
Salmonella sp.	±	Intestinal epithelium (varies)
Salmonella typhi	+	Intestinal lymphoid tissue, liver, biliary tract
Cryptosporidium	+	Intestinal epithelium
Giardia lamblia	+	Attached to intestinal epithelium
Entamoeba histolytica	+	Invasion of intestinal epithelium

[a] Described for pigs, foals, calves, sheep, dogs, mice, man and turkeys; maximum susceptibility in the first few weeks of life.
[b] Strain ETEC remains in the lumen; EIEC is similar to *Shigella*, EHEC reaches subepithelial tissues.

intermediate stages between initial infection, through clinical diarrhoea to recovery from infection. We either do not know or can only infer what the intermediate stages are for the other examples alluded to – signified by broken arrows (Fig. 8.20) – leading to a return to normal in those cases in which disease is self-limiting.

Campylobacter jejuni does not figure in our treatment so far despite the fact that *C. jejuni* and related species are the most common bacterial cause of diarrhoea in many industrialised countries. This is because of a severe lack of relevant 'mechanistic' information due to the lack of good experimental models; hence we know very little about the detailed mechanisms of pathogenicity of this hugely important pathogen. The clinical picture of the pathogenesis of *C. jejuni* infection may be summarised as follows. In developing countries the most common clinical presentation is mild watery diarrhoea, whereas in developed countries disease often manifests as a severe inflammatory diarrhoea. No evidence has yet been found to suggest that the watery type and severe bloody type of diarrhoeas can be explained in terms of a *C. jejuni* equivalent of the ETEC and EHEC mechanisms described above. Current thinking proposes that the different disease patterns reflect the immunological status of the host. Those with full immunity experience no clinical disease, whereas those with no pre-immunity experience the full-blown bloody diarrhoea and those with partial

Table 8.8. Types of intestinal infection

Types of infection	Microorganism	Disease
Microorganism attaches to epithelium of small intestine, rarely penetrates and causes disease (diarrhoea) often by forming a toxin(s) which induces fluid loss from epithelial cells	*Vibrio cholerae* *E. coli* (certain strains) *Giardia lamblia*	Cholera Infantile gastroenteritis (certain types) or mild cholera-like disease in adults (travellers' diarrhoea) Calf diarrhoea Giardiasis
Microorganism attaches to and penetrates epithelium of large intestine (*Shigella*) or ileum (*Salmonella*), causing disease by shedding/killing epithelial cells (exotoxin?) and inducing diarrhoea. Subepithelial penetration uncommon	*Shigella* spp. *Salmonella* (certain species)[a] *E. coli* (certain strains) *Campylobacter jejuni* Human diarrhoea viruses *Eimeria* spp. *Entamoeba histolytica*	Bacillary dysentery Salmonellosis Coliform enteritis or dysentery Piglet diarrhoea Diarrhoea, enteritis in man[b] Gastroenteritis Coccidiosis in domestic animals (may cause diarrhoea and blood loss) Amoebic dysentery
Microorganism attaches to and penetrates intestinal wall. Also invades subepithelial tissues, sometimes (typhoid, hepatitis A) spreading systemically	*Salmonella typhi* and *paratyphi* *Salmonella* (certain species) *E. coli* (certain strains) Hepatitis A virus Reoviruses, enteroviruses	Enteric fever (typhoid) Salmonellosis (severe form) Calf enteritis Varied Hepatitis

[a] There are more than 1000 serotypes of *Salmonella*, distinct from *Salmonella typhi* and *Salmonella paratyphi*. They are primarily parasites of animals, ranging from pythons to elephants, and their importance for man is their great tendency to colonise domestic animals. Pigs and poultry are commonly affected, and human disease follows the consumption of contaminated meat or eggs.
[b] Other campylobacters cause sepsis, abortion and enteritis in animals.

immunity, watery diarrhoea. The incubation period can range from 1 to 7 days and acute diarrhoea can last for 1–2 days with abdominal pain which may persist after diarrhoea has stopped. Diarrhoeal stools often contain fresh blood, mucus and an inflammatory exudate with leucocytes; bacteremia may also occur though it is rarely reported. Infected mucosae may be oedematous and hyperaemic with petechial haemorrhages. The disease, even its severe form, tends to be self-limiting, despite the fact that organisms may be isolated for several weeks after resolution of the symptoms. We do, however, know that there is a strong correlation between infection with *C. jejuni* and Guillain–Barré syndrome which is the most notable complication of *C. jejuni* infection. Guillain–Barré syndrome is a peripheral neuropathy, and one possible cause may be an autoimmune phenomenon arising from molecular

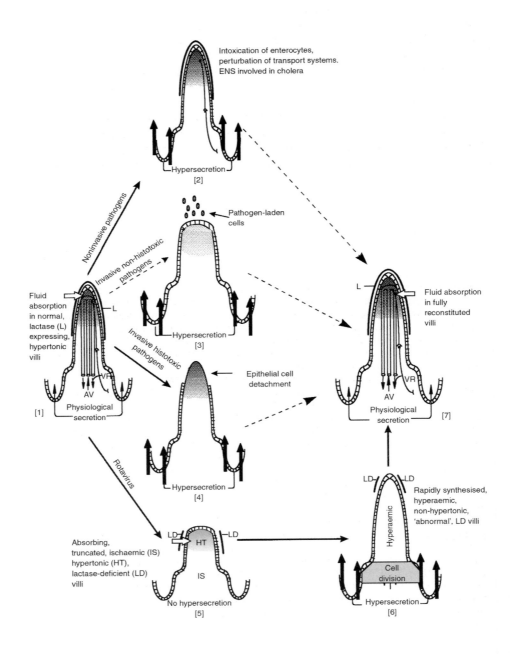

mimicry between the polysaccharide side chains of *C. jejuni* and neural gangliosides.*

While there are reasonable models for studying colonisation and initial invasion, there is a problem regarding experimental animal models in which to reproduce the extreme form of bloody diarrhoea seen in humans. However, the situation concerning *C. jejuni* is probably about to change dramatically. New strategies based on the use of the new technology of 'microarrays'† are now being used. By this means, and by reference to the genomic atlas, it is theoretically possible to identify which genes are expressed under different sets of experimental conditions including those which mimic the infection environment. Doubtless a plethora of new data is about to be generated from which we hope to learn more of the disease-conferring attributes of *C. jejuni* and related species.

Rotaviruses are known to invade intestinal epithelial cells and cause diarrhoea in man, foals, dogs, pigs, mice, etc. Extensive multiplication takes place and very large amounts of virus (10^{11} particles g^{-1}) are shed in faeces. The conventional wisdom is that tips of villi especially are

---

* Guillain–Barré syndrome is also associated with certain virus infections, and 'flu vaccination (see Ch. 12).

† Microarrays: see Ch. 1.

Fig. 8.20 Diarrhoeal mechanisms: initial stages and (for rotavirus) some intermediate stages in disease progression. This represents a schematic summary of the text on diarrhoeal mechanisms. In all cases, broken arrows indicate uncertainty about the number and nature of intermediate steps in the return to normality of affected villi in self-limiting diarrhoeal disease. For clarity, the blood supply in [2] and both blood supply and enteric nervous system (ENS) in [3], [4], [5] and [6] have been omitted.
[1] represents a normal villus; the shading intensity (as in Fig. 8.19) represents the magnitude of osmolarity. [2] Intoxication of villi by noninvasive pathogens such as *V. cholerae* and ETEC. The main diarrhoeal determinant is CT in *V. cholerae* and LT and ST in ETEC. However, as discussed in the text, toxins are not the whole story, hence the broken arrows. [3] Represents disease caused by invasive pathogens such as nonhistotoxic *S. typhimurium* and rotavirus. Villi are shortened with presumed loss of absorption and observed increase in secretion. Again the mechanistic pathway for return to normality is not known for bacterial infections. [4] Loss of epithelia due to a histotoxin seen in some strains of *S. typhimurium*. Clearly loss of enterocytes will affect absorption and open up other routes for progressive invasion. Again note the broken arrow. [5] A more complete experimentally based understanding of the pathophysiological mechanisms is possible in rotavirus infection of neonatal mice ([5], [6] and [7]). The main point is that conventional wisdom is not sustained: maximum diarrhoea occurred during the resynthesis of truncated villi and villus shortening was preceded/caused by ischaemia. Prolongation of diarrhoea coincided with non-hypertonic villi; diarrhoea ceased on reconstitution of hypertonic villus tip regions. It is possible to infer that some of these intermediate steps take place in other gut infections.

affected, leading to reduced absorption of fluid from the lumen. In addition destruction of enterocytes leads to a loss in lactase resulting in an accumulation of lactose in the gut causing an osmotic flux of fluid into the intestine. A major study of rotavirus-induced diarrhoea in neonatal mice provides a different model of this important disease of children. The main features of this model are summarised in Fig. 8.20. Oral infection of the gut induces ischaemia in villi, followed by hypoxia, enterocyte damage, and shortening of villi. The perception is that it is the induction of ischaemia and not viral replication *per se* that results in these changes. It is during rapid resynthesis of the atrophied villi that maximum diarrhoea occurs due to the transient accumulation of excess NaCl in dividing cells. Prolongation of diarrhoea is seen to be due to the hyperaemic state of the newly reconstructed villi which reduces the hypertonicity of villi. Resolution of the diarrhoea occurs when microcirculation is restored to normal with concomitant restoration of hypertonic tip zones in villi.

The preceding description of the self-limiting diarrhoea induced by rotavirus in neonatal mice is that of a basic response probably applicable to many diarrhoeas since the features of the post-peak phase have often been reported or can be inferred in other infections. However, the observed pathology will be different according to age, host species, or the inducing pathogen. For example, in rotavirus-infected lambs, villus atrophy and crypt hypertrophy occur (the latter indicative of crypt cell division) but as in mice, infected lambs are not lactose intolerant. In rotavirus-infected swine piglets, crypt hypertrophy occurs but villus atrophy is severe, the animals are lactose intolerant and mortality is high; a similar situation exists for the coronavirus, transmissible gastroenteritis (TGE) virus of swine. The latter has often been used as the model for infantile diarrhoea but the question is whether human infants are more like piglets or lambs. Clinical studies have shown that recovery from *mild*, acute gastroenteritis of rotavirus origin occurs within 2 weeks irrespective of the carbohydrate ingested. Clearly, the severity of disease and the clinical outcome will depend on the extent of 'vertical' villus/crypt involvement and the regions of intestine infected. When villus erosion is severe, then lactose may cause an 'osmotic' purge or be fermented by intestinal bacteria to short-chain fatty acids which stimulate secretion in the colon. Astroviruses, Norwalk virus, caliciviruses and certain adenoviruses all cause gastroenteritic disease by infecting enterocytes. However, parvoviruses cause severe intestinal disease in dogs by virtue of their predilection for the mitotically active crypt cells which is the cause of the near-complete erosion of villi similar to that seen after exposure to sublethal doses of irradiation.

Can we be more specific about the viral determinants responsible for triggering these complex host reactions? It has recently been shown that a non-structural rotavirus protein, NSP4, induces diarrhoea in mice when introduced into the ileum, by causing increased Cl$^-$ secretion. An apparent exception to the 'rule' that viruses do not form toxins!

Entamoeba histolytica causes lysis of target cells apparently by direct contact with the cell membrane. This pathogen produces under *in vitro* conditions a spectacular array of potential (but as yet unproven) virulence determinants including: proteases that round up cells, pore-forming proteins, collagenases and oligosaccharidases and neurotransmitter-like compounds; the latter can induce intestinal fluid secretion. Some of these factors have been implicated as the determinants responsible for liver abscess formation.

Although much research has been focused on toxins, their mode of action, and their role in disease, it is useful to compare different types of intestinal infection and to refer to the concept of *food poisoning*. Types of intestinal infection are set out in Table 8.8. Food poisoning is a loosely used term, and usually refers to illnesses caused by preformed toxins in food, or sometimes to illnesses that come on within a day or so after eating contaminated food. Food may be contaminated with plant poisons, fungal poisons (e.g. poisoning due to *Amanita phalloides*), fish poisons,* heavy metals, as well as with bacterial toxins or bacteria.

References

Alouf, J. E. and Freer, J. (eds) (1999). 'The Comprehensive Sourcebook of Bacterial Protein Toxins'. Academic Press, London. (This excellent book contains up to date treatments on many of the protein toxin topics alluded to in Ch. 8 of this book.)

Borriello, S. P. (1998). Pathogenesis of Clostridium difficile infection. *J. Antimicrob. Chemother.* **41**, 13–19.

Buchmeier, M. J. *et al.* (1980). The virology and immunology of lymphocytic choriomeningitis virus infection. *Adv. Immunol.* **30**, 275–331.

Burke, B. and Desselberger, U. (1996). Rotavirus pathogenicity. *Virology* **218**, 299–305.

Casali, P. and Oldstone, M. B. A. (1983). Immune complexes in viral infection. *Curr. Topics Microb. Immunol.* **104**, 7–48.

Dale, J. B. and Beachey, E. H. (1985). Epitopes of streptococcal M proteins shared with cardiac myosin. *J. Exp. Med.* **162**, 583–591.

Fitzgerald, T. J. (1981). Pathogenesis and immunology of *Treponema pallidum*. *Annu. Rev. Microbiol.* **35**, 29–54.

Fleischer, B. and Hartwig, U. (1992). T-lymphocyte stimulation by microbial antigens. *In* 'Biological Significance of Superantigens' (B. Fleischer, ed.). *Chem. Immunol.* Vol. 55, pp. 36–64. Karger, Basel.

---

* Ingestion of scombroid fish (mackerel, etc.) containing large amounts of histamine or similar substances leads to headache, flushing, nausea and vomiting within an hour.

Fontaine, A., Arondel, J. and Sansonetti, P. J. (1988). Role of Shiga toxin in the pathogenesis of bacillary dysentery, studied by using a Tox-mutant of *Shigella dysenteriae*-1. *Infect. Immun.* **56**, 3099–3109.

Hamilton, P. J. *et al.* (1977). Disseminated intravascular coagulation: A review. *J. Clin. Path.* **31**, 609–619.

Hirst, T. R. (1999). Cholera toxin and *Escherichia coli* heat labile enterotoxin. *In* 'The Comprehensive Sourcebook of Bacterial Protein Toxins' (J. E. Alouf and J. Freer, eds), pp. 104–129. Academic Press, London.

Hormaeche, C. E., Penn, C. W. and Smyth, C. J. (eds) (1992). 'Molecular Biology of Bacterial Infection. Current Status and Future Perspectives'. Soc. Gen. Microbiology Symposium 49, Cambridge University Press, Cambridge.

Hornick, R. B. *et al.* (1970). Typhoid fever: pathogenesis and immuno-logic control. *N. Engl. J. Med.* **283**, 739.

Kaper, J. B., Morris, J. G. and Levine, M. M. (1995) Cholera. *Clin. Microbiol. Rev.* **8**, 48–86.

Karaolis, D. K. R., Somara, S., Maneval, D. R., Johnson, J. A. and Kaper, J. B. (1999). A bacteriophage encoding a pathogenicity island, a type-IV pilus and a phage receptor in cholera bacteria. *Nature* **399**, 375–379.

Ketley, J. M. (1997). Pathogenesis of enteric infection by *Campylobacter*. *Microbiology* **143**, 5–21.

Khan, S. A. *et al.* (1998). A lethal role for lipid A in *Salmonella* infec-tions. *Molec. Microbiol.* **29**, 571–579.

Laforce, F. M. (1994). Anthrax. *Clin. Infect. Dis.* **19**, 1009–1114.

Levin, J., van Deventer, S. J. H., van der Poll, T. and Sturk, A. (eds) (1994). 'Bacterial Endotoxins. Basic Science to Anti-Sepsis Strategies'. Progress in Clinical and Biological Research. Vol. 388, John Wiley & Sons Inc., New York.

Levin, J., Alving, C. R., Munford, R. S. and Redl, H. (eds). (1995). 'Bacterial Endotoxins. Lipopolysaccharides From Genes to Therapy'. Progress in Clinical and Biological Research, Vol. 392, John Wiley & Sons Inc., New York.

Lencer, W. I., Hirst, T. R. and Holmes, R. K. (1999). Membrane traffic and the cellular uptake of cholera toxin. *Biochim. Biophys. Acta Molec. Cell Res.* **1450**, 177–190.

Lodge, J. M., Bolton, A. J., Martin, G. D., Osborne, M. P., Ketley, J. M. and Stephen, J. (1999). A histotoxin produced by *Salmonella*. *J. Med. Microbiol.* **48**, 811–818.

Lundgren, O. and Jodal, M. (1997). The enteric nervous system and cholera toxin-induced secretion. *Comp. Biochem. Physiol. A* **118**, 319–327.

Mathan, M. M., Chandy, G. and Mathan, V. I. (1995). Ultrastructural changes in the upper small intestinal mucosa in patients with cholera. *Gastroenterology* **109**, 422–430.

McGee, Z. A. *et al.* (1981). Pathogenic mechanism of *Neisseria gonor-*

rhoeae: observations on damage to human fallopian tubes in organ cultures by gonococci of colony Type I or Type 4. *J. Infect. Dis.* **143**, 413–422; 432–439.

Mims, C. A. (1957). Rift Valley Fever virus in mice VI: Histological changes in the liver in relation to virus multiplication. *Austral. J. Exp. Biol. Med. Sci.* **35**, 595.

Mims, C. A. (1985). Viral aetiology of diseases of obscure origin. *Brit. Med. Bull.* **41**, 63–69.

Nataro, J. P. and Kaper, J. B. (1998). Diarrheagenic *Escherichia coli*. *Clin. Microbiol. Rev.* **11**, 142–201.

Olsnes, S., Wesche, J. and Falnes, P. Ø. (1999). Binding, uptake, routing and translocation of toxins with intracellular sites of action. *In* 'The Comprehensive Sourcebook of Bacterial Protein Toxins' (J. E. Alouf and J. Freer, eds), pp. 73–93. Academic Press, London.

Osborne, M. P., Haddon, S. J., Worton, K.J., Spencer, A. J., Starkey, W. G., Thornber, D. and Stephen, J. (1991). Rotavirus-induced changes in the microcirculation of intestinal villi of neonatal mice in relation to the induction and persistence of diarrhea. *J. Pediatr. Gastroenterol. Nutr.* **12**, 111–120.

Poewe, W., Schelosky, L., Kleedorfer, B., Heinen, F., Wagner, M. and Deuschl, G. (1992). Treatment of spasmodic torticollis with local injections of botulinum toxin. *J. Neurol.* **239**, 21–25.

Raudin, J. I. (1986). Pathogenesis of diseases caused by *Entamoeba histolytica*: studies of adherence, secreted toxins and contact-dependent cytolysis. *Rev. Infect. Dis.* **8**, 247–260.

Rodriguez, M., von Wedel, R. J., Garrett, R. S. *et al.* (1983). Pituitary dwarfism in mice persistently infected with lymphocytic choriomeningitis virus. *Lab. Invest.* **49**, 48.

Schiavo, G., Benfenati, F., Poulain, B., Rossetto, O., de Laureto, P. P., DasGupta, B. R. and Montecucco, C. (1992). Tetanus and botulinum-B neurotoxins block neurotransmitter release by proteolytic cleavage of synaptobrevin. *Nature* **359**, 832–833.

Schiavo, G., Poulain, B., Rossetto, O., Benfenati, F., Tauc, L. and Montecucco, C. (1992). Tetanus toxin is a zinc protein and its inhibition of neurotransmitter release and protease activity depend on zinc. *EMBO J.* **11**, 3577–3583.

Silva, T. M. J., Schleupner, M. A., Tacket, C. O., Steiner, T. S., Kaper, J. B., Edelman, R. and Guerrant, R. L. (1996). New evidence for an inflammatory component in diarrhea caused by selected new, live attenuated cholera vaccines and by El Tor and O139 *Vibrio cholerae*. *Infect. Immun.* **64**, 2362–2364.

Spencer, A. J., Osborne, M. P., Haddon, S. J., Collins, J., Starkey, W. G., Candy, D. C. A. and Stephen, J. (1990). X-ray-microanalysis of rotavirus-infected mouse intestine – a new concept of diarrheal secretion. *J. Pediatr. Gastroenterol. Nutr.* **10**, 516–529.

Stephen, J. (2000). Pathogenesis of infectious diarrhea: A minireview. *Can. J. Gastroenterol.* (in press).

Svanborg, C., Godaly, G. and Hedlund, M. (1999). Cytokine responses during mucosal infections: role in disease pathogenesis and host defence. *Curr. Opin. Microbiol.* **2**, 99–105.

Uchida, H., Kiyokawa, N., Horie, H., Fujimoto, J. and Takeda, T. (1999). The detection of Shiga toxins in the kidney of a patient with hemolytic uremic syndrome. *Pediatr. Res.* **45**, 133–137.

VanderSpeck, J. C. and Murphy, J. R. (1999). Diphtheria toxin-based interlukin-2 fusion proteins. *In* 'The Comprehensive Sourcebook of Bacterial Protein Toxins' (J. E. Alouf, and J. Freer, eds), pp. 682–690. Academic Press, London.

Welliver, R. C. *et al.* (1981). The development of respiratory syncytial virus-specific IgE and the release of histamine in naso-pharyngeal secretions after infection. *N. Engl. J. Med.* **305**, 841–845.

Williams, R. C. (1981). Immune complexes in human diseases. *Annu. Rev. Med.* **32**, 13–28.

Yuki, N. (1999). Pathogenesis of Guillain-Barre and Miller Fisher syndromes subsequent to *Campylobacter jejuni* enteritis. *Jap. J. Infect. Dis.* **52**, 99–105.

9

Recovery from Infection

If there is to be recovery from an infection, it is first necessary that the multiplication of the infectious agent is brought under control. The microbe must decrease in numbers and cease to spread through the body or cause progressive damage. This is accomplished by immunological and other factors whose action is now to be described. The average multiplication rate of various microorganisms in the infected host as shown by doubling times (Table 8.2), is nearly always longer than in artificial culture under optimal conditions. This in itself reflects the operation of antimicrobial forces. In the process of recovery from an infectious disease, damaged tissues must of course be repaired and reconstituted. Sometimes the microorganism is completely destroyed and tissues sterilised, but often this fails to take place and the microorganism persists in the body, in some instances continuing to cause minor pathological changes. The individual is nevertheless said to have recovered from the acute infection and is usually resistant to re-infection with the same microorganism. Persistent infections are dealt with in Ch. 10.

Immunological Factors in Recovery

The mechanisms of recovery from a primary infection are not necessarily the same as those responsible for resistance to re-infection (see below). For instance, antibody to measles is of prime importance in resistance to re-infection and susceptible children can be passively

307

protected by the antibody present in pooled normal human serum. But, compared with cell-mediated immunity (CMI), antibody plays only a small part in the recovery from initial infection with measles virus. Antibody, T cells, NK cells, complement, phagocytes and interferon are involved in the response to nearly all infections and, without any doubt, are together responsible for recovery. They constitute a mighty antimicrobial force, whose action is illustrated diagramatically in Fig. 9.1. In only a few instances, however, have the different components of this mighty force been dissected out and separately evaluated. All components normally operate together, and to some extent the attempt to make separate evaluations is rather like deciding on the relative importance of 2, 3 and 4 in producing the product 24. For instance, polymorphs play a vital role in recovery from many bacterial infections, but they necessarily operate in conjunction with antibody and complement. Macrophages play a vital role both in the induction and in the expression of CMI responses (see below). There are naturally occurring diseases in which one component is deficient or absent, and sometimes a component can be eliminated experimentally. More commonly the deficiency diseases are mixed in type, and experimentally it is often impossible to eliminate one component without also affecting the others. Indeed, when one component is eliminated its job may be taken over by others.

One major difficulty in assessing the importance of immune

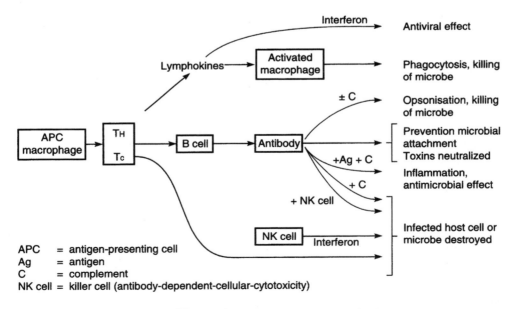

Fig. 9.1 Immune responses in infection.

responses is that nearly all microorganisms are very complex, with large numbers of antigens. Various tests for antibody and T cells are carried out, but it is not always possible to test the response to a defined antigen, or to know precisely which antigens are important for infection and pathogenicity.

Antibody

The different types of antibody and the ways in which they have an antimicrobial action are listed in Ch. 6. Antibody actions against microorganisms are further discussed at the end of this chapter under 'Resistance to Re-infection'.

In some infections antibody plays a major part in the process of recovery (see Table 9.1). For instance, viruses producing systemic disease, with a plasma viraemia (see p. 130), are controlled primarily by circulating antibody. This seems to be so in yellow fever or poliomyelitis virus infections. Children with severe hypogammaglobulinaemia are unable to form antibodies to poliovirus, and are about 10 000 times more likely than normal individuals to develop paralytic disease (which is generally of a chronic type) after live virus vaccina-

Table 9.1. Antibody and CMI in resistance to systemic infections[a]

Type of resistance	Antibody	CMI

Recovery from primary infection	Yellow fever
	Polioviruses
	Coxsackie viruses
		Poxviruses
		  e.g. ectromelia (mice),
		    vaccinia (man)
		Herpes simplex
		Varicella-zoster
		Cytomegalovirus

	Streptococci Staphylococci *Neisseria meningitidis* *Haemophilus influenzae* Malaria? *Candida* spp. *Giardia lamblia*	LCM virus (mice) Measles Tuberculosis Leprosy Typhoid Systemic fungal infections Chronic mucocutaneous candidiasis?
Resistance to re-infection	Nearly all viruses including measles Most bacteria	Tuberculosis Leprosy
Resistance to reactivation of latent infection	Herpes simplex?	Varicella-zoster Cytomegalovirus Tuberculosis *Pneumocystis carinii*

[a] Either antibody or CMI is known to be the major factor in the examples given. But in many other infections there is no information, and sometimes both types of immunity are important.

tion.* They have normal CMI and interferon responses, normal phagocytic cells and complement, but lack the specific antibody which must be produced if virus multiplication and spread to the central nervous system (CNS) are to be inhibited.

Antibody on its own can neutralise virus infectivity. Such neutralising antibodies attach to specific neutralisation sites on surface proteins (poliovirus has four and influenza virus has five neutralisation sites) and only through these sites is the virus neutralised. Antibodies do attach to other sites but these are non-neutralising. Neutralising antibodies do not need to coat the virion and rarely inhibit attachment of the virion to the host cell receptor. In fact most kinetics of neutralisation are single hit, meaning that one virus particle is neutralised by one molecule of antibody. How can this be when an antibody (IgG) molecule (molecular wt 150 000) is less than the size of one of the hundreds of surface proteins of an enveloped virus like influenza? It is thought that antibodies usually act by interfering with uncoating, either by triggering a stage of uncoating prematurely or by preventing uncoating by cross-linking surface structures on the virion. For example, some antibodies inhibit fusion (the first stage of uncoating). Still others inhibit later stages of uncoating from taking place. How they do so is not understood. Neutralising antibodies can be of the IgG, IgA or IgM immunoglobulin isotypes.

Antibody also promotes the uptake and digestion of virus by phagocytic cells, so that the virus–antibody complex is finally taken up and disposed of. Antibodies also act against viruses by clumping them, by destroying them with the help of complement, or by inducing inflammatory responses following their interaction with viral antigens (see Ch. 6).

Various bacteria have been shown to make specific attachments to epithelial surfaces and here secretory IgA antibodies are significant. IgA antibodies are formed in most infections of mucosal surfaces whether bacterial, viral or due to other microorganisms. They tend to prevent re-infection, but if formed early enough in the primary infection they could block the attachment of the microorganism to susceptible cells or cell surfaces (see Table 2.1) and thus interfere with the spread of infection. Their actual function in recovery, however, is doubtful. As was pointed out earlier, virus infections that are limited to epithelial surfaces and do not have a time-consuming spread of infection through the body, have incubation periods of no more than a few days. There is little opportunity for the slowly evolving immune response to play an important role in recovery, and virus replication is

---

* Agammaglobulinaemics are also susceptible to pneumococcal infections. Theoretically, opsonisation of these bacteria should occur after activation of the alternative complement pathway (Ch. 6), but antibody appears to be needed for optimal uptake and killing by phagocytes. Antibody may also be needed for lysis of virus-infected host cells after complement pathway activation (see pp. 323–4).

often inhibited before there has been a detectable IgA response. On the other hand, it must be remembered that antibodies (IgG or IgA) can be produced locally within 2 days after experimental respiratory tract infections, for instance, and they would not be detected routinely when bound to viral antigens at this stage. But interferon is produced by the first infected cell, and is likely to have an important local antiviral action. If the process of infection takes longer, then secretory IgA antibodies have more opportunity to aid recovery. When the intestinal protozoan *Giardia lamblia* causes symptoms, these are not seen until 6–15 days after infection. A role for secretory IgA antibodies is indicated because patients with a shortage of these antibodies show troublesome and persistent giardial infection.

Quite clearly, as discussed in Ch. 4, the antibody response to streptococci, staphylococci and various encapsulated bacteria such as the pneumococcus is of particular importance. These are the common pyogenic (pus-forming) infections. For its antibacterial function, antibody needs to operate together with phagocytic cells and complement and, if either of these are missing, resistance to pyogenic infections is impaired. Children with agammaglobulinaemia suffer repeated infections with pyogenic bacteria.* The spleen is an important site of antibody formation, and when the spleen has been removed surgically, or rendered incompetent in children with sickle cell disease (see p. 367), there is increased susceptibility to such infections. On the other hand, many bacterial infections (tuberculosis, syphilis, typhoid, gonorrhoea) can persist or can re-infect in spite of the presence of large amounts of antibody. This is discussed more fully in Ch. 7, and it is a reminder of the frequent inability of antibodies to ensure recovery.

Antibodies are vital in recovery from diseases caused by toxins, such as diphtheria and tetanus. As soon as antibodies have been formed to neutralise the powerful toxins and prevent further tissue damage, recovery is possible; without antibodies the other antibacterial forces may operate in vain. In diphtheria the patient often recovers and is immune to the toxin without having controlled the infection itself, and remains a carrier.

Circulating antibodies are probably important in the recovery from infection with certain protozoa such as malaria. Here, in particular, antibody must be directed against the relevant stage of the microorganism (especially the merozoite) and also against the relevant antigen on the microorganism. Merozoites are the forms that specifically absorb to red blood cells and parasitise them, and protective antibodies coat the merozoite surface and inhibit this absorption, at the same time promoting phagocytosis by the reticuloendothelial system.

---

* Infants with congenital (inherited) forms of agammaglobulinaemia remain well until about 9 months of age because the gift of maternal IgG via the placenta gives passive protection during this period.

Host defences against fungi are less clearly defined but there are indications that CMI is more important than antibody. Disseminated infection with certain fungi (*Coccidioides*, *Histoplasma*) occurs even in the presence of high antibody titres, and in such cases there is usually no CMI demonstrable by skin tests (p. 170), suggesting that T-cell responses matter most. Local infections with fungi elicit good CMI responses but poor antibody responses, and the patient recovers. Severe mucocutaneous candidiasis is seen in those with defective CMI, in spite of normal antibody production.

Small microorganisms such as viruses may have no more than one (human immunodeficiency virus; HIV) or two different proteins on their surface. The surface of influenza virus, for instance, consists of 500 or more haemagglutinin trimers, interlaced with about 100 neuraminidase tetramers. Antibodies to either antigen protect against infection, although the haemagglutinin is the major neutralisation antigen. Antibody to the neuraminidase inhibits its enzymic activity. It does not prevent infection of the cell, but prevents the dissemination of newly formed virus, and thus hinders the spread of infection. This occurs because neuraminidase is required to digest sialic acid receptors on the cell from which it has just emerged, and thus prevent the virus from attaching to the cell from which it has just emerged. One can begin to work out the mechanisms of antibody protection in a relatively simple microorganism of this sort.* Larger microorganisms, however, generally have many different proteins and carbohydrates on their surface. Some of these will be concerned with vital steps in the process of infection, and antibodies to specific neutralisation sites on these will be protective. Antibodies to other antigenic sites on these structures and even to some complete structures will not be protective, and when they are attached to the microbial surface may even physically interfere with (block) the action of protective antibodies. In addition, a large assortment of irrelevant antibodies are produced to internal components of the microorganism. Antibodies themselves differ in the firmness of the combination they make with antigens and may be of high or low avidity (see Glossary). Thus the quality of the antibody also matters. Protection by antibody is therefore a complicated matter, and if there is no protection in spite of the presence of large amounts of antibody, one has to ask first what components of the microbe these antibodies are combining with, and whether these antibodies have the relevant specificity for the job. Second, one needs to ask whether the antibody itself is of sufficient quality, and of the appropriate isotype.

---

* The haemagglutinin of influenza virus was the first envelope protein for which a three-dimensional structure was determined. Analysis with monoclonal antibodies defined five antigenic sites which mediate neutralisation, and their physical location. These are also the sites which undergo antigenic drift (p. 209).

Cell-mediated immunity

There is good evidence that T-cell-mediated immunity is of supreme importance in recovery from a variety of microbial infections. These tend to be infections in which the microorganism replicates intracellularly (see Table 9.1). Tissue responses in the host bear the hallmarks of T-cell involvement, the infiltrating cells consisting primarily of lymphocytes and macrophages. Macrophages are often infected. Infections of this nature include tuberculosis, brucellosis, listeriosis, tularaemia, syphilis, tuberculoid leprosy (p. 320) and leishmaniasis. In *Leishmania* infection, recovery is associated with the development of a Th1 response. This is orchestrated by the production of interleukin-12 (IL-12) from infected macrophages which acts on either natural killer (NK) cells or CD4 T cells to produce interferon-γ (IFN-γ) and tumour necrosis factor (TNF), which in turn feeds back on macrophages to induce nitric oxide, an important molecule in controlling this parasite. The blockade of IL-12 activity *in vivo*, either by neutralising antibodies or the use of IL-12-deficient mice, leads to the development of a Th2 response which fails to protect the host from a generalised parasite infection, through a lack of nitric oxide production. Similar mechanisms operate in recovery from *Listeria* infection, illustrating the central role of IL-12 and IFN-γ in the evolution of Th1-protective immune responses. In some situations persistence of antigen, as with *M. tuberculosis*, can lead to protracted Th1 responses resulting in chronic inflammation. These responses are characteristic of delayed-type hypersensitivity which can be demonstrated in a specific manner by skin testing.

As pointed out earlier, CMI develops in many other infections but is not very clearly associated with recovery. On infection with *Streptococcus pyogenes*, for instance, delayed hypersensitivity to the streptococcal products streptokinase and streptodornase develops, but it is less important than antibody in recovery from infection. An interesting distinction can be made between different types of infectious agent and the immune strategy most likely to be effective (Table 9.2).

The clearest picture about CMI in recovery comes from certain virus infections, particularly herpes viruses, poxviruses, influenza virus, and it is first necessary to refer to the salient features of these infections which make the CMI response important. Antibodies neutralise free virus particles liberated from cells, but, despite the help of complement and of phagocytes with Fc receptors, often fail to influence events in infected cells. Action on the infected cell seems necessary for recovery from the above virus infections. The destruction of cells infected with viruses takes place in various ways, but depends on the mechanism of virus maturation in the cell. Many viruses, such as poliovirus or papilloma viruses, replicate and produce fully infectious particles inside the cytoplasm. These particles are nucleocapsids, consisting of the nucleic acid with its protein coat (capsid), and are exposed to antibody when

Table 9.2. Immune defences appropriate to different types of infection

Type of infectious agent	Primary immune defence	Immune mechanism	Additional immune defences	Examples
1. Multiplies inside tissue cells	Antibody rarely inhibits virus attachment but prevents entry or a later stage of infection. Entry of other microbes inhibited by coating with antibody and/or complement	Antibody production (IgG, IgA, IgM)	Kill infected cell[a]	Most viruses Rickettsias, malarial merozoites
2. Multiplies inside phagocytes	Activate phagocytes and thus render them resistant to infection	T cells generate cytokines	Kill infected[a] phagocyte	Certain viruses *Mycobacterium tuberculosis Leishmania*, trypanosomes (see Table 4.2)
3. Multiplies outside cells	Kill microbe extracellularly or intracellularly	Complement-mediated lysis[b] Opsonised phagocytosis and killing	Neutralise microbial toxins	Most bacteria Trypanosomes
4. Multiplies outside cells, but attachment to body surface necessary for invasion	Prevent attachment by coating microbial surface with specific antibody	Antibody production (mainly secretory IgA)	As under 3	Streptococci *Neisseriae E. coli*, etc. (see Table 2.1, pp. 14–18)

[a] Mechanisms are antibody complement, antibody-dependent cell cytotoxicity (ADCC), cytotoxic T cell, or natural killer cell (see Ch. 6).
[b] NK cells (e.g. in *Toxoplasma gondii* infection) or T cells may also have a role.

liberated from the cell (Fig. 9.2). Other viruses do not have to wait for cell disruption, but are liberated by a process of budding from the cell membrane. The viral nucleocapsid in the cytoplasm becomes closely associated with the cell membrane and acts as an *initiation* point for the viral envelope proteins (Fig. 9.2). The virus particle finally matures by budding through the altered cell membrane, acquiring an envelope as it does so. Such viruses are referred to as 'enveloped' and include HIV, herpes viruses, myxo- and paramyxo-viruses, etc. (pp. 422–423). There are two important consequences of this mechanism of virus maturation. First, virus can be released even though the cell remains

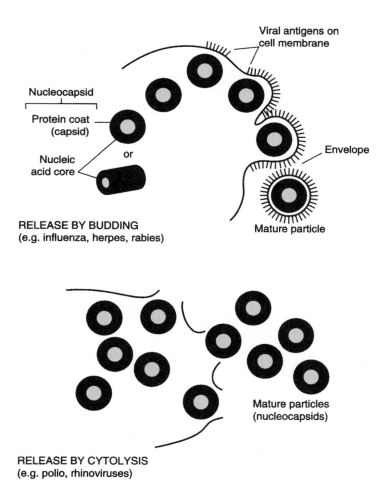

Fig. 9.2 Diagram to illustrate mechanisms of virus release from infected cell. Nucleocapsids may be spherical (herpes viruses) or tubular (influenza). Budding may also take place from nuclear membrane (herpes viruses) or from the membrane lining cytoplasmic vacuoles (coronaviruses, flaviviruses).

alive and intact. Second, the foreign viral antigens appearing on the cell surface are recognised by the host and an immune response is generated with the infected cell as the target. The significance of this is that the infected cell can be destroyed before virus has been liberated,* and can also be destroyed in oncogenic and other virus infections in which virus is liberated from the cell over long periods.

Viral antigens are also formed on the cell surface during the replication of certain nonenveloped viruses such as adenoviruses. The surface antigens are not incorporated into the virus particle, but the infected cell bearing the antigens can be recognised and destroyed by immune mechanisms. Also, although destruction of infected host cells has long been considered a feature of viral rather than other infections, there is now evidence for this occurring with other infections. Host cells infected *in vitro* with protozoa (*Plasmodia* and *Theileria*), with rickettsia (*Coxiella burnetii*) and with certain bacteria (*Listeria*) express microbial antigens on their surface and are thus vulnerable to immune lysis. This is achieved as a natural consequence of degradation of intracellular antigens by the proteolytic machinery of the infected cell and the presentation of antigenic peptides by MHC class I. Consequently, whether a cell is infected by an enveloped or nonenveloped virus, a bacterium or a parasite, antigens can be recognised by cytotoxic T cells.

The immune mechanisms for the destruction of cells bearing foreign antigens on their surface can be summarised as 'burns', 'pores' or 'poisons' and are as follows:

1. As mentioned above (Ch. 6) T cell receptors only recognise peptides in association with MHC class I proteins (CD8 T cells) or class II proteins (CD4 T cells) on the target cell surface. Any viral protein can be processed in this way, and usually internal virion proteins or nonvirion (e.g. nuclear-transcription factors) proteins provide the major target for T cells. In order to destroy a target cell, T cells must become activated. Once this is achieved the activated cytotoxic T cell (CTL) makes intimate contact with the target cell membrane and delivers a lethal hit. The T cell then disengages and moves to another target. Killing of target cells occurs by one of two mechanisms. One mechanism involves releasing the contents of cytotoxic granules containing perforin,† which deposits 'pores' in the

---

* Viral antigens often appear on the cell surface very early in the replication process, many hours before progeny virus particles have been formed. These antigens can be identified on staining with fluorescent antibody. As many as ten distinct viral antigens (glycoproteins) appear on the cell surface in the case of a large virus such as herpes simplex.

† Perforin is homologous to C9, the pore-forming component of complement. Both proteins polymerise on contact with the target cell breaching the membrane and producing pores through which electrolytes and other molecules flow causing cell damage.

membrane of the target cell, and granzyme B, which enters the cell through the pores to act as a 'poison' in triggering apoptosis. In contrast to apoptosis, cellular necrosis may occur (possibly as a result of large amounts of perforin being deposited) in which there is a leakage of cell components and K$^+$ ions, an influx of water and Na$^+$ ions, and the target cell swells up and dies. Natural killer cells also use the perforin lytic mechanism to kill target cells. A second method that triggers cell death is the interaction of Fas (a TNF-like receptor on target cells) with the Fas ligand (on T cells), said to be the 'kiss of death', due to the activation of the 'death' domain in the cytoplasmic tail of Fas. This results in the initiation of a cascade of cellular proteases leading to apoptosis.

2. Macrophages, polymorphs and NK cells have the ability to destroy target cells with the assistance of specific antibody (ADCC). Antibody combines with antigen on the infected cell surface, and the killer cell attaches to the antibody-coated cells via the Fc receptor. The process is enhanced when complement is activated, the C3b molecules deposited on the cell surface being recognised by mononuclear and phagocytic cells that bear C3b receptors. The final killing mechanism is not clear, but the killer cell releases oxygen radicals and hydrogen peroxide (see p. 92) which 'burn' the target cell. The Fc receptors for IgG and IgE on eosinophils enable them to kill multicellular parasites such as schistosomes (see pp. 88–89) after adhering in large numbers to the antibody-coated surface of these parasites. This involves releasing toxic proteins (e.g. major basic protein and eosinophil cationic protein) directly onto the parasite surface to 'burn' holes in the tegument enabling eosinophils to enter the parasite to deliver the *coup de grace*. These proteins are so toxic they can also damage mammalian cells, so the eosinophil carefully seals the area on the parasite where the proteins are delivered. Why the eosinophil is not destroyed is a mystery.

Destruction of infected cells is not the only mechanism available to T cells in controlling a virus infection. In hepatitis B virus infection of the liver and in herpes simplex virus infection of neurons, CD8 T cells prefer to 'cure' the infection rather than kill the cells. This has been demonstrated in a mouse model of hepatitis B virus infection in which every hepatocyte becomes infected. By delivering immune T cells to these mice the infection is readily controlled, but widespread destruction of hepatocytes is not observed. The key protective mechanism is IFN-γ, released by the activated CD8 T cells, which blocks virus replication and rids the cells of the viral genome. A similar mechanism operates in hepatitis B virus (HBV) infected chimpanzees, the other natural host for this virus. In neurons infected with herpes simplex virus, it is likely that similar CD8 T cell control mechanisms occur, since neurons expressing late virus proteins (an indicator of the late stages of virus replication) can be prevented from cell death and 'cured' of this productive infection. However, the virus may persist in a latent form. This strategy benefits the virus in terms of its survival as a latent infection and also the host in retaining the function of these irreplaceable cells. A decision on whether HBV or herpes virus infected cells are killed or cured could be related to the amount of MHC class I expression on the infected cell. When MHC expression is high these cells can be targeted by cytotoxic T cells and killed. When MHC expression is low, cytotoxic T cells may have difficulty directly engaging the target cell, but can still influence virus replication through the local release of IFN-γ. This further illustrates the diversity of anti-viral mechanisms at the disposal of the host.

The sequence of events with herpes virus, poxvirus and measles virus infections appears to be as follows (as outlined in Ch. 6). At sites of virus multiplication, T lymphocytes, in the course of their normal movements through the body encounter viral peptides that are complexed with MHC proteins on the surface of a dendritic cell or other antigen-presenting cell. When a T cell encounters the antigen to which it is specifically sensitised, it becomes activated and divides to give fresh supplies of specifically sensitised T cells. These can react with any cell presenting the relevant peptide in association with a MHC molecule. Cytokines are liberated to attract macrophages and other leucocytes and focus them onto the site of infection. Infected cells are destroyed or cured by cytotoxic T lymphocytes and other cells, and virus material and cell debris is phagocytosed and disposed of by activated macrophages. Similar events occur in lymph nodes to which virus or virus antigens have been brought by lymphatic drainage.

The best way of discovering the function of a bodily mechanism or organ is to see what happens when it is removed (see also agammaglobulinaemia, p. 310). In experimental infections, CMI can be inactivated without affecting antibody or interferon responses, and changes in the disease are then studied.

A defined depletion of T cells can be achieved by treatment with monoclonal antibodies specific for CD4 or CD8 proteins. This powerful approach enables T cells or other cells to be depleted at any stage in the immune response to infective agent. An alternative method is to use transgenic 'knockout' mice in which the gene encoding the protein of interest is inactivated at the DNA level. This powerful technology enables selected defects in host defence to occur, resulting in deficiencies of e.g. IFN-γ, TNF, IL-2, IL-4, IL-10, or their receptors; CD8 and CD4 T-cell function (disrupt CD8 or CD4 genes); and B-cell function (disrupt expression of IgM). Most of these gene knockout mice develop normally and remain well, but show increased susceptibility to intracellular infections caused by various viruses, bacteria or protozoa. But the picture is complex. Deleting one cytokine or cell function upsets a delicate network of antimicrobial forces. Often a different defence mechanism takes over the function of the one that has been deleted. In other words, there is a redundancy in host defence mechanisms, as might be expected as an evolutionary response to infectious agents that often evade or interfere with these mechanisms.

At the clinical level, albeit without the precise focusing achieved in the knockout mice, evidence for the importance of CMI in the control of infections comes from studies on patients with defective CMI. Very rarely, infants are born with an absent or poorly developed thymus gland (thymic aplasia or hypoplasia). Their T lymphocytes fail to differentiate and develop, giving rise to severe CMI deficiency. Although their T-cell-dependent antibody response is also defective, they make a normal T-cell-independent antibody response (mainly IgM) (see Ch. 6 and Glossary). Thymic aplasia, although so rare,* is a 'pure' deficit and gives some insight into the importance of CMI in infectious diseases. Affected infants show normal ability to control most bacterial infections, but a greatly increased susceptibility to infections with various viruses and certain other intracellular microorganisms. After measles infection, for instance, there is no rash, but an uncontrolled and progressive growth of virus in the respiratory tract, leading to fatal giant cell pneumonia. Evidently the CMI response controls the infectious process and at the same time plays a vital role in the development of skin lesions. In the days when affected children were vaccinated against smallpox with vaccinia virus, the virus grew as usual in epidermal cells at the inoculation site to give an increasing zone of skin destruction. In normal infants there was an inflammatory response at the edges of the lesion after 6–8 days and this led to inhibition of virus growth, then scabbing and healing of the lesion. The infant with thymic aplasia, however, did not show this response and the destructive skin lesion continued to enlarge, occupying an ever-increasing area of the arm and shoulder. The infection could be controlled by local injection of immune lymphocytes from a closely related donor, but not by antibody. Infants with this type of immune deficiency also tend to suffer severe generalised infections with herpes simplex virus. In addition they show increased susceptibility to other intracellular microorganisms. When they are vaccinated against tuberculosis with live BCG vaccine, the attenuated bacteria, instead of undergoing limited growth with induction of a good CMI response,

---

* Mixed antibody and CMI deficiencies are commoner, with defects in CMI, antibody response and phagocytic cells. The infants suffer from superficial *Candida albicans* infections, and commonly die with *Pneumocystis carinii* (see Glossary), pneumonia or generalised infections with vaccinia, varicella or measles. Immune deficiencies (mostly CMI) are seen in adults with Hodgkin's disease, or after immunosuppression for organ transplants. In these patients, who have encountered the common infections in early life, there is reactivation of persistent infections such as varicella-zoster, tuberculosis, herpes simplex, cytomegalovirus, warts and *Pneumocystis carinii* infection. Gram-negative bacterial pneumonia may also occur. In many immunodeficiencies the primary defect is unknown. Patients lacking adenosine deaminase, an enzyme in the purine salvage pathway, develop a severe combined immunodeficiency (SCID). Their lymphocytes fail to mature, macrophage activation is defective, and they die early as a result of infection unless given a bone marrow transplant.

multiply in an uncontrolled fashion and may eventually kill the patient. The CMI response is therefore necessary for the control of infection with intracellular bacteria of this type.

There are two more clinical examples of the importance of CMI in recovery from nonviral intracellular infections. Leprosy, caused by *Mycobacterium leprae*, exists in a spectrum of clinical forms. At one end of the spectrum is tuberculoid leprosy. Here, the infection is kept under some degree of control, with infiltrations of lymphocytes and macrophages into infected areas such as the nasal mucous membranes. In the lesions there are very few bacteria and all the signs of a strong CMI response. Injection of lepromin (leprosy antigens) into the skin of the infected patient gives a rather slowly evolving but strong delayed hypersensitivity reaction. This type of leprosy is called tuberculoid leprosy because the host response is similar to that in tuberculosis. At the other end of the leprosy spectrum* (lepromatous leprosy) there are very few lymphocytes or macrophages in the lesions and large numbers of extracellular bacteria. Associated with these appearances indicating less effective control of the infection, there is a weak or absent skin response to lepromin. Lepromin is a crude bacterial extract and the exact antigens to which the lepromatous patient fails to respond are not known.† Antibodies are formed in larger amounts than in tuberculoid patients and indeed may give rise to immune complex phenomena in lepromatous patients (see Ch. 8), but these antibodies fail to control the infection. The second clinical example of the importance of CMI in recovery concerns the disease chronic mucocutaneous candidiasis. Children with immunodeficiency disease sometimes develop severe and generalised skin lesions caused by the normally harmless fungus *Candida albicans*. Antibody to *Candida* is formed but the CMI response is often inadequate, and these patients can be cured by supplying the missing CMI response. This can be given in the form of repeated injections of a mysterious 'transfer factor' (see Glossary). The CMI response may have additional antimicrobial effects in chronic infections with certain intracellular organisms. When the microorganism persists as a source of antigenic stimulation and the CMI-induced influx of mononuclear cells continues, a granuloma may be formed (see below). The focus of infection tends to be walled off, and this is often associated with the inhibition of microbial growth. Granulomas are a feature of respiratory tuberculosis, contributing to pulmonary fibrosis. Granulomas, however, can result from chronic accumulation of immune complexes as well as from chronic local CMI reactions (see Ch. 8).

* To some extent there is also a clinical spectrum in tuberculosis, according to the type of immune response, some (more susceptible) patients showing strong antibody responses and weak cell-mediated immune responses.

† Lymphocytes from patients show normal transformation responses to other mycobacteria such as BCG and *Mycobacterium lepraemurium* (see Ch. 7).

Phagocytosis

Phagocytes play a central role in resistance to and recovery from infectious diseases. In the old days physicians saw the formation of pus (see p. 94) as a valiant attempt to control infection, and referred to it as 'laudable pus', especially when it was thick and creamy. An account of the antimicrobial functions of phagocytes and the consequences of phagocyte defects is given in Ch. 4.

Inflammation

Inflammation, whether induced by immunological reactions, tissue damage or microbial products plays a vital role in recovery from infection (see also Chs 3 and 6). Inflammation is necessary for the proper functioning of the immune defences because it focuses all circulating antimicrobial factors onto the site of infection. The circulating antimicrobial forces that arrive in tissues include polymorphs, macrophages, lymphocytes, antibodies, activated complement components, and materials like fibrin that play a part in certain infections.* The increased blood supply and temperature in inflamed tissues favour maximal metabolic activity on the part of leucocytes, and the slight lowering of pH tends to inhibit the multiplication of many extracellular microorganisms. The prompt increase in circulating polymorphs during pyogenic infections is caused in the first place by the release of cells held in reserve in the bone marrow, but there is also an increase in the rate of production. Monocyte release and production is controlled independently. At least four colony-stimulating factors, all glycoproteins, control the mitosis of polymorph and macrophage precursors, and their final differentiation and activity. They are present in increased amounts in serum during infection, and in animals the serum levels are dramatically raised by the injection of endotoxin.

Circulating polymorphs show increased functional activity during pyogenic infections and readily take up and reduce a certain yellow dye (nitroblue-tetrazolium), forming dark blue deposits in the cytoplasm. An increase in the proportion of polymorphs showing this reaction reflects their increased activity, but the test is of no value in the diagnosis of pyogenic infections because of false-positive and false-negative

---

* There have been numerous reports of antimicrobial factors present in normal serum. Doubtless some of these involved alternative pathway activation of complement, as in the case of pathogenic *Neisseria* (see below). Trypanocidal factors in normal human serum may be related to natural resistance to trypanosomiasis. It has been known since 1902 that *Trypanosoma brucei*, which is not infectious for man, is lysed by something present in normal human serum, whereas *Trypanosoma rhodesiensi* and *Trypanosoma gambiensi*, which infect man and cause sleeping sickness, are relatively resistant. The trypanocidal factor has been shown to be a high-density lipoprotein.

results. In any case, increased reduction of the dye is not necessarily associated with increased bactericidal activity.

When inflammation becomes severe or widespread, there is a general body response with the appearance of acute-phase proteins in the blood (see p. 78). As a result two classical changes can be detected in the blood. The first is an increase in the erythrocyte sedimentation rate (ESR), and this is a clinically useful indication that inflammation or tissue destruction is occurring somewhere in the body. The exact mechanism of the increase is not understood. The second change is the appearance in the blood of increased quantities of a β-globulin synthesised in the liver and detected by its precipitation after the addition of the C carbohydrate of the pneumococcus. It is therefore called C-reactive protein.* Very small amounts are present in the blood of normal individuals, but there is a 1000-fold increase within 24 h of the onset of inflammation. After binding to substances derived from microorganisms and from damaged host cells, it activates the complement system, acts as an opsonin, and possibly serves a useful function. Both the ESR and C-reactive protein changes are nonspecific sequelae to inflammation of any sort, whether infectious or noninfectious.

When the infection is persistent, inflammation may become chronic, lasting weeks or months. Infections do not generally last for long periods if they induce acute polymorphonuclear inflammation; the battle between host and microbe is decided at an early stage.† Chronic inflammation depends on a constant leakage of microbial products and antigens from the site of infection. The type of infection that persists and causes chronic inflammation is generally an intracellular bacterial or fungal or chlamydial infection. In these infections there is a chronic CMI response, with proliferation of lymphocytes and fibroblasts in infected areas, a steady influx of macrophages, and the formation of giant and epithelioid cells.‡ Episodes of tissue necrosis alternate with repair and the formation of granulation tissue, then fibrous tissue. It is a ding-dong battle between microorganisms and host antimicrobial forces. The resulting granuloma (see also above) can be regarded as an attempt to wall off the infected area. Chronic infections with chronic inflammation and granuloma formation include tuberculosis, syphilis, actinomycosis, leprosy, lymphogranuloma inguinale and coccidiodomycosis. Chronic viral infections are not associated with chronic inflammatory responses, probably because virus growth is often defective and no more than minute amounts of antigen are liberated.

---

* C-reactive protein and other acute-phase proteins are formed as a result of the action of mediators such as IL-1, IL-6 and TNF.

† Occasionally this is not so, and there is continued polymorph infiltration, as for instance in chronic osteomyelitis or a pilonidal sinus.

‡ Epithelioid cells are poorly phagocytic, highly secretory, and about 20 μm in diameter, whereas giant cells, formed by the fusion of macrophages, are up to 300 μm in diameter, containing up to 30 nuclei.

Complement

Complement has been discussed and invoked on many occasions in Chs 6–8 and in this chapter. It should be remembered that some of the complement components are quite large molecules, and do not readily leave the circulation except where there is local inflammation. Complement can carry out antimicrobial activities in the following ways.

1. *Complement lysis*. Complement reacts with antibody (IgG and IgM) that has attached to the surface of infected cells or to the surface of certain microorganisms, and destroys the cell or microorganism after making holes in the surface membrane. Gram-negative bacteria are killed in this way, and also enveloped viruses such as rubella and parainfluenza (although as mentioned above lack of complement does not exacerbate these virus infections). Because of the amplification occurring in the complement system (see Ch. 6), especially when the alternative pathway is also activated, antibody attached to the surface of a microorganism is more likely to induce complement lysis than it is to neutralise it. Complement lysis is therefore perhaps more important when antibody molecules are in short supply, early in the immune response. Bacteria with surface polysaccharide components can activate complement without the need for antibody (see 5, below), as can host cells infected with viruses such as measles. In the latter case alternative pathway activation by itself does not do enough damage to kill the cell. Presumably, less severe membrane lesions can be repaired; antibody as well as complement must be present for lysis.

2. *Complement opsonisation*. Complement reacts with antibody attached to the surface of microorganisms, providing additional receptor sites for phagocytosis by cells bearing the appropriate complement receptors, like polymorphs or macrophages. Phagocytosis is also promoted by antibody attached to the microorganism because of the Fc receptors on phagocytes, but when complement is activated there are many more molecules of C3b present as a result of the amplification phenomenon. Therefore complement often has a more pronounced opsonising effect than antibody alone and for some bacteria, such as the pneumococcus, opsonisation actually depends on complement. Complement opsonisation is important when the antibody is IgM, because human phagocytes do not have receptors for the Fc region of IgM. Complement can also act as an opsonin though not always so effectively, in the absence of antibody (see 5, below).

3. *Complement-mediated inflammation*. Specific antibodies react with microbial antigens that are either free or on the surface of microorganisms. Following this antigen–antibody reaction, complement is activated, with generation of inflammatory and chemotactic factors (C3a and C5a). These substances focus antimicrobial serum factors and leucocytes onto the site of infection.

4. *Complement-assisted neutralisation of viruses*. In the case of
viruses coated with antibody, complement adds to the mass of
molecules on the virus surface and may hinder attachment of virus
to susceptible cells. In some situations complement can mediate
neutralisation of virus coated with a non-neutralising antibody.
This will depend on the antibody isotype and presumably the
density of antibody on the surface of the virus. Some viruses
(murine leukaemia virus, Sindbis virus) can directly activate the
complement system by interaction of virion envelope proteins with
C1q or C3, resulting in the neutralisation of infectivity.

5. *Complement-assisted cell lysis*. C3b deposition on infected host cells
not only opsonises (see above) but also augments cell-mediated
cytotoxicity (ADCC, see p. 165) and antibody-dependent lysis of
cells. The latter is generally regarded as an inefficient process
requiring around 10^5 antibody molecules to kill measles virus
infected cells.

6. *Complement opsonisation via alternative pathway*. Complement
reacting with endotoxin on the surface of Gram-negative bacilli,
with capsular polysaccharide of pneumococci, etc., or with *Candida*,
is activated via the alternative pathway (see p. 177) and C3b-medi-
ated opsonisation takes place. It seems likely that this is important
in natural resistance to infection.

Unfortunately, there is little direct evidence that the above antimicrobial activities of complement are in fact important in the body. The rare
patients with C3 deficiency develop repeated pyogenic infections, and
C3-deficient mice show increased susceptibility to plague and to
staphylococcal infections. Mice with C5 deficiency (controlled by a
single gene) are more susceptible to *Candida* infection, probably
because of inadequate opsonisation. Patients with C5–C8 deficiencies,
however, are often particularly susceptible to disseminated or recurrent neisserial infection. In this case the bactericidal rather than the
opsonising action of complement seems important. But observations on
complement deficiencies are probably too limited to draw firm conclusions and there have been few clearly defined deficiencies. The system
is a highly complex one, with alternative pathways, positive feedback
amplification and multiple inhibitors. A similar complement system
occurs in a wide range of vertebrates and it must be assumed that such
a complex, powerful system confers some biological advantage,
presumably by giving resistance to microbial infections.

Interferons

The interferons are cytokines, members of a family of cell-regulatory
proteins produced by all vertebrates. There are three types of inter-

feron: alpha (α), beta (β) and gamma (γ). Alpha and beta interferons are very similar, and are made by nearly all cells in the body, including epithelial cells, neurons, muscle cells, etc. in response to viral and other infections (bacteria, mycoplasma, protozoa). There are 12 human alpha interferon genes and one beta interferon gene encoded on the short arm of chromosome 9. Gamma interferon is produced by NK cells and by T lymphocytes following antigen-specific stimulation. Only one gene exists for gamma interferon, encoded on chromosome 12. All are cytokines with immunoregulatory functions as well as the antimicrobial action described below.

Viruses are the most important inducers of interferon-(INF) α and -β, the stimulus to the cell being the double-stranded RNA formed during virus replication (Fig. 9.3). Interferons act on uninfected cells, binding to a cell surface receptor and activating a number of genes involved in immunity to viruses. Some of these gene products (2′5′A synthetase/RNAseL, Mx protein) target viral messenger RNA and others

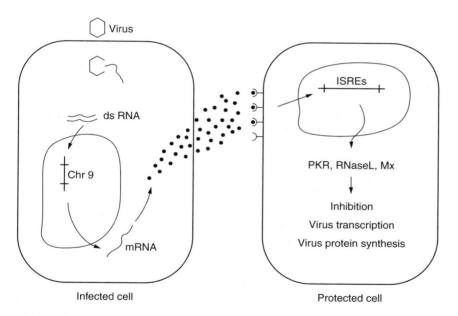

Fig. 9.3 Mechanism of induction and expression of α and β interferon. Virus infection results in dsRNA intermediates initiating interferon gene expression on chromosome 9 (man). Interferon produced attaches to a receptor on neighbouring uninfected cells and initiates gene expression via transcription factors binding to interferon response elements (ISREs – nucleotide elements associated with a number genes, including immune response genes) leading to expression of PKR, RNaseL, Mx, etc. These proteins inhibit any virus replication by attacking virus RNA or protein synthesis machinery. Interferons also increase NK cell activity. Interferon-γ produced by T cells and NK cells functions in a similar way in protecting uninfected cells.

(PKR) inactivate polypeptide chain elongation, blocking viral protein synthesis. Interferons are exceedingly potent in vitro, being active at about 10^{-15}M. They have no direct action on virus itself and do not interfere with viral entry into the susceptible cell. The interferons produced by different species of animals are to a large extent species specific in their action. Interferon liberated from infected cells can reach other cells in the vicinity by diffusion and establishes an antiviral state which protects them from infection.* A cell is thus protected from infection with all viruses for a period of up to 24 h. It would seem inevitable that interferon is important in recovery from virus infections, whether on epithelial surfaces or in solid tissues. Interferons, moreover, have other effects on host resistance. They activate NK cells and control T-cell activity by upregulating the expression of MHC proteins and thus the concentration of available peptide antigen.

It is now clear that interferon plays a central role in limiting virus infection *in vivo*. Evidence for this comes from various experimental approaches. The most dramatic are mice lacking the receptor for IFN-α/β (deleted by transgenic 'knockout' technology) which means they are unable to respond to interferon produced during a virus infection. Such animals are highly susceptible to many virus infections, including herpes simplex, MHV-68, Semliki forest virus. Alpha/beta interferon can be selectively inhibited in mice by treatment with antibody to interferon. When this is done, enhanced susceptibility to certain virus infections is observed. Interferon has also been given passively to experimental animals and can be effectively induced by the administration of a synthetic ds-RNA preparation (poly I : poly C). Antiviral results are demonstrable in experimental infections, and are most clearly seen in infections of epithelial surfaces such as the conjunctiva or respiratory tract, and when treatment is begun before rather than after infection.

In humans, naturally occurring deficiencies in IFN are rare, partly because, for IFN-α at least, there are so many different genes involved. A study of 30 children who suffered from recurrent respiratory tract infections identified four with impaired interferon production. When these particular children were infected with common cold viruses, IFN-α could not be detected in nasal washings. Their peripheral blood leucocytes also failed to produce IFN-α on repeated testing *in vitro*, although INF-γ production was normal.

Interferon would seem to be the ideal antiviral chemotherapeutic agent for use in man, being produced naturally by human cells, nonimmunogenic and active against a broad spectrum of viruses. However, it does cause influenza-like symptoms. So far, results in human patients have not been dramatic. For instance, volunteers infected intranasally

---

* Interferon is also induced by nonviral agents such as rickettsiae and certain bacteria, and will protect cells from various nonviral intracellular microorganisms.

with rhinoviruses and other respiratory viruses have been given either poly I : poly C or repeated very large doses of purified human interferon by the same route, but with only slight protection. However, it has proved useful in clearing up some cases of chronic hepatitis B infection, and is being used to treat hepatitis caused by the flavivirus, hepatitis C. HBV downregulates the expression of MHC class I proteins on infected hepatocytes, thus preventing CD8 T cells from destroying infected cells. By treating with IFN-γ or IFN-α, expression of MHC class I proteins is upregulated and CTLs can act. Initially the patient may become ill from the effects of interferon, but eventually, a virus-free liver regenerates. Here interferon is exercising its regulatory function as well as its antiviral effect. A strain of mice can be created in which the IFN-γ gene or its receptor has been inactivated or 'knocked-out' transgenically. In the absence of pathogens, mice developed normally, but they were more susceptible to the intracellular bacteria *Mycobacterium bovis* and *Listeria monocytogenes* and to vaccinia virus (but not to influenza virus). The multiplicity of effects of this interferon was demonstrated by impairment in these mice of the functions of macrophages and NK cells, reduction of macrophage MHC class II proteins, uncontrolled proliferation of splenocytes, and a reduction in the amount of antigen-specific IgG2a. Further support for the importance of interferons in antimicrobial defences comes from the discovery that certain viruses have gene products that interfere with the antiviral action of interferon. Hepatitis B virus (via a domain in its polymerase) and adenovirus (via the E1A protein) block interferon-induced signalling, and adenoviruses and Epstein–Barr virus block activation of the interferon-induced RNA-dependent protein kinase. Poxviruses such as myxoma and vaccinia virus do it by encoding proteins that act as interferon receptors and thus neutralising interferon.

Multimechanistic Recovery: an Example

Although the host factors responsible for recovery have been described separately, they generally act together. Recovery is multimechanistic. As an example, Hormaeche's group in Newcastle have made extensive studies of the mechanisms involved in controlling *Salmonella* infections caused by *S. typhimurium* in mice, the most widely used model for typhoid-like disease caused by *S. typhi* in man. This embraces many of the features already dealt with in this chapter and anticipates some dealt with in Ch. 11. The system involves intravenous injection* of mice with organisms and, over a period of several days, estimation of

---

* The events after oral infection are more complex, and less well understood. There are the extra features of bacterial entry into gut epithelial cells and transit through Peyer's patches, mesenteric lymph nodes and lymphatics, before the blood is invaded.

the bacterial populations present in liver and spleen (the principal relevant components of the reticuloendothelial system (RES; see Glossary), whose mononuclear cells represent the main battleground in this infection). Sublethal infection proceeds in at least four distinct phases, schematically depicted in Fig. 9.4, and with some *Salmonella* this confers solid immunity to rechallenge.

Phase 1: initial inactivation of the inoculum. This is a constant finding representing the transition from the *in vitro* to the *in vivo* phenotype. The decline is due to immunologically nonspecific uptake and destruction in macrophages of the RES. It is enhanced when animals are pretreated with opsonising antibody, as would be expected (see Fig. 5.4).

Phase 2: exponential growth in the RES. This occurs during the first week with an estimated doubling time for *Salmonella* of *ca.* 2–5 h; killing rates are also slow. Three factors can affect phase 2. (i) Inoculum dose. By increasing the dose the pattern of phases 1 and 2 remains the same but raised to a higher level. When the inoculum reaches LD_{50} or higher (see Glossary) no phase 3 is observed; phase 2 continues till lethal numbers (10^8–10^9) are reached. With very high doses, the slope

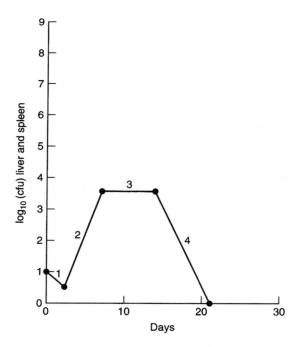

Fig. 9.4 The four phases of a sublethal *Salmonella* infection in mice. Phase 1: initial inactivation of a large fraction of the challenge inoculum. Phase 2: exponential growth in the RES over the first week. Phase 3: plateau phase in which growth is suppressed. Phase 4: clearance of the organisms from the RES. cfu, colony-forming units.

of phase steepens and the time to death shortens. (ii) Virulence of the bacteria. Increase in the slope of phase 2 is also a function of the virulence of the strain. (iii) Innate resistance of the host. A gene *ity* (immunity to typhimurium), expressed through macrophages controls phase 2. A similar situation exists for *Leishmania donovani, Mycobacterium tuberculosis* BCG, and *Mycobacterium lepraemurium*.

Phase 3 is essential for the host to survive. It is not mediated by T cells but requires continued production of TNF-α, which stimulates the production of IFN-γ. Studies with the Listeria model indicate that TNF-α is produced by macrophages which stimulate NK cells to release IFN-γ, which in turn activates newly recruited macrophages. It is of interest that, during the plateau phase, mice show a manifest macrophage-mediated immunosuppression towards other antigens.

Phase 4 is the clearance phase which does require the presence of T cells, causing macrophage activation. CD8 as well as CD4 T cells are involved. Again host genes play an important role in this phase.

Temperature

In man the mean daily body temperature is 36.8°C with a daily variation of only 1.3°C, the maximum being at about 18.00 h, the minimum at about 03.00 h. This almost constant body temperature, like the almost constant level of blood sugar, illustrates Claude Bernard's dictum that '*La fixité du milieu intérieur est la condition de la vie libre*'. If the individual is to function steadily in spite of changes in the external environment, the internal environment must remain constant. The brain is one of the most sensitive parts of the body to departures from normality. At temperatures below 27.7–30°C people become unconscious, at 40.5°C or above they become disoriented and may be maniacal; above 43.3°C they are comatose. A rise in body temperature is one of the most frequent and familiar responses to infection, whether the infection is largely restricted to body surfaces (common cold, influenza) or is obviously generalised (measles, typhoid, malaria). During fever the appetite is often lost and headache may result from dilation of meningeal blood vessels. The temperature rise is largely due to an increase in heat production, and the raised metabolic rate, together with reduced food intake, results in a high excretion of nitrogen in the urine. There is rapid wasting of body fat and muscles if the fever is prolonged.

In infectious diseases there is a common mediator of the febrile response known as endogenous pyrogen (interleukin-1).* It is present

---

* In addition, IL-6 causes fever by acting on the hypothalamus, whereas TNF (in the LPS fever model in rats) tends to reduce an already elevated temperature

in inflammatory exudates and in the plasma during fever, and acts on the temperature-regulating centre in the anterior hypothalamus, resetting the body thermostat. Endogenous pyrogen is produced by macrophages and certain other cells, and as little as 30–50 ng causes fever in rabbits. It can be induced by immunological mechanisms. Fever is a common accompaniment of generalised antigen–antibody reactions. For instance, rabbits immunised with bovine serum albumin develop fever when injected with this antigen. Systemic virus infections such as the exanthems (see Glossary) are characterised by an asymptomatic incubation period during which virus replicates and spreads through the body, followed by a sudden onset of illness with fever. The febrile reaction is mainly due to the immune response to the virus; hence its relatively sudden onset a week or two after infection. The CMI response (see below) as well as the antibody response is involved. Antigen–antibody reactions, in addition to causing fever, can also give rashes, joint swelling and pain, even glomerulonephritis (see Ch. 8). The first signs of illness in hepatitis B, before jaundice, are often 'allergic' in nature and mediated by antigen–antibody interactions, with fever, joint pains and fleeting rashes.

The generalised CMI response in the infected host is also a cause of fever, e.g. in tuberculosis, brucellosis and perhaps staphylococcal and cryptococcal infections. Tuberculin added to alveolar macrophages from an immunised animal generates endogenous pyrogen (interleukin-1). Also patients with chronic brucellosis develop fever when injected with 10 g of purified brucella.

Certain bacterial products are pyrogenic. The peptidoglycan in the cell wall of staphylococci causes monocytes to liberate endogenous pyrogen. More importantly, endotoxins from Gram-negative bacteria also have this effect, as little as 2 ng of *Salmonella* endotoxin kg^{-1} causing fever in man. Endotoxin is present in the circulation during systemic infection with Gram-negative bacteria, but tolerance to endotoxin-induced fever develops quite rapidly, and endotoxin itself probably makes no more than a partial contribution to the febrile response, even in infections such as typhoid and dysentery. There is no good evidence that other microbial products or toxins cause fever other than by immunological mechanisms. In the old days before penicillin, pneumococcal pneumonia used to give one of the highest fevers known in man with dramatic and severe onset, the temperature often rising to 40°C within 12 h. These bacteria, however, have no endotoxin or other pyrogens and the mechanisms were presumably immunological.

When human volunteers were infected with influenza virus, those with the most IL-6 and IFN-γ in nasal fluids had higher temperatures, as well as more virus, more mucous production and more symptoms. In the case of influenza virus infection of ferrets, the best animal model for human influenza, there is a direct correlation between virulence and viral pyrogenicity. Endogenous pyrogen is released locally in the respiratory tract as a result of virus–phagocyte interaction. The haemagglutinin and/or neuraminidase surface glycoproteins of the virion are responsible.

Since fever is such a constant sequel of infection, it is natural to suppose that it has some antimicrobial function. Thomas Sydenham in the seventeenth century wrote that 'Fever is a mighty engine which nature brings into the world for the conquest of her enemies'. Bodily functions are profoundly disturbed by fever. Metabolic activity is increased in phagocytic cells, and studies *in vitro* show that there are large increases in T-cell proliferation and in antibody production at febrile temperatures. The evidence, however, is disappointing. Temperature-sensitive mutants of certain viruses are often less virulent, and experimental virus infections can sometimes be made more severe by preventing fever with antipyretic drugs. When fever is induced in infected animals by raising the environmental temperature, there are also other complex physiological changes, making it difficult to interpret such experiments. In two bacterial infections, gonorrhoea and syphilis, the microbes themselves are actually killed by febrile temperatures, but in the natural disease these temperatures are rarely reached. Before the introduction of antibiotics, patients with these two diseases were infected with malaria in order to induce body temperatures high enough to eradicate the infection (following which the malaria was treated with quinine).

If fever is of value to the host, one might expect microbes to attempt to prevent it. Vaccinia virus, which normally fails to cause fever in mice, produces a soluble receptor for IL-1b (the fever mediator) and virus strains lacking the gene for this receptor do cause fever. We may ask whether T pallidum actively inhibits the fever response.

Fever is costly in energy and is an ancient bodily response, having evolved with the vertebrates over hundreds of millions of years. Perhaps one day some more convincing evidence will emerge to give substance to Sydenham's eloquent convictions.

Tissue Repair

Once the multiplication of the infecting microorganism has been controlled, and the microorganism itself perhaps eliminated from the body, the next step in the process of recovery is to tidy up the debris and repair the damaged tissues. In other words pathogenesis is followed by 'pathoexodus'. Four examples will be given, in the skin, respiratory tract, liver and the foetus. At the molecular level a profusion of mediators are involved. Cytokines, because of their effects on cell growth and differentiation, play a part at all stages in the repair process.

In the skin

During recovery from a boil, for instance, the sequence of events is as follows. Superficial tissue debris, including necrotic epidermis, inflammatory cells and plasma exudate, dries off as a scab. This gives mechanical protection, acts as a barrier to further infection, and can be shed to the exterior after repair is completed. Below the scab, phagocytic cells clear up the debris and fibroblasts move in, multiply, and lay down a mucopolysaccharide matrix over the underlying intact tissues. New blood vessels are formed, and later on lymphatics, by sprouting of the endothelial cells of neighbouring vessels into the fibroblast matrix. The newly formed capillaries advance into the damaged zone at 0.1–0.6 mm a day. They are fragile and leaky, and there is a continuous extravasation of polymorphs, macrophages and fibroblasts into the matrix. As seen from the surface, each collection of capillary loops in the fibroblast matrix looks like a small red granule and this soft vascular material is therefore called granulation tissue. It bleeds easily and with its rich blood supply and abundant phagocytic cells is well protected against infection. Meanwhile, epidermal cells at the edges of the gap have been multiplying. The newly formed layer of cells creeps over the granulation tissue, and the epidermis is thus reconstituted. Fibroblasts in the granulation tissue lay down reticulum fibres, and later collagen. If the area of epidermal cell destruction is large, and when underlying sebaceous glands, hair follicles, etc. are destroyed, a great deal of collagenous fibrous tissue is formed to repair the gap. The newly formed collagen in fibrous tissue contracts and tends to bring the skin edges together. Contracting collagen can strangle an organ like the liver, but in the skin it merely forms a scar. A scar is a characteristic sequel to vaccination with BCG, or to a bacterial infection involving sebaceous glands, as seen in severe acne.

In the respiratory tract

After infection with a rhinovirus or influenza virus, there are large areas where the epithelial cells are destroyed, mucociliary transport is defective (see Ch. 2) and the underlying cells vulnerable to secondary bacterial infection. Phagocytic cells must now ingest and dispose of tissue debris, and the epithelial surface must be reconstituted by a burst of mitotic activity in adjacent epithelial cells. To some extent pre-existing cells can slide across the gap but repair depends on mitosis in cells at the edges. The process of repair takes several days, and the mechanism is the same whether the damage is caused by viruses, bacteria or chemicals. Epithelial regeneration is particularly rapid in respiratory epithelium, and also in conjunctiva, oropharynx and muco-cutaneous junctions, but it is delayed if the infection continues. After chronic bacterial or chemical damage there is an increase in mucus-producing goblet cells in the respiratory epithelium, and sometimes impairment of mucociliary mechanisms, resulting in the condition called chronic bronchitis. As a rule, however, recovery is complete.

In the liver

During recovery from focal hepatitis, polymorphs and macrophages are active in areas of tissue damage, phagocytosing dead and damaged hepatic cells, Kupffer cells, biliary epithelial cells, inflammatory cells and microorganisms. As this proceeds, neighbouring hepatic cells and bile duct epithelial cells divide to replace missing cells. This, together with cell movement and rearrangement, leads to remodelling of the lobules and the restoration of normal appearances. If supporting tissues have been significantly damaged, and particularly if there are repeated episodes of necrosis, healing involves scar formation. When this is widespread it is referred to as cirrhosis, the bands of fibrous tissue dividing up the organ into irregular islands. The regenerating islands enlarge to form nodules, the fibrous tissue thickens and contracts and there is obvious distortion of structure, with circulatory impairment, biliary obstruction and liver dysfunction.

In most tissues, repair with restoration of structural integrity can be achieved by fibrous tissue formation. Recovery of function depends more on the ability of differentiated cells in damaged tissues to increase their numbers again and thus restore functional integrity. Liver cells or epithelial cells have a great capacity for mitosis, and the intestinal epithelium, respiratory epithelium or liver can be restored to normal without great difficulty. In the case of cardiac muscle, striated muscle or brain, the differentiated cells show little if any mitotic capacity and destruction in these tissues results in a permanent deficit in the number of cells. This may be of no consequence in a muscle as long as firm scar tissue repairs the damage, but it may be important in the central nervous system. Anterior horn cells destroyed by poliovirus cannot be replaced, and if enough are destroyed there will be a permanent paralysis, although some restoration of function takes place by learning to use muscles more effectively and by the recovery of damaged anterior horn cells.

In the foetus

Tissue repair in the foetus is in some ways easier and in others more difficult. In general there is a very great capacity for repair and reconstitution of damaged tissues. Primitive mitotic cells abound, organs are in a state of plasticity, and in the developmental process itself tissue destruction and repair accompanies mitosis and construction. On the other hand, at critical times in foetal life, there is a programmed cell division and differentiation in the course of constructing certain major organs. If one of these organs is damaged at this critical time, the developmental process is upset and the organ is malformed. This is what happens when rubella virus infects the human foetus during the first three months of pregnancy. Depending on the exact organ system being formed at the time of foetal infection, there may be damage to the heart, eyes, ears or brain, resulting in congenital heart disease, cataract, deafness or mental retardation in the infant. Other infections (see Table 5.3) affect particularly the central nervous system of the foetus (toxoplasmosis, cytomegalovirus, syphilis) and sometimes bones and teeth (syphilis).

If the foetal infection is severe, as is the rule with vaccinia virus or with most bacteria, foetal death and abortion is the inevitable consequence. There are only a small number of microorganisms that infect the foetus and interfere with development without proving fatal. This type of nicely balanced pathogenicity is needed if the infected foetus is to survive and be born with a malformation. Even the infections that cause malformations (teratogenic infections) are sometimes severe enough to kill the foetus. In most congenital infections the microorganism remains present and is detectable in the newborn infant (cytomegalovirus, rubella, syphilis, etc.), often persisting for many years. It is a striking feature of most teratogenic foetal infections (rubella, cytomegalovirus, toxoplasmosis) that the mother suffers a very mild or completely inapparent infection.

Certain microorganisms infect the foetus and damage developing organs, but are then eliminated from the body. The damaged organs are formed as best as possible, and at birth there are no signs that the malformation was caused by a microorganism. Tissues are sterile, and no inflammatory responses are visible histologically. Thus when a pregnant hamster is infected with K virus (a polyomavirus), there is infection of the dividing cells that are to form the molecular layers constituting the bulk of the cerebellum. These cells are destroyed, the cerebellum therefore fails to develop normally, and the newborn hamster shows severe signs of cerebellar dysfunction, although it is perfectly well in every other way. The affected cerebellum is small and greatly depleted of cells, but there is no evidence of past microbial infection.

Resistance to Re-infection

Resistance to re-infection depends on the immune response generated during primary infection. Passive immunisation with antibody is known to protect humans against measles, hepatitis A, hepatitis B, rabies, etc., and the passively acquired (maternal) immunity of the newborn child or calf to a great variety of infections is another example of the resistance conferred by specific antibody. Most resistance to re-infection is antibody-mediated (Table 9.1). IgG antibodies generally continue to be formed in the body for many years following the initial infection; IgA antibodies are less persistent than IgG antibodies. Even if antibody levels have sunk to undetectable levels, memory cells from the initial infection are often present in large enough numbers to give an accelerated (anamnestic) response within a few days of re-infection. This is especially important in infectious diseases with incubation periods measured in weeks because there is time enough for the anamnestic response to operate and terminate the infection during the incubation period, before production of clinical disease. Sometimes resistance to re-infection is maintained by repeated subclinical infections, each of which boosts the immune response. For instance, children catching rubella at school can re-infect their immune parents subclinically and this is detected by a rise in antibody levels. Resistance to rubella, diphtheria and perhaps other infectious diseases is maintained in this way.

Antibodies protect against infection in a number of ways (see pp. 165–166). For instance, they attach to the microbial surface and promote its uptake by phagocytic cells, acting as opsonins. Other antibodies protect against reinfection by combining with the microbial surface and blocking attachment to susceptible cells or body surfaces. Microorganisms that need to make specific attachments are listed in Table 2.1. However, circulating IgG or IgM antibodies coat polioviruses, coxsackie viruses or adenoviruses, and act by interfering with viral uncoating (see above) rather than by blocking attachment to susceptible cells. Secretory IgA antibodies are particularly important because they can act on the microorganism before its attachment to a body surface. They do not act as opsonins; they do not lyse microorganisms because there is no complement on body surfaces, and in any case they fix complement poorly. But by preventing the attachment of microorganisms such as *Vibrio cholerae* to intestinal epithelium, the gonococcus to urethral epithelium, or *Chlamydia* to the conjunctiva, IgA antibodies can ensure that these microorganisms are carried away in fluid secretions rather than initiate infection. Acquired resistance to infection of the surface of the body is often of short duration. For instance, resistance to gonorrhoea or parainfluenza viruses following natural infection seems to last only for a month or so, and in childhood repeated infections with respiratory syncytial virus and *Mycoplasma pneumoniae* are common. Presumably the IgA antibodies that mediate resistance are short lived and IgA memory cells do not generate a good enough or rapid enough secondary response.

Resistance to re-infection, since it is immunological in nature, refers especially to the antigenic nature of the original infecting microorganism. Resistance to measles or mumps means resistance to measles or mumps wherever or whenever they occur, because these viruses are of only one type (monotypic) immunologically.* Resistance to the disease influenza or poliomyelitis, however, depends on the separate acquisition of resistance to a number of distinct antigenic types of influenza or polio viruses. Resistance to streptococci depends on the acquisition of antibodies to the M protein in the bacterial cell wall, and since there are at least 10 types of M protein that circulate quite commonly in communities (40–50 types of M protein altogether), repeated infections with *Streptococcus pyogenes* occur as antibodies are gradually developed against the various types. Often, however, different serological types of a given microorganism show some overlap so that antibodies to one type can confer partial resistance to another.

When resistance to a disease appears not to develop, the possibility of multiple antigenic types must be considered. There are multiple antigenically distinct types of gonococcus, for instance, as discussed on pp. 207–208, a fact that helps account for successive attacks of gonorrhoea. Numerous attacks of nonspecific urethritis are to be expected because of the variety of microorganisms that cause this condition. In one study, 40% of attacks were due to *Chlamydia*, but there are 12 known antigenic types.

Resistance to re-infection can also be mediated by CMI. The CMI response generated on primary infection lasts weeks or perhaps months rather than years, and there is an accelerated CMI response on re-infection, although less vigorous than in the case of antibodies. Nearly always a persistent infection is needed to give continued CMI resistance and infections showing this are usually intracellular in nature. For instance, resistance to re-infection with tuberculosis, syphilis and possibly malaria, depends on the active presence of the microorganism in the body, with continuous stimulation of the antibody and CMI responses. In most of these instances, resistance to re-infection is CMI-mediated. There are a few examples, however, such as measles, in which recovery from primary infection is largely due to CMI, but resistance to re-infection is attributable to antibody.

---

* Differences between geographical strains may be demonstrable on genetic analysis, but antibody to one strain generally neutralises other strains. If these differences become more marked, they may be significant, and we may need to know, for instance, whether current vaccines protect against all strains of hepatitis B virus or rabies virus. Still smaller differences are detectable between viruses isolated from different individuals in a given geographical region. The differences are minor and generally insignificant, but can be useful as markers for tracing the source of epidemics. Ultimately, because viruses are so mutable, even the progeny in a given infected host may not be uniform, and should perhaps be regarded as a virus population. This is particularly true in HIV infections.

References

Alluwaimi, A. M., Smith, H. and Sweet, C. (1994). Role of surface glycoproteins in influenza virus pyrogenicity. *J. Gen. Virol.* **74**, 2835–2840.

Baumann, H. and Gauldie, J. (1994). The acute phase response. *Immunol. Today* **15**, 74–80.

Bogdan, C., Gessner, A., Solbach, W. and Rollinghoff, M. (1996). Invasion, control and persistence of Leishmania parasites. *Curr. Opin. Immunol.* **8**, 517–525.

Chisari, F. V. and Ferrari, C. (1995) Immunopathogenesis of viral hepatitis. *Annu. Rev. Immunol.* **13**, 29–60.

Dimmock, N. J. (1993). Neutralization of animal viruses. *Curr. Topics Microbiol. Immunol.* **183**, 1–149.

Dinarello, C. A. (1996). Biologic basis for interleukin-1 in disease. *Blood* **87**, 2095–2147.

Hormaeche, C. E., Villarreal, B., Mastroeni, P., Dougan, G. and Chatfield, S. N. (1993). Immunity mechanisms in experimental salmonellosis. *In* 'Biology of *Salmonella*' (F. C. Cabello, C. E. Hormaeche, P. Mastroeni and L. Bonina, eds), pp. 223–235. NATO ASI Series, Series A: Life Sciences 245.

Jones, B. D. and Falkow, S. (1996). Salmonellosis: Host immune responses and bacterial determinants. *Annu. Rev. Immunol.* **14**, 533–561.

Kaufmann, S. H. E. and Ladel, C. H. (1994). Application of knockout mice to the experimental analysis of infection with bacteria and protozoa. *Trends Microbiol.* **2**, 235.

Mabruk, M. J. E. M. F., Flack, A. M., Glasgow, G. M., Smyth, J. M., Folan, J. C., Bannigan, J. G., O'Sullivan, M. A., Sheahan, B. J. and Atkins, G. J. (1988). Teratogenicity of the Semliki Forest virus mutant *ts*22 for the foetal mouse: induction of skeletal and skin defects. *J. Gen. Virol.* **69**, 2755–2762.

Morgan, B. P. and Walport, M. J. (1991). Complement deficiency and disease. *Immunol. Today* **12**, 301–306.

Ravetch, J. V. and Clynes, R. A. (1998). Divergent roles for Rc receptors and complement *in vivo*. *Annu. Rev. Immunol.* **16**, 421–432.

Review (1991). The biology of complement. *Immunol. Today* **12**, 291–342.

Reviews (1992). Cytokines in infectious diseases. *Immunol. Revs.* **127**.

Roberts, N. J. (1991). Impact of temperature elevation on immunological defences. *Rev. Infect. Dis.* **13**, 462.

Rogers, T. J. and Balish, E. (1980). Immunity to *Candida albicans*. *Microbiol. Rev.* **44**, 660–682.

Samuel, C. E. (1991). Antiviral actions of interferon: interferon-regulated cellular proteins and their surprisingly selective antiviral activities. *Virol.* **183**, 1–11.

Thomas, H. C. (1990). Management of chronic hepatitis virus infection.

In 'Control of Virus Diseases' (N. J. Dimmock, P. D. Griffiths and C. R. Madeley, eds), pp. 243–259. Soc. Gen. Microbiol. Symp. **45**, Cambridge University Press, Cambridge.

Unanue, E. R. (1997). Inter-relationship among macrophages, natural killer cells and neutrophils in early stages of *Listeria* resistance. *Curr. Opin. Immunol.* **9**, 35–43.

Van den Broek, M. F., Muller, U., Huang, S., Zinkernagel, R. M. and Aguet, M. (1995). Immune defence in mice lacking type I and/or type II interferon receptors. *Immunol. Rev.* **148**, 5–18.

Vilcek, J. and Sen, G. C. (1996). Interferons and other cytokines. In 'Fields Virology', 3rd edn (B. N. Fields, D. N. Knipe, P. M. Howley *et al.*, eds.) pp. 375–399. Lippincott-Raven, Philadelphia.

Wilson, I. A. and Cox, N. J. (1990). Structural basis of immune recognition of influenza virus hemagglutinin. *Annu. Rev. Immunol.* **8**, 737–771.

Wyatt, H. V. (1973). Poliomyelitis in hypogammaglobulinaemics. *J. Infect. Dis.* **128**, 802.

10

Failure to Eliminate Microbe

There are many infections in which the microorganism is not eliminated from the body, but persists in the host for months, years or a lifetime. Examples of persistent infections are given in Table 10.1. One way of looking at persistent infections is to regard them as failures of the host defence mechanisms which are designed to eliminate invading microorganisms from tissues. There are various ways in which the host defence mechanisms can fail and various methods by which the microbes can overcome them. Microbial adaptations to the encounter with the phagocytic cell are described in Ch. 4 and the ways in which the immune responses are by-passed are described in Ch. 7. Persistent infections usually represent a secondary event, following on from an initial acute infection. Primary infection with herpes simplex virus causes an acute stomatitis with herpetic (literally meaning creeping, and hence snake-like) lesions on the tongue which gives the virus its name, and this is followed by the persistent infection described below. Persistent infections are not usually significant causes of acute illness, but they are particularly important for five reasons:

1. They enable the infectious agent to persist in the community (see below);
2. They can be activated in immunosuppressed patients, and sometimes (e.g. herpes simplex) in normal people (latency);
3. Some are associated with immunopathological disease (see pp. 283–284);
4. Some are associated with neoplasms (see Table 8.1);

Table 10.1. Examples of persistent infections (mainly human)

Microorganism	Site of persistence	Consequence	Infectiousness of persistent microorganism	Shedding of microorganism to exterior
Viruses				
Herpes simplex	Dorsal root ganglia	Activation, cold sore	−	+
Varicella-zoster	Dorsal root ganglia	Activation, zoster	−	+
Cytomegalovirus	Lymphoid tissue	Activation ± disease	−	+
	Salivary glands	None known	+	+
HHV 6	Lymphoid system	None known	±	±
Epstein–Barr virus	Lymphoid tissue	Lymphoid tumour	−	−
	Epithelium	Nasopharyngeal carcinoma	++	−
	Salivary glands	None known	+	+
Hepatitis B	Liver (virus shed into blood)	Chronic hepatitis; liver cancer	+	+
Adenoviruses	Lymphoid tissue	None known	−	+
Polyomavirus (mice)	Kidney tubules	None known	+	+
Polyomaviruses	Kidney	Activation (pregnancy, immunosuppression)	−	+
Leukaemia viruses	Lymphoid and other tissues	Late leukaemia	±	−
Measles	Brain	Subacute sclerosing panencephalitis	±	−
HIV	Lymphocytes, macrophages	Chronic disease	+	+
Chlamydia				
Trachoma	Conjunctiva	Chronic disease and blindness	+	?
Psittacosis	Lung (rarely in man)	None known	?	−
	Spleen (of bird)	Activation	±	+
Rickettsia				
Rickettsia prowazeki	Lymph node	Activation	?	+
Rickettsia burneti (sheep)	Spleen ?	Activation; source of human Q fever	−	+

Bacteria

Salmonella typhi	Gall bladder	+	Intermittent shedding in urine, faeces	+
Mycobacterium tuberculosis	Lung or lymph node (macrophages?)	?	Activation, tuberculosis in middle age	+
Treponema pallidum	Disseminated	±	Chronic disease	−

Protozoa

Plasmodium vivax	Liver	?	Activation, clinical malaria	+
Toxoplasma gondii	Lymphoid tissue, muscle, brain	±	Activation, neurological disease	−
Trypanosoma cruzi	Blood, macrophages	±	Chronic disease	−

5. Some are immunosuppressive (human immunodeficiency virus; HIV) and permit disease caused by other normally harmless persistent microorganisms.

Persistent infections cannot by definition be acutely lethal; in fact they tend to cause only mild tissue damage or disease in the host. The mild diseases caused by persistent infections with adenoviruses, herpes viruses, typhoid or malaria can be contrasted with the serious diseases caused by the nonpersistent microorganisms of plague, cholera, yellow fever or paralytic poliomyelitis.

In certain acute infections the patient appears to recover, but there is later a relapse. Following typhoid, for instance, 8–10% of patients suffer relapses, although usually mild. In such instances the infection is not strictly persistent, but it seems as if the host's immune forces need repeated stimulation before there is complete elimination of the infectious agent.

Latency

Any form of persistent infection endows a microorganism with a greatly enhanced ability to remain in the host population as well as in the infected individual and latency, which means that the latent object is present but not apparent, represents an extreme manifestation of persistence. By definition, the microorganism is not detectable in infectious form during the latent period of infection.

The significance of persistence in latent form becomes clear when measles is compared with chickenpox. Measles is not normally a persistent infection, and after an individual has been infected and suffered the characteristic illness and rash, the immune response controls the infection and eliminates the virus from the body. Immunity to re-infection is lifelong, and a continued supply of fresh susceptible hosts must be found if the virus is to persist in the community. The virus does not survive for long outside the body, and therefore cannot persist in a community unless there is, at all times, someone actually infected with measles. Before the general use of measles vaccine, measles used to come to towns and cities every few years, infecting the susceptible children who had appeared since the last epidemic, and then disappearing again. The virus had to be re-introduced into the community at intervals because there were not enough susceptible children appearing to keep the infection going all the time. From studies of island communities it has been shown that the minimum population to maintain measles without introduction from the outside is about 500 000. Chickenpox, in contrast to measles, causes a persistent infection (see below). During childhood infection with chickenpox the virus ascends to the dorsal root ganglia of sensory nerves supplying the affected skin areas, and stays there in an essentially noninfectious state after recovery and elimination of virus from the rest of the body (Fig. 10.1). The disease, chickenpox, disappears temporarily from the community. Virus in the dorsal root ganglia is kept under control by CMI, but the strength of this CMI response weakens as individuals age, and there is an increasing likelihood in older people that latency will be broken and one of the ganglia will produce enough infectious virus to spread down the peripheral nerve to the skin. This disease is called zoster (shingles) and is characterised by a crop of vesicles which is restricted to the distribution of that particular nerve. The vesicles are rich in virus which is capable of causing chickenpox in any susceptible children who have appeared in the community. Studies of island communities have shown that chickenpox can maintain itself indefinitely in a community of less than 1000 individuals.* Chickenpox is described as a persistent infection characterised by latency; in fact the sequence of events is an acute infection, apparent recovery from the original infection (the latent phase), and later in life a second acute infection with disease as the virus reappears and is once again shed to the exterior.

Herpes simplex virus gives rise to an exactly comparable latent infection. Infection normally occurs during infancy or early childhood, and causes a mild acute illness with stomatitis and slight fever. The virus travels up the axons of sensory nerves to the trigeminal ganglion supplying the mouth and related areas, and after apparent recovery from the initial infection, virus remains latent in neurons in the ganglion (Fig. 10.2). At intervals later in life virus can be activated in the ganglion, travel down the nerve and cause a vesicular eruption, usually round the lips or nostrils. This is called a cold sore, and virus from the cold sore can infect a susceptible individual. Certain individuals (about 10% of the population in the UK) are particularly prone to cold sores. Factors that activate virus in the ganglion include colds and other fevers, menstruation and psychological factors. Their mode of action is not understood. Sunlight is also well known to activate cold sores. It probably acts on the skin around lips or nostrils, causing inflammation and stimulating sensory nerves and thus the latent virus in the ganglion, or by causing a subclinical, spontaneously reactivating lesion to become an overt lesion. The eruption is restricted to the area supplied by the particular sensory ganglion that was involved during the original childhood infection. Herpes simplex may infect other areas of the body. A small child falls and hurts its knee, the knee is kissed

---

* The viruses that maintain themselves in small, completely isolated Indian communities in the Amazon basin are therefore persistent viruses (see Table 10.1) rather than nonpersistent viruses such as polio, influenza or measles. This was shown from antibody surveys carried out shortly after first contact of these communities with the outside world. Other infections which can be maintained in small populations are those in which there is persistent shedding of the microorganism (typhoid, tuberculosis, see below) or in which there is a reservoir of infection in some other host species (yellow fever, plague).

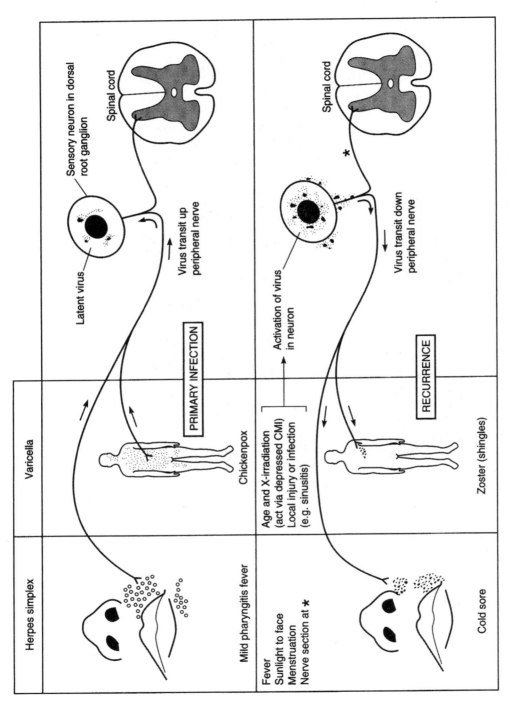

Fig. 10.1 Mechanism of latent herpes simplex and varicella-zoster virus infection in man.

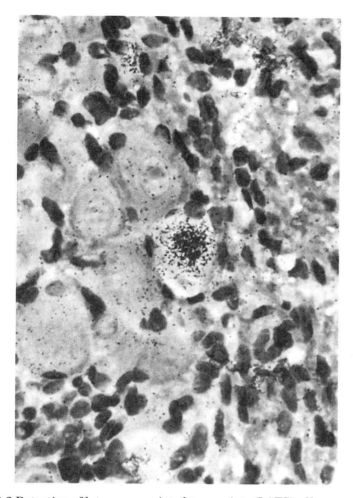

Fig. 10.2 Detection of latency-associated transcripts (LATS) of herpes simplex virus using the technique of *in situ* hybridisation. The picture shows a paraffin section of a dorsal root (sensory) ganglion from a mouse infected with virus 50 days earlier. At this time there is no infectious virus present and no viral proteins are expressed in the neuron. During this period of virus latency, the only viral activity is the expression of LATS, viral RNA of unknown function. The LATS are detected by *in situ* labelling with a ^{35}S-labelled DNA probe of complementary sequence. A positive reaction, seen by autoradiography as silver grains, is present over one neuron. Haematoxylin eosin stain ($\times$ 520). (By kind permission of Dr S. Wharton, Department of Pathology, University of Cambridge.)

better by an aunt with a cold sore, and the knee is now the primary site of infection. Recurrent 'cold sores' in this individual involve the knee. Likewise venereally transmitted herpes simplex, causing primary infection of the penis or cervix, will give recurrent lesions in these areas if reactivated later in life.

During the latent stage, herpes simplex virus DNA is present in neurons in dorsal root ganglia. The latent state seems to be delicately balanced, because the virus is activated merely by exposing without touching the trigeminal ganglion at operation. We know something about the mechanism of latency and reactivation of herpes simplex virus. In an acute infection, activation of transcription of the virion genome depends on the interaction of a virion protein, VP16, with a cellular protein(s). Latency arises either when the activating cellular proteins are absent, or repressor proteins bind to VP16 and/or bind to the activating cellular proteins, with the result that normal transcription of virus genes is inhibited. During latency there is very limited transcription of a specific class of viral RNAs known as latency-associated transcripts (LATS) (see Fig. 10.2). The function of LATS is still uncertain, since virus mutants lacking LAT sequences can still establish a latent infection and undergo reactivation of virus. Neurons harbour multiple copies of the herpes virus genome, with a range of copy numbers in individual cells. The viral genome is not integrated with host DNA and it is not in its usual linear form. Instead it circularises, and exists in the cell in a free episomal form. The amount of viral DNA in sensory ganglia does appear to correlate with the frequency of reactivation. Furthermore, reactivation only occurs in a small fraction of the neurons that contain viral DNA.

It has been suggested from experimental work with mice that reactivation is quite common in ganglion cells, but immune responses generally suppress virus replication before the full pathogenic sequence can be enacted. The virus must first travel down the nerve (apparently in the axon and at about $9\,\mathrm{mm\,h^{-1}}$), then infect dermal cells and finally epidermal cells, before a lesion is produced. Looked at this way, each clinical lesion represents a failure to control the growth and spread of reactivating virus. In other words, reactivation is a two-stage process, and reactivation of cytomegalovirus and papillomavirus can be interpreted in the same way (Fig. 10.3). In the case of herpes simplex virus and varicella-zoster the first stage takes place in the ganglion and occurs spontaneously, while the second requires the

Fig 10.3 Two stages in reactivation of latent viruses: herpes simplex (HSV), varicella-zoster (VZV), cytomegalovirus (CMV), wart viruses, polyomaviruses.

spread of virus from nerve endings to dermal and epidermal cells, and is subject to immune control. After virus has reactivated in sensory neurons, sensations such as itching are generated in the areas supplied by affected neurons, perhaps before virus actually reaches the skin. When pseudorabies, a similar virus, multiplies in sensory neurons in pigs, these phenomena are prominent enough to give the condition the name 'mad itch'. With reactivation of varicella-zoster virus in man the skin lesions may be small (or even absent) in comparison with the area affected by pain or paraesthesia. Herpes simplex reactivations can occur without visible skin or mucosal lesions. At times, therefore, immune forces seem to control the infection before skin lesions can be produced. Whatever the stage of reactivation that is under immune control, it is clear that CMI is involved. For unknown reasons the CMI response to varicella-zoster is selectively depressed in the elderly and in patients with lymphomas (see below), whereas responses to other persistent infections such as herpes simplex and cytomegalovirus are unaffected. Thus varicella-zoster is the persistent infection that commonly reactivates to cause disease in these individuals.

Brill–Zinsser disease is a rickettsial example of latency. Following complete clinical recovery from typhus, the rickettsias sometimes become latent in lymph nodes or the reticuloendothelial system. After 10 or more years, unknown influences cause the latent infection to be activated and the individual suffers a mild illness, less severe than the original typhus but with the rickettsias once again present in the blood. If the human body louse is present, it can acquire the infection following a blood feed and transmits it as typhus to susceptible individuals. Sheep become latently infected with *Rickettsia burneti*. The infection may reactivate in late pregnancy, and very large numbers of organisms are then shed in urine, faeces, amniotic fluid and placenta. A stable infectious aerosol is formed which can cause Q fever in susceptible farmers or veterinary surgeons.

Malaria provides a classic example of protozoal latency. After clinical recovery, particularly from *vivax* malaria, the parasite disappears from the blood circulation and enters a state of latency in the liver.* The specialised liver form is called a hypnozoite (sleeping animalcule).† Subsequently, often after many years, the parasite in the liver re-infects red blood cells to give a fresh clinical attack of malaria. Malarial

---

* In latency due to non-*vivax* types of human malaria, the parasite stays in red blood cells that have been sequestered in circulatory backwaters.

† In the case of the subspecies *Plasmodium vivax hibernans*, all the sporozoites injected by the mosquito turn into hypnozoites, and the red blood cell stage does not begin for several months. This is useful to the parasite in its northern range, enabling it to survive during the cooler winter when mosquitoes are not available, and produce the blood forms when mosquitoes have reappeared. It was this subspecies that at one time caused malaria in the east of England and in The Netherlands.

latency is particularly striking in those from temperate climates who become infected in the tropics, return home, and suffer an attack of malaria many years later.

Tuberculosis sometimes gives a type of latent infection. The bacteria remain dormant in the body after the initial infection and recovery, and can later be reactivated to give clinical disease. In the old days before BCG vaccination, many town and city dwellers were infected in early life, but in most cases the infection in the lung or lymph node was controlled and remained subclinical giving rise to a healed primary focus. After the age of about 40, perhaps with the general age-related decrease in the strength of CMI, there is an increasing likelihood that the bacteria in a primary focus will become active again and cause clinical disease. Respiratory tuberculosis in the middle-aged patient generally arises in this way.

Viable but noncultivable forms

Study of bacterial persistence and latency is complicated by the occurrence of viable but noncultivable forms. This is an interesting phenomenon exhibited by certain Gram-negative non-spore-forming and non-cyst-forming bacteria (including *Campylobacter jejuni*, *Vibrio cholerae*, *Vibrio fulnificus* (an estuarine bacterium) and *Salmonella enteritidis*). Such organisms are triggered by a variety of different environmental conditions to assume peculiar morphologies and are then extremely difficult (or to date, impossible) to cultivate *in vitro*. Failure to recover them reflects our ignorance of the necessary culture conditions, rather than indicating that they are dead! In fact, success in reverting some 'noncultivable forms' to normal forms has been reported. Long-term persistence of *Chlamydia psittaci* and *C. trachomatis* is reported not only *in vitro* but also in healthy birds and humans. The mechanism is not clear, but it may be important in conditions such as trachoma, Reiter's disease and chronic pelvic disease with infertility in women. Noncultivable *Chlamydia* are sometimes detectable *in vivo* by specific antibody and DNA probes. Chlamydial reticulate bodies (RBs) replicate but are not in themselves infectious. Their replication is complex and is particularly susceptible to the availability of nutrients; for example, replication is retarded by omission of cysteine from the medium, and completely inhibited by an imbalance in the concentration of structurally related amino acids. In addition, restriction of the supply of amino acids also leads to the production of abnormal forms similar to those induced by penicillin. It could be that failure to cultivate *Chlamydia* from sites *in vivo*, where their presence can be demonstrated, is due to the very low infectivity for cell cultures of abnormal forms induced by the unfavourable nutritional status of certain environments.

Persistent Infection with Shedding

Persistence is the state where microorganisms can be found continuously in the individual, and in this particular category they are shed more or less continuously, often for many years, without causing further disease. After recovery from typhoid, for instance, bacteria sometimes persist for long periods in the gall bladder. Scarred, avascular areas of the gall bladder are colonised, where the bacteria enjoy a certain freedom from the blood-borne antimicrobial forces of the host. Typhoid bacilli are discharged intermittently into the bile and thus the faeces. Two to five per cent of typhoid cases become faecal excretors and nearly all of them are women, because gall bladder damage and scarring is commoner in women. Such carriers of typhoid are important sources of infection as they are apparently healthy. 'Typhoid Mary' was a carrier who was employed as a cook in the USA, and moved from one place to another, cooking for eight different families and causing more than 200 cases of typhoid before she was finally caught and pensioned off. Typhoid can also persist in the urinary tract, giving urinary spread of disease from the carrier. One serious outbreak of typhoid in Croydon, London, in 1937 was traced to a carrier who had been employed during work on water pipes supplying the affected area, and had urinated on nearby ground and contaminated the water supply. There were 310 cases, with 43 deaths.

A carrier state is also seen in certain bacterial infections of the body surfaces. After recovery from diphtheria, scarlet fever or whooping cough, the bacteria often persist in the nasopharynx for many months, serving as a source of infection for susceptible individuals. The mechanisms of persistence are not understood, but it must be remembered that the normal resident bacteria of the nasopharynx by definition are also persistent, and often include potentially pathogenic bacteria such as the meningococcus, the pneumococcus and pathogenic strains of group A streptococcus or *Staphylococcus aureus*. It is uncommon for these bacteria to give trouble; for every person suffering from meningococcal disease there are about 1000 unaffected carriers. The fact that bacteria capable of causing diphtheria, scarlet fever or whooping cough can also persist in this site is perhaps not surprising.

Entamoeba histolytica often causes a persistent infection, and cysts can be shed in the faeces for many years after recovery from amoebic dysentery or after subclinical infection. The cysts are highly resistant and infectious. *Entamoeba* persistence is not unexpected because several other *Entamoeba* species are regular human commensals. Persistent asymptomatic carriage also occurs with other intestinal protozoal pathogens, including *Cryptosporidium*, *Giardia* and *Blastocystis*.

Persistent virus infections include Epstein–Barr virus and herpes simplex virus infections, and these are shed in saliva, often for long periods after the initial infection. Epstein–Barr virus is detectable in throat washings for at least several years. Herpes simplex virus reappears in the mouth later in life if there are cold sores, and repeated tests on given individuals have shown that the virus is also intermittently present in oral secretions at other times, though this may simply reflect the asymptomatic reactivation of latent virus. Hepatitis B virus persists in the blood for long periods, and perhaps for life in up to 10% of adults and 90% of neonates. About 0.1% of individuals in northern Europe and North America are carriers, many of them apparently normal and healthy, and the incidence is much higher in India and SE Asia (3–5%) and Oceanic Islands (10–15%). The blood of a carrier is infectious and can be transmitted to susceptible individuals via blood transfusions, the contaminated syringes of drug addicts, and the tattooist's, acupuncturist's or ear-piercer's needle.* In the case of hepatitis C, the commonest cause of post-transfusion hepatitis, up to 90% of infected adults become carriers in spite of apparently vigorous antibody and CMI responses.

Other viruses are shed in urine. Polyomavirus, for instance, causes a natural infection of mice and establishes foci of infection in kidney tubules, whence it is discharged into the urine. Mice remain perfectly well, but infection is persistent and urine is the major vehicle for the spread of infection between individuals.† When the host animal is in contact with humans or invades their dwellings, there are opportunities for human infection. Important human diseases acquired from animal urine in this way include leptospirosis (see Ch. 2), Lassa fever (see Ch. 1), Bolivian haemorrhagic fever (see Ch. 11), and hantavirus infections.‡ Cytomegalovirus is present in the urine of about 10% of children under the age of 5 in London, but it is unlikely that this is important in the spread of infection. Certain viruses, such as mammary tumour virus of mice and cytomegalovirus in man are shed persistently in milk, and among the bacteria both *Brucella* and tubercle bacilli are present in the milk of persistently infected cows.

There are several human infections in which the microorganism often persists in tissues for long periods and at the same time causes

---

* Hepatitis B infection also often occurs in individuals who have not been transfused with blood or injected with contaminated needles. Evidently there are other important mechanisms for the transmission of this infection. Vertical transmission from mother to infant is the most important one. Infection is common in male homosexuals, occurs also in heterosexual females, and in some carriers the virus is detectable in saliva and semen.

† The human polyomaviruses (JC and BK), in contrast, show latency rather than persistence with shedding, in the kidneys of many adults. Virus is reactivated and shed in urine during normal pregnancy, old age, and also in the immunosuppressed kidney transplant patient.

‡ Hantaviruses excreted by rodents in Korea infected about 3000 US servicemen, causing haemorrhagic fever and renal disease, during the Korean war. More recently (1993) an outbreak of a mysterious pulmonary disease in New Mexico, USA, was shown to be due to another hantavirus shed from infected dormice.

chronic disease. These include tuberculosis, leprosy, syphilis, brucellosis and trachoma. In these infections the disease is the result of a long drawn out battle between the microorganism and the immune and tissue defences of the host, sometimes one and sometimes the other gaining the upper hand. In the case of leprosy and tuberculosis, bacteria continue to be shed to the exterior and infect others. Each of these infections is a tribute to the ability of the microorganism to survive and multiply in the face of host defence responses, and the progressive tissue damage is partly a direct result of bacterial activity, but largely attributable to the host responses (see Ch. 8). The protozoa of malaria and trypanosomiasis also give rise to chronic infections, and the ways in which these microorganisms evade host immune responses are discussed in Ch. 7.

An unusual group of infectious agents persists in the body after infection, and gives rise to progressive and fatal neurological disease after prolonged incubation periods. This comprises scrapie, bovine spongiform encephalopathy (BSE), transmissible mink encephalopathy, kuru and Creutzfeld–Jacob disease. Scrapie is a naturally occurring disease of sheep that has been present in Europe and the UK for about 300 years. The brain is involved, and the disease is so called because affected sheep itch, and scrape themselves against posts and fences to relieve this symptom. In the laboratory, scrapie is transmissible to mice and other animals, and the feeding of infected sheep's heads to mink on mink farms in the USA has given rise to the disease called transmissible mink encephalopathy. BSE appeared mainly in dairy cattle in the UK in the 1980s and, like the mink disease, resulted from the adaptation of scrapie to a new host species. Investigations suggest that it arose from inadvertently incorporating scrapie-infected sheep residues from abbatoirs into artificial cattle food. This coincided with a change in practice in abbatoirs in which residues were no longer extracted with organic solvents at 70°C for 8 h followed by treatment with superheated steam to remove the solvent. It is likely that scrapie infectivity, which is notoriously resistant to inactivation, had until then been destroyed or reduced to a noninfectious level by this treatment. A similar disease, presumably transmitted in the same way, has been reported in domestic cats and in certain zoo antelopes. The agent responsible for BSE has been passed to mice and indeed resembles scrapie, but importantly it is not identical and has presumably undergone adaptation to grow in cattle. Unfortunately BSE, unlike scrapie, can be transmitted by the oral route to humans, in whom it causes a fatal CJD-like disease ('new variant CJD'). Nearly all the 50 or so cases have been in England, the home of BSE in cattle, and resulted from consumption of infected neural or lymphoid material before the control measures became effective. Altogether there have been nearly 200 000 cases of BSE in cattle, but none in those born since 1996.

Kuru was a fatal neurological disease of humans in Papua New Guinea, spread by ritual cannibalism. It was restricted to the Fore tribes, and a total of 3700 cases occurred in a population of 35 000. The condition was not communicated directly from person to person and none of the hundreds of children born and suckled by mothers with kuru developed the disease. Those dying with kuru were eaten by relatives as a mark of respect. The women and children ate the brain and acquired the disease, whereas the men either did not participate or preferred extraneural tissues and consequently were not often affected. Infection may have taken place via abrasions on fingers and mouth rather than in the gastrointestinal tract. At one time up to half the women in affected villages were suffering from kuru. Cannibalism in New Guinea has now died out and so, therefore, has the disease. The last person with kuru died in 1998. A suggested origin of kuru was from the cannibalistic consumption of a missionary who died from Creutzfeld–Jacob disease.

Creutzfeld–Jacob disease is another rare neurological disease of humans, occurring sporadically all over the world, caused by infectious agents similar to those of scrapie and kuru, but with an unknown mode of transmission.* About 10% of cases occur in certain families (see below).

In all these diseases, the incubation period represents a large fraction of the life span of the host – 6 months with mouse scrapie, 2 years with sheep scrapie, 4–6 years with BSE in cattle, and 4–20 years with kuru in humans. During this time the microbial agent steadily increases in amount, first in lymphoid tissues and then in the brain. Cells infected are neurons, follicular dendritic cells, and probably astrocytes and B cells. The process of infection and production of pathological changes is slow but proceeds inexorably and no antibody is formed. One prominent pathological feature is a fine vacuolation in the brain and this group of diseases was therefore called the 'spongiform encephalopathies'. These agents are not typical viruses, their exact mode of replication is still a mystery, and they have not been shown to contain either DNA or RNA. Indeed the spontaneous occurrence of these diseases in those with mutations in the PrP gene (see below) adds a new dimension to our concept of infection.

Molecular biology, however, has made spectacular contributions to our understanding. The pathological changes in the brain are associated with conversion of a host-coded prion protein (PrP^c) present in neurons in the normal brain, into a modified form, the scrapie prion protein (PrP^{sc}). This is resistant to normal proteolytic processing, accumulates in synaptic structures and results in death of neurons, possibly by apoptosis. The presence of PrP^{sc} in the brain is a diagnostic

---

* Transmission from patient to patient has been recorded following the use of neurosurgical instruments, after injection of growth hormone or chorionic gonadotrophic hormone preparations prepared from large numbers of human pituitary glands, a few of which were infected, and also after transplantation of cornea or dura mater from a donor who turned out to be infected.

feature of this group of diseases. It seems likely that PrP^{sc} is the actual infectious agent. However, it has not been explained how the different incubation periods which are so characteristic of the 15–20 different strains of scrapie agent in mice can be attributed to, and coded for, by a normal host protein. Either there is an as-yet unrecognised nucleic acid component, or the PrP protein exists in multiple conformations. Transgenic 'knockout' mice which lack the PrP gene seem normal in all respects except that they cannot be infected with scrapie. The discovery of the PrP gene raised the possibility that mutations in this gene would account for the 10% of cases of CJD that are familial and do not follow exposure to infection. So far 18 different mutations have been recorded, especially at codons 53, 129, 178 and 200, which are associated with familial CJD. Presumably the mutant PrP undergoes spontaneous conversion to the pathogenic PrP^{sc} form, thus causing the disease.

Lastly there are two inherited forms of human spongiform encephalopathy in which, as in familial cases of CJD, there are mutations in the PrP gene. One is called Gerstmann–Straussler–Schenker (GSS) syndrome, and PrP knockout mice made transgenic for the mutant GSS syndrome PrP gene spontaneously develop typical spongiform encephalopathy. The other is fatal familial insomnia, a rare disease with lesions in the thalamus, which has been shown to be associated with a mutation in codon 178 of the PrP gene.

Epidemiological Significance of Persistent Infection with Shedding

There are obvious advantages to a microorganism if it persists in the host and is shed from the body for long periods after the initial infection. Maintenance of the infection in a host community is made easier, and herpes simplex, varicella-zoster, tuberculosis, typhoid and other conditions have already been discussed from this point of view.

The epidemiological advantages of prolonged shedding of microorganisms to the exterior are well illustrated in the cases of myxomatosis and cholera. They are not truly persistent infections, but show the results of an increased period of shedding during the acute disease and convalescence. Each provides an excellent example of the natural evolution of an infectious disease. Myxomatosis is a virus disease of rabbits, spread mechanically by biting arthropods. When introduced into Australia in 1950, it caused nearly 100% mortality in the rabbit population. But rabbits were never eliminated from Australia because within the next 5 years or so a new and more stable host–microbe balance evolved. There was a change in both the virus and in the host species. First, rabbits with a genetically based susceptibility to myxomatosis were weeded out, leaving a rabbit population that was by nature more resistant, only 25% of them dying after infection with virulent virus. Second, the virus changed. Infected rabbits develop virus-rich swellings on the ears and face, and these skin lesions serve as sources of infection for the mosquitoes that carry virus to other rabbits. In the early stages of the Australian epidemic, when the infection was very severe, rabbits died a few days after developing these swellings. Later, however, a strain of virus emerged which was much less lethal and allowed the rabbit with the virus-rich swellings to live for a week or two, even to survive, and this gave greatly increased opportunities for virus spread by mosquitoes. The less lethal strain of virus therefore replaced the original virulent strain in the rabbit population.

Cholera is an intestinal infection caused by *Vibrio cholerae* and is transmitted by faecal contamination of water supplies, and it is always spread with great rapidity and efficiency in crowded human communities in the absence of satisfactory sanitary arrangements. This was so in the nineteenth century in London, and nowadays in India or parts of the Middle East, and is all too familiar in refugee camps throughout the world. Classically there is a very short illness, characterised by vomiting, diarrhoea, dehydration and shock, which is often lethal within 24 h. The bacteria persist in water for a week or so and convalescent patients may continue to excrete bacteria in faeces for a few weeks. Classical cholera, however, has been replaced in many regions by the other biotype, El Tor.* The great majority of infections with strains of El Tor are symptomless, but these people excrete bacteria for longer periods than in classical cholera. El Tor is also less readily inactivated, and can thus spread from person to person by contact, feeding utensils, etc., as well as by faecal contamination of drinking water. The El Tor strain has largely replaced classical strains of cholera, mainly because it is shed from the patient for a longer period and does not depend for its transmission on a contaminated water supply.

* The two biotypes can only be distinguished by biochemical tests, and they belong to the O1 (O = somatic antigen) serovar. The El Tor biotype was isolated in 1905 from dead pilgrims from Mecca who had been quarantined in a camp in Sinai. They had shown no sign of cholera. In general El Tor causes disease in 1 in 50 infections compared with 1 in 6 with the classical strain. However, virulence is difficult to compare since there are now known to be harmless strains of classical *V. cholerae*, which cannot be distinguished by normal typing methods. The avirulent strains are even being considered as a live vaccine.

Non-O1 *V. cholerae* are widely prevalent in the aquatic environment, and are not usually very pathogenic, producing only small amounts of toxin. But a non-O1 strain appeared in India in 1992 (the 0139 Bengal strain), causing severe cholera, and this strain has spread across the world; 01-type (whole bacterial cell) vaccines give no protection, and the World Health Organisation (WHO) no longer recommends their use.

Persistent Infection without Shedding

A large proportion of the microorganisms that persist in the body are rarely if ever shed to the exterior. Their importance is for the individual rather than for the community. Most of them give rise to no ill effects, but one or two may cause trouble if the immune responses are weakened, and one or two can ultimately cause cancer. Most of them are viruses, and viruses have a unique ability to persist and multiply in cells, often in a defective (noninfectious) form (see Ch. 7). Many adenoviruses, for instance, persist in lymphoid tissues after initial infection, causing no disease, but are still recoverable from normal adenoids or tonsils. There is little or no infectious virus in these tissues because of effective control by immune or other mechanisms, but when the tissue is removed and placed in culture where the controls are no longer present, the infectious virus appears. Adenoviruses are recoverable from one-third of all adenoids and tonsils removed during the first decade of life, and they must be regarded as part of the normal microbial flora of man. Certain herpes viruses, including Epstein–Barr virus, cytomegalovirus, human herpes virus type 6 (HHV6) and Marek's disease virus in chickens, also show persistent infection of lymphoid tissue. In the case of Epstein–Barr virus, circulating B cells contain viral DNA but no infectious virus, whereas T cells are the target in HHV6. Cytomegalovirus is present in monocytes of about 5% of normal people (e.g. healthy blood donors), so that infection can occur as a result of blood transfusions.

In the early days of tissue culture, when normal monkey and human kidney cells were used for the propagation of polio and other viruses, a number of viruses were isolated from the kidneys of normal individuals. These included reoviruses, measles virus, cytomegaloviruses, and a papovavirus (SV40) in monkey kidneys (see p. 406). Most human kidneys contain the polyomaviruses BK and JC, with sporadic appearance of virus in urine. The form in which these viruses exist in the normal kidney is not clear. There is a final group of persistent viral infections that are nearly always completely harmless to the host, but sometimes, often a very long period after initial infection, they cause malignant change. These are the retroviruses that cause mammary tumours in mice, and leukaemias (sometimes other types of tumour) in mice, humans and other mammals. Retroviruses are RNA viruses that contain a reverse transcriptase enzyme, which transcribes viral RNA into cDNA as a necessary part of the replication cycle. This DNA is integrated into the genome of the infected cell and becomes part of it. Thus if the egg is infected (or the sperm in the case of mouse mammary tumour virus), the viral genome is present in all embryonic cells and is transferred from one generation to the next, via the offspring. This is an example of vertical transmission (see p. 3), and these viruses are endogenous retrovirus.

They start off as infectious, but soon accumulate mutations and deletions, and eventually lose infectivity, remaining as DNA in the host genome, no more than 'retroviral fossils' protected, replicated and handed down the generations as if they were the host's own genes. This surely represents the ultimate, the final logical step in parasitism!* It becomes difficult to determine which is host and which is parasite, and the word infection loses much of its meaning. Indeed if the host benefited from the presence of the viral genome, the association could be classed as symbiotic.

Why are the retrovirus sequences so common, and why are there often multiple copies of them? For example, the human genome contains 50–100 copies of human endogenous retrovirus-W (HERV-W). For one thing, they are not pathogenic, and can confer protection against related exogenous retroviruses, for instance by coding for molecules that block virus adsorption to cells, or perhaps they have a role in foetal development. On the other hand, it could be that they are such successful parasites it is just too difficult to keep them out of, or to get rid of them from host DNA.

For an infection transmitted exclusively vertically, via the egg or sperm, there is of course no need for the virus ever to mature into an infectious particle. Its continued presence in the descendent host generations is ensured. Some transmission between individuals sometimes occurs, however, after birth, as with the mammary tumour virus of mice transmitted via milk to the offspring or leukaemia virus of cats transmitted horizontally between individuals, and in these circumstances infectious virus must be produced.

Other retroviruses more regularly undergo a full cycle of replication in cells throughout the body, and are shed in saliva, milk, blood, etc., to infect other individuals. These can be transmitted horizontally and are called exogenous retroviruses. Cats infected with feline leukaemia virus for instance excrete in their saliva up to 10^6 infectious doses ml^{-1}.

The various mouse leukaemia viruses are present from birth in all individuals of all known strains of mice, but leukaemia is a relatively uncommon and late consequence of infection. From the virus point of view, leukaemia is an irrelevant result because it does not help viral persistence in the individual nor transmission to fresh individuals. The incidence and type of leukaemia depends on the virus and on the genetic constitution of the mouse, but the mechanism of leukaemia induction is not completely understood. In man, HTLV1 (p. 221) infects

---

* Tests for endogenous viral nucleic acid sequences show that they are very common in the host genome of vertebrates. In humans about 30% of the total DNA is retroviral in origin, if all retroviral elements bounded by long-term repeats are included. Evidently, these viruses have repeatedly colonised vertebrates during evolution. A few (e.g. HERV-W) have the promoters, enhancers, repressors, to give them tissue-specific expression, for instance in placenta or foetal tissues.

T cells, increasing the density of interleukin-2 (IL-2) receptor, on these cells and thus ensuring their continued growth, which sometimes leads to malignant change.

The lentiviruses ('slow' viruses) are another group of retroviruses, which includes visna of sheep and goats, equine infectious anaemia of horses, and HIV (see p. 191) of humans. All are persistent infections, causing chronic disease and showing antigenic variation in the infected host (see pp. 207–208). As is so often the case in persistent virus infections, many of which are listed in Table 7.1, macrophages and lymphocytes are infected.

The phenomenon of integration of viral genome into host cell genome is not unfamiliar to microbiologists. It is a feature of the so-called temperate (nonlysogenic) infection of bacteria with bacteriophages (viruses of bacteria). The infection is harmless, and the bacteriophage genome expresses only those few proteins which are necessary for the maintenance of the lysogenic state. The diphtheria toxin of *Corynebacterium diphtheriae* is a protein encoded by a lysogenic phage. Following treatment with certain inducing agents or occasionally spontaneously, control processes are upset and the infection reverts to a lytic one in which infectious phage is produced and the host bacterium is destroyed. Integration also occurs experimentally when certain papovaviruses such as SV40 in monkeys or polyomavirus in mice enter the right kind of host cell. The cell is transformed, developing new viral surface antigens and malignant properties. These papovaviruses, however, are not known to cause malignant tumours under natural conditions.

Significance for the Individual of Persistent Infections

Persistent infections that are normally held in check by immune defences can be activated when immune defences are weakened. This is a major feature in AIDS. It also occurs when patients for kidney transplantation are given immunosuppressive drugs, and the persistent but normally harmless cytomegalovirus for instance is activated in most patients within a month or two, often giving rise to fever, pneumonitis or hepatitis. Even warts are activated and appear sometimes in large numbers. The CMI response is depressed in patients with certain tumours of lymphoid tissues, such as Hodgkin's disease, and these patients may suffer from activation of persistent tuberculosis, varicella-zoster, or cytomegalovirus infections. Not all persistent infections are activated, and there is no evidence for an increased incidence of herpes simplex cold sores because CMI impairment is specific for varicella-zoster rather than herpes simplex. The reason for the characteristic spectrum of reactivating persistent infections in acquired immunodeficiency syndrome (AIDS), Hodgkin's disease, transplant patients, etc., is not known. However, if reactivation with renewed shedding and transmission is to be advantageous to the virus, it should take place during the normal life span rather than as a result of disease. Accordingly, reactivation and shedding occurs in humans during normal pregnancy (BK and JC polyomaviruses) and also in old age (varicella-zoster virus, as described earlier in this chapter).

Persistent infections induce persistent immune responses, and these, although failing to eliminate the microorganism, sometimes cause pathological changes. The continued immune response to infections such as tuberculosis, syphilis, etc., leads to chronic disease, as mentioned above. In many cases the granuloma is the characteristic lesion formed around persistent foci of infection, making a major contribution to the disease itself. Persistent infections are often associated with persistence of microorganisms or microbial antigens in the blood. Circulating immune complexes are formed under these circumstances (see Ch. 8), and can give rise to a number of pathological changes, including glomerulonephritis. If the lesions at the sites of microbial persistence are trivial, immune complex formation is sometimes the major disease process. This seems to be the case in some types of chronic glomerulonephritis in man.

A persistent infection is associated with some cases of post-viral or chronic fatigue syndrome (also known as 'ME' or 'yuppie flu') in which patients suffer extreme fatigue after moderate exertion. Almost certainly there are a variety of causes, and the contribution of psychiatric (depressive) factors can be difficult to disentangle. However, certain investigations of muscle biopsies demonstrated that a significant proportion (53%) contained enterovirus RNA sequences, and a smaller proportion contained Epstein–Barr virus DNA. Conceivably, persistent virus infections are responsible for at least some cases of the post-viral fatigue syndrome.

One important consequence of persistent infection of major significance for the individual is that persistent microorganisms may eventually induce tumour formation. The viral leukaemias, sarcomas, mammary carcinomas and leukoses of mice, cats, chickens and other animals are caused by persistent RNA tumour viruses, when present in individuals of suitable age and genetic constitution. In man (see Table 8.1) certain types of leukaemia are caused by human T-cell leukaemia viruses 1 and 2 (HTLV1 and 2). The human wart is a benign tumour caused by a persistent virus, and cancer of the cervix is now very closely associated with a few of the sexually transmitted papilloma (wart) viruses. Burkitt's lymphoma and nasopharyngeal carcinoma are caused by Epstein–Barr virus, and liver cancer by hepatitis B virus. Very few people infected with these viruses develop a malignant tumour, and various cofactors are probably necessary, but have not yet been identified.

Conclusions

Microbial persistence, in summary, is a common sequel to viral, chlamydial and intracellular bacterial infections. Many of the severe infections causing illness and death in communities (poliomyelitis, plague, yellow fever, cholera) are not persistent and the micro-organisms are eliminated from the body after recovery. Persistent infections are often important from the microbe's point of view, enabling it to be maintained in small or isolated host communities. Persistent infections also generally present problems in the development of vaccines (see Ch. 12). They are becoming relatively more important, both for the individual and for the community, as the nonpersistent infections are eliminated by public health measures and by vaccination. Not only may they reactivate and cause troublesome infections in immunocompromised or immunosuppressed patients, but some of them can cause malignant tumours.

References

Beatty, W. L., Byrne, G. I. and Morrison, R. P. (1994). Repeated and persistent infection with chlamydia and the development of chronic inflammation and disease. *Trends Microbiol.* **2**, 94–98.

Bock, G. R. and Whelan, J. (eds) (1993). 'Chronic Fatigue Syndrome'. Ciba Foundation Symposium 173. John Wiley and Sons, Chichester.

Büeler, H., Aguzzi, A., Sailer, A., Greiner, R.-A., Autenreid, P., Aguet, M. and Weissman, C. (1993). Mice devoid of PrP are resistant to scrapie. *Cell* **73**, 1339–1347.

Campo, M. S. (1992). Cell transformation by animal papillomaviruses. *J. Gen. Virol.* **73**, 217–222.

Carr, K. (1993). Prion diseases. *Nature* (Lond.) **365**, 386.

Coles, A. M., Reynolds, D. J., Harper, A., Devitt, A. and Pearce, J. H. (1993). Low-nutrient induction of abnormal chlamydial development: A novel component of chlamydial pathogenesis? *FEMS Microbiol. Lett.* **106**, 193–200.

Dalgleish, A. G. (1991). Viruses and cancer. *Brit. Med. Bull.* **47**, 21–46.

Garcia-Blanco, M. A. and Cullen, B. R. (1991). Molecular basis of latency in pathogenic human viruses. *Science* **254**, 815–820.

Horn, T. M., Huebner, K., Croce, C. and Callaban, R (1986). Chromosomal locations of members of a family of novel endogenous human retroviral genomes. *J. Virol.* **58**, 955–959.

Howard, C. R. (1986). The biology of hepadnaviruses. *J. Gen. Virol.* **67**, 1215–1235.

Johnson, R. T. and Gibbs, C. J. (1998). Creutzfeld–Jakob disease and related transmissible spongiform encephalopathies. *N. Engl. J. Med.* **33**, 339–359; 1994–2004.

Krotoski, W. A. (1985). Discovery of the hypnozite and a new theory of malarial relapse. *Trans. Roy. Soc. Trop. Med. Hyg.* **79**, 1–11.

Levy, J. A. (1993). Pathogenesis of human immunodeficiency virus infection. *Microbiol. Rev.* **57**, 183–289.

Manson, J. C. (1999) Understanding transmission of the prion diseases. *Trends Microbiol.* **7**, 465–467.

Marrack, P. and Kappler, J. (1994). Subversion of the immune system by pathogens. *Cell* **76**, 323–332.

Miller, A. E. (1980). Selective decline in cellular immune response to varicella-zoster in the elderly. *Neurol.* **30**, 582–587.

Mims, C. A. (1981). Vertical transmission of viruses. *Microbiol. Rev.* **45**, 267–286.

Penn, C. W. (1992). Chronic infections, latency, and the carrier state. *In* 'Molecular Biology of Bacterial Infection. Current Status and Future Perspectives' (C. E. Hormaeche, C. W. Penn and C. J. Smyth, eds), pp. 107–126. Society for General Microbiology Symposium 49. Cambridge University Press, Cambridge.

Preston, C. M. (2000). Repression of viral transcription during herpes simplex virus latency. *J. Gen. Virol.* **81**, 1–19

Rickinson, A. B. and Kieff, E. (1996). Epstein–Barr virus. *In* 'Fields Virology' (B. N. Fields, D. M. Knipe and P. M. Howley, eds), pp. 2397–2446. Lippincott-Raven, Philadelphia.

Rouse, B. T. (1992). Herpes simplex virus: pathogenesis, immuno-biology and control. *Curr. Topics Microbiol. Immunol.* **179**, 1–179.

Stevens, J. G. (1991). Herpes simplex virus:neuroinvasiveness, neuro-virulence and latency. *Sem. Neurosci.* **3**, 141–147.

Teich, N. (ed.) (1991a). Viral oncogenes. Part I. *Sem. Virol.* **2**(5).

Teich, N. (ed.) (1991b). Viral oncogenes. Part II. *Sem. Virol.* **2**(6).

Yoffe, B. and Noonan, C. A. (1993). Progress and perspectives in human hepatitis B virus research. *Prog. Med. Virol.* **40**, 107–140.

11

Host and Microbial Factors Influencing Susceptibility

A host may be susceptible to infection by a given microorganism but rarely suffers harmful effects. In the old days, everyone was susceptible to infection with polioviruses or tubercle bacilli, but relatively few became paralysed or developed pulmonary tuberculosis. Not only this, but host susceptibility to infection often varies independently of susceptibility to disease. From the microorganism's point of view, infectiousness or transmissibility is not the same as pathogenicity. Transmissibility in fact depends on the extent of shedding of microorganisms from the infected individual, on the stability of microorganisms outside the host, and on the ease with which infection is established in new hosts. Each of these factors shows great variation. Variations in the ease with which infection is established are illustrated in Table 11.1. It can be seen that the dose required to produce infection, disease or death, depends on the microorganism, the route of infection, the host, and on other factors.

The word virulence is sometimes used to refer to the infectiousness or transmissibility of a microorganism, but the word as used here will refer instead to its pathogenicity, or ability to cause damage and disease in the host. An infection can be totally harmless and asymptomatic or lead to a lethal disease, depending on the results of the encounter between microorganism and host. The characteristics of both the microbe ('seed') and the host ('soil') contribute to the outcome of an infection, and either can exercise a determining influence. To put it as a platitude, it takes two (microbe and host) to make an infection or a disease. Some of the host and microbial factors influencing susceptibility to disease are discussed in this chapter.

Table 11.1.   Examples of variations in the dose of microorganisms required to produce infection, disease or death in the host

Microorganism	Host	Routes of infection	(ID_{50}), Minimally infectious disease producing (DD_{50}) or lethal dose $(LD_{50})^a$
Rhinovirus	Man	Nasal cavity	1 $TCID_{50}^b$ (DD_{50})
		Conjunctiva	16 $TCID_{50}$ (DD_{50})
		Posterior pharyngeal wall	200 $TCID_{50}$ (DD_{50})
Salmonella typhi	Man	Oral	$\leq 10^5$ bacteria (DD_{50})
Shigella dysenteriae	Man	Oral	10 bacteria (DD_{50})
Vibrio cholerae	Man	Oral	10^8 bacteria (DD_{50})
		Oral (together with bicarbonate, see Ch. 2)	10^4 bacteria (DD_{50})
Giardia lamblia	Man	Oral	10 cysts (ID_{50})
Mycobacterium tuberculosis	Man	Inhalation	1–10 bacteria (ID_{50})
Ectromelia (mousepox) virus: virulent strain	Mouse (C57BL or WEHI strain)	Footpad	1–2 virus particles[c] (ID_{50})
	Mouse (WEHI strain)	Footpad	25 virus particles (LD_{50})
	Mouse (C57BL strain)	Footpad	10^7 virus particles (LD_{50})

Examples from man, together with one example from an experimental animal (mouse) to show host genetic effects (mouse strain differences).
[a] '50' means that this dose produces the effect in 50% of those inoculated.
[b] Tissue culture infectious doses.
[c] By electron microscopy.

Genetic Factors in the Microorganism

The microorganism's ability to infect a given host is genetically determined and many microorganisms infect only one particular host species. For instance, measles, trachoma, typhoid (*Salmonella typhi*) and our warts are exclusively human infections. Others are less specific, rabies and anthrax seemingly capable of infecting all mammals. Pathogenicity or virulence is also a function of the microbial genome. Virulence is being increasingly recognised to depend on coordinated expression of numerous genes, whose products mediate adherence, antiphagocytic activity, immune evasion, production of toxins, etc. Environmental stimuli inform the bacterium about its changing surroundings, and for a pathogen these may vary enormously according to the stage of the infection process. Such changes may be sensed in physical terms (pH, temperature, osmolarity), nutrient availability, or in terms of hostile host defences. Some examples have already been given earlier: *Shigella* invasion (Ch. 2) and *Salmonella* intraphagocytic survival (Ch. 4).* The latter is an example of a growing number of systems which belong to the family of 'histidine protein kinase/response regulators', two component signal transduction systems. The first component is a sensor protein – a protein kinase – which is sensitive to fluctuations in one or more environmental parameters. In response to the environmental signal, it autophosphorylates at a conserved histidine residue. The second component is a regulatory protein which is phosphorylated at a conserved aspartic acid residue by the activated kinase, and this alters some cellular function, usually at the level of transcription. When the genes which respond to the stimulus are co-regulated by a common protein, they are described as a regulon. A group of regulons which responds independently to the same stimulus is called a stimulon. Some examples are given in Table 11.2.

Table 11.2. Examples of two component systems for regulating the expression of bacterial virulence components

Genes *Acronym*; function controlled	Initial signal	Pathogens
ompR/envZ Outer membrane proteins	Osmotic stress	*Salmonella typhimurium* *Shigella flexneri*
phoP/phoQ Phosphatase; intraphagocytic survival	Multifactorial	*S. typhimurium*
bvg *Bordetella* virulence gene; adhesins and toxins	Temperature	*Bordetella* spp.
algR Alginate synthesis in the cystic fibrosis lung	Osmotic stress	*Pseudomonas aeruginosa*
pilA/pilB Adhesion pili	?	*Neisseria gonorrhoeae*
agr Accessory gene regulator; expression of range of toxins	?	*Staphylococcus aureus*
toxR Single protein system with both sensing and kinase activity Expression of cholera toxin and toxin co-regulated pili	Osmotic stress, temperature, pH, amino acid availability	*Vibrio cholerae*
mry Product activates gene encoding M protein	CO_2	*Streptococcus pyogenes*

* A detailed treatment of this subject is beyond the scope of this book but the serious student is referred to Dorman (1994).

Changes in virulence may arise from apparently trivial changes in the genome. Minor changes in the M protein that coats group A streptococci can lead to major changes in bacterial virulence, and a single amino acid change in the haemagglutinin of influenza virus can convert a relatively avirulent strain into a lethal strain for mice. The differences between variola minor and variola major used to be a matter of life and death when human beings were infected, but are only detectable with difficulty in the laboratory. A fresh pandemic strain of influenza A virus is able to spread readily in the community because one of the surface proteins of the virus shows a major difference from pre-existing strains and therefore no one has immunity to infection (see Ch. 7). The change in surface protein must be associated with a high level of transmissibility, but may or may not be associated with an increase in virulence. The Hong Kong strain (1968) of influenza A virus, for instance, was avirulent in comparison with the devastating strain of 1918. Artificial influenza viruses can be made in the laboratory; the surface components of new strains and the avirulence characters of other strains are welded together by a process of genetic recombination to give hybrid viruses that have potential as vaccines.

The pathogenicity for a given host is often dramatically altered following the repeated growth of a microorganism under unfamiliar circumstances outside the body. The laboratory passage of pathogenic viruses in cultured cells often leads to great reductions in pathogenicity (attenuation) in the original host, and this has been a standard procedure for the production of live virus vaccines (see poliovirus, below, and Ch. 12). The attenuated strain breeds true and is a genetic variant. A similar phenomenon is seen with bacteria. For instance, BCG (Bacille Calmette–Guérin) vaccine consists of a strain of bovine tubercle bacillus, highly attenuated after 350 subcultures over the course of 15 years in glycerin–bile–potato culture medium. Also, gonococci cultivated in artificial media after isolation from the human urethra show rapid change to an almost nonpathogenic form.

Until recently, little was known of the genetics of animal viruses because it was not easy to find good genetic markers and use classical approaches such as recombination. The reoviruses, however, have a genome consisting of ten discrete segments (genes) of dsRNA and, when cells are infected with more than one type of reovirus, the different genes undergo reassortment in the progeny virus. Painstaking studies of the properties of strains of virus produced in this way have enabled scientists to identify genes and gene products that control the virulence of reoviruses in mice. For instance, there is a viral surface (capsid) polypeptide that binds the virus to neurons which is obviously important in the production of encephalitis, and a surface polypeptide that confers resistance to intestinal proteases gives the virus the capacity to infect via the alimentary canal. When the genes for both these polypeptides are present in a virus strain, it can infect by mouth and cause encephalitis. The genetic analysis of virus virulence has now changed as a result of powerful new research methods, such as the cloning of DNA sequences, and their application to viruses (and also to bacteria) is transforming microbial genetics. For instance, by introducing segments of DNA from virulent microbial strains into avirulent strains, the virulence genes can be identified. Viruses have a smaller, simpler genome than other microorganisms, and for many viruses the entire genome has been sequenced, and functions are slowly being assigned to specific sequences. But usually it is not possible to account for virulence in terms of a single specific gene product because virulence is often multifactorial. Also, even when a gene product is closely associated with virulence, it is likely to be a perplexing further step to say how it operates *in vivo*. Nevertheless, recombinant DNA technology is greatly accelerating progress, and advancing our understanding of virulence.

One example which illustrates both the power and limitations of recombinant DNA technology is the live oral poliovirus vaccine. Virulent virus of each of the three serotypes was attenuated empirically by passage at low temperature (33°C) by Albert Sabin working in the USA in the 1950s. The vaccine is excellent as most people in the world know, but the type 3 vaccine strain reverts to virulence at a very low frequency (one case per 10^8 doses of vaccine) and causes paralytic poliomyelitis. Virus isolated from patients has been completely sequenced and compared with the sequences of the attenuated vaccine strain and of the virulent wild-type virus. This showed that there was only one nucleotide change common to the acquisition of avirulence and the subsequent reversion to virulence. The problem is that the mutation is in a *noncoding* region of the genome, and it is not understood how this affects virulence.

Studies of chimeric polioviruses artificially produced from vaccine and virulent strains have identified critical point mutations with marked effect on neurovirulence or neuroinvasiveness. But their mode of action is not clear and, in any case, virulence in the natural disease depends on invasion not only of the nervous system but also of intestine, lymph nodes and blood. In the case of influenza, it would be of immense importance to identify the mutation(s) that conferred such virulence on the 1918 strain, but studies of the 1918 haemagglutinin have so far failed to give an answer (see Table 7.2).

All viruses are equally mutable, and the selection of mutant viruses is the mechanism for the changes that take place during attenuation. However, RNA viruses accumulate more mutations than DNA viruses as they have no 'proofreading and error correction' mechanism to detect and put right any mistakes. Changes in the pathogenicity of myxomavirus (a DNA virus) during the evolution of the virus in wild rabbit populations have been referred to in Ch. 10, but nothing is known of the mechanisms or genetic basis of these changes. Sometimes the mechanism of the change in pathogenicity of a virus variant has been elucidated. Mousepox is an infectious disease of mice, comparable to smallpox in man, caused by ectromelia, another DNA virus. There is a virulent strain that kills all infected mice following extensive growth of virus in the liver. After repeated growth of this strain in an unnatural laboratory host, the chick embryo, a variant virus strain emerged that is well adapted for growth in the chick embryo, but has a greatly reduced pathogenicity and is nonlethal for mice. It has been shown that the virulent strain of virus grows readily in liver macrophages (Kupffer cells) and subsequently infects hepatic cells to cause extensive liver necrosis, whereas the avirulent variant infects liver macrophages with difficulty and therefore infects no more than the occasional hepatic cell. In this instance, decreased virulence is due to decreased ability to infect macrophages (pp. 132–5).

For bacteria, mutation is also one of the important types of change in genetic constitution. The progeny of a single bacterial cell are not genetically homogeneous; a small proportion of them are mutants. The mutation rate for a given genetic change varies between 1 in 10^7 and 1 in 10^{10}, and the spontaneous mutants only replace the original type if they are favoured (selected) by the environment. There are about 6000 genes in a bacterium such as *E. coli*, and mutations may involve changes in structure, biochemical activity, antigenic properties, ability to produce toxins, etc., any of which could lead to changes in pathogenicity. Much of bacteriology consists of a study of mutants, especially those that acquire new antigens or toxins.* Smooth–rough variation is an important type of mutation affecting pathogenicity. Certain bacteria owe their pathogenicity to a surface component or capsule that interferes with phagocytosis by polymorphs and macrophages (see Ch. 4). This surface material often gives the bacterial colonies formed on artificial media a 'smooth' appearance. The surface material is lost during long periods of growth of bacteria in the laboratory, and in the host this leads to more efficient phagocytosis and decreased pathogenicity. The colonies now have a 'rough' appearance. Smooth–rough variations are common in salmonellas, shigellas and pneumococci, and represent bacterial mutations. Genetic changes in bacteria are frequently due to extrachromosomal genetic elements called plasmids (see Glossary). Plasmids also often carry determinants for toxin production (e.g. *E. coli* enterotoxin, *S. aureus* exfoliative toxin), or for colonisation (pili of enteropathogenic *E. coli*) or invasiveness (*Shigella flexneri*). Plasmids are transferred between bacteria and are important in the transfer of antibiotic resistance between intestinal bacteria (p. 275).

We understand much less about the genetic control of pathogenicity in protozoa. An individual with trypanosomiasis is persistently

---

* Remember that transmissibility, if it goes with virulence, is a property of supreme importance. The recently described (Oshkosh) strain of *M. tuberculosis* is about three times as transmissible as other strains as, well as virulent, and the genetic basis for this will doubtless soon be revealed.

infected because the parasite undergoes periodic changes in surface coat proteins. These are programmed from the trypanosome genome and enable it to evade host immune defences. Twenty-two different types of *Entamoeba histolytica* can be distinguished by isoenzyme electrophoresis, and 12 appear to be nonpathogenic, failing to invade tissues and cause disease (see p. 303). But the basis for amoebic pathogenicity is not understood, and it is not clear whether these types are genetically stable.

Genetic Factors in the Host

Susceptibility to infectious disease is always influenced and is sometimes determined by the genetic constitution of the host. An impressive example of individual differences in susceptibility that are assumed to be to a large extent genetic in origin was provided in Lubeck in 1926. Living virulent tubercle bacilli instead of vaccine was inadvertently given to 249 babies. There were 76 deaths, but the rest developed minor lesions, survived, and were alive and well 12 years later. The infecting material and the dose was in each case identical. Overstating the point, one could say that, for a given microorganism, there are almost as many different diseases as there are susceptible individuals.*

Sometimes the mechanism of genetic susceptibility can be defined with some precision. People with the sickle cell trait show a markedly decreased susceptibility to malaria. Malarial merozoites parasitise red blood cells and metabolise haemoglobin, freeing haem and utilising globin as a source of amino acids. A single gene present in these individuals causes a substitution of the amino acid valine for glutamic acid at one point in the β-polypeptide chain of the haemoglobin molecule. The new haemoglobin (haemoglobin S) becomes insoluble when reduced, and precipitates inside the red cell envelope, distorting the cell into the shape of a sickle. In the homozygote there are two of these genes and the individual suffers from the disease sickle cell anaemia, but in heterozygous form (sickle cell trait) the gene is less harmful, and provides a resistance to severe forms of falciparum malaria that ensures its selection in endemic malarial regions. Parasitised red cells readily sickle following utilisation of oxygen by the developing parasite. Perhaps the haemoglobin S crystals kill the parasite, and in any

---

* Another unfortunate example occurred in 1942 when more than 45 000 US military personnel were vaccinated against yellow fever but were inadvertently injected at the same time with hepatitis B virus which was present as a containment in the human serum used to stabilise the vaccine. There were 914 clinical cases, of which 580 were mild, 301 moderate and 33 severe. Even with a given vaccine lot, the incubation period varied from 10 to 20 weeks. Serological tests were not then available, so the number of subclinical infections is not known. In this case, physiological as well as genetic differences in susceptibility presumably played a part.

case such cells are removed from the circulation at an early stage by the reticuloendothelial system. This probably accounts for the resistance to malaria of those with the sickle cell trait. The gene would be eliminated from populations in 10–20 generations unless it conferred some advantage, and restriction endonuclease analysis of the gene in peoples in India and West Africa show that it has arisen independently in these malarial countries. It is not often that genetic susceptibility to infectious disease of man has been defined in this way, both genetically and at the biochemical level in the infected individual. A similar protection against malaria is conferred on those with another type of abnormal haemoglobin called haemoglobin C, also produced by a single amino acid change in the polypeptide chain. Both are hereditary conditions and are found in malarious countries. The Duffy antigen provides another example of powerful genetic selection by an important pathogen. Red blood cells from most West Africans, in contrast to those from Europeans, lack the Duffy antigen, which acts as an attachment site for *Plasmodium vivax*. This type of malaria is accordingly almost unknown in West Africa.

People with the sickle cell trait (heterozygotes) do not suffer clinically, but the homozygotes often die during childhood. They not only develop anaemia (their red blood cells are more fragile because they sickle and unsickle under normal circumstances while circulating), but also show increased susceptibility to infection. Pneumococci in particular cause trouble, probably a sequel to spleen dysfunction (p. 311) which in turn is a result of the repeated infarcts that occur in the spleen, as well as in other organs.

Human susceptibility to diseases such as tuberculosis and rheumatic fever is influenced by the genetic constitution of the host, as indicated in familial studies (rheumatic fever) and differences in racial susceptibility (tuberculosis). During the great ravages of pulmonary tuberculosis in European countries in the seventeenth, eighteenth and nineteenth centuries, genetically susceptible individuals were weeded out. As recently as 1850, mortality rates in Boston, New York, London, Paris and Berlin were higher than 500 per 100 000. With improvements in living conditions, these fell to 180 per 100 000 by 1900, and they have fallen greatly since then, but it appears that the present populations have a certain amount of genetic resistance to the disease (see also Ch. 1). Previously unexposed inhabitants of Africa, the Pacific Islands and elsewhere show greater susceptibility. Extensive lung involvement is still common in infected Africans,* and in the Plains

* It is estimated that about one-third of the present world's population have been infected, mostly in developing countries, and although only about 10% of them become ill, tuberculosis was responsible for 1.5 million deaths worldwide in 1988. Even in Europe, this ancient and preventable disease continues at a scandalous level. About 60 000 died in Europe in 1998, most of them from low-income families. If it were typhoid or leprosy, this would be regarded as intolerable.

Indians living in the Qu'Appelle Valley reservation in Saskatchewan, Canada in 1886, the disease spread through the body affecting glands, bones, joints, meninges, to give a death rate of 9000 per 100 000. Yellow fever, on the other hand, appears to have originated in Africa, and African people show greater resistance to the disease than Europeans. Yellow fever was transported to the Americas (together with the transmitting mosquito *Aedes aegypti*) on the slave ships, and the first American cases were recorded in Yucatan in 1640. The disease was more lethal in the previously unexposed and genetically more susceptible people of Central and South America.

Genetic influences in man are often difficult to dissociate from nutrition and other socioeconomic factors. Measles, for example, is largely a mild disease in developed countries but can have a mortality rate of up to 50% in severely malnourished children in the Third World. The genetic effect is clearly shown to be distinct from environmental effects in studies of the occurrence of diseases such as tuberculosis in identical twins who have lived apart. In one classical study with tuberculosis, it was shown that 87% of identical twins also had the disease, whereas only 26% of nonidentical twins were affected. The identical twins, moreover, showed a similar type of clinical disease, which has also been shown for leprosy. Naturally occurring changes in the genetic resistance of rabbits to myxomavirus are referred to on pp. 353–4.

The picture is much clearer in certain experimental infections in animals. For instance, the susceptibility of mice to certain viruses and to enteric bacteria such as *Salmonella* is under genetic control, and susceptible and resistant strains have been developed by breeding. Susceptibility of mice to *Salmonella* infections is under the control of many genes. In one instance, that of susceptibility to the lethal effect of intracerebrally injected yellow fever virus, resistance is inherited as a single dominant genetic factor. The basis of resistance in the brain is not understood, but it presumably involves the susceptibility of neurons to infection with yellow fever virus. Resistance to mouse hepatitis virus is also under simple genetic control and this seems to operate by restricting virus growth in liver macrophages, and thus preventing infection of liver cells (see also mousepox, p. 366). Genetic factors presumably control the behaviour and characteristics of macrophages, and macrophages play a central role as determinants of pathogenicity in many viral, bacterial and other infections.

Species differences in susceptibility to infection, of course, are also genetically determined, and in some instances the mechanisms have been identified. Guinea-pigs are resistant to South American strains of *Yersinia pestis*, and this is because asparagine, a bacterial growth requirement, is missing from guinea-pig serum, which contains the enzyme asparaginase. Bacterial growth is accordingly slower, giving time for the immune response to control the infection. Susceptibility of the bovine placenta to *Brucella abortus* is associated with the presence of a bacterial growth stimulant, erythritol. This is not present in the human placenta which is therefore resistant to infection. Accordingly, cows but not people abort when infected with *Brucella abortus*.

Genetic susceptibility of the host to infectious disease may operate at the level of the immune response and there is increasing evidence that this is an important phenomenon. The development of immune responsiveness in general is under genetic control, as illustrated by the failure of development in congenital agammaglobulinaemia and thymic aplasia (see Ch. 9), but genetic control also operates at a more specific level because immune responses to antigens are under the control of specific immune response genes. There are many genes for each of the MHC class I and class II proteins (HLA-A, B, C and HLA-DP, DQ, DR in man; and H-2 K, D and L and H-2 I in mice). The MHC proteins control immune responses by interacting with and presenting foreign peptides to T cells, individual MHC proteins reacting or not reacting with any one peptide according to its sequence (see p. 152). If a foreign protein gave no peptide reactive with a MHC protein, there would be no T-cell response to that protein. Individuals with a gene conferring a poor immune response* to a given microbial antigen, especially a surface antigen, are likely to have difficulty controlling infection with that particular microorganism. On the other hand, those with a poor response are less likely to suffer any immunopathological consequences of the infection.

Tissue typing for the various histocompatibility or transplantation antigens is a sophisticated and commonly used test, performed on a person's blood lymphocytes. HLA type correlates with susceptibility to certain types of infections. The association is often with the type and severity of disease rather than with the incidence of disease, which is presumably influenced by non-MHC genes. For instance, HLA-DR2 and DQ1 are associated with lepromatous leprosy, and HLA-DR3 with tuberculoid leprosy. HLA-DR2 is also associated with susceptibility to pulmonary tuberculosis. This means that people with these MHC proteins have a statistically increased chance of contracting a particular type of disease. In terms of individual genes, mutations in the NRAMP-1 (natural resistance-associated macrophage protein-1) gene in mice and men, or mutations in a human gene on chromosome 6q controlling the receptor for gamma interferon, are associated with increased susceptibility to tuberculosis.

Most associations are slight but there seems to be the clearest association when there is a strong autoimmune component, and the most striking example is ankylosing spondylitis, where 92% of patients are HLA-B27, compared with 9% in the general population. The mechanism is not clear, but it has been suggested that, when HLA-B27 indi-

---

* In the case of antibody, a poor immune response may mean production of antibody too slowly, in small amounts, of the wrong class, of low avidity, or against less relevant microbial antigens (see p. 198).

viduals make an immune response to antigens of *Klebsiella** or possibly other gut bacteria, there is cross-reaction with antigens on their own lymphocytes and in joint tissues. This leads to a crippling autoimmune disease. Also, in rheumatoid arthritis, a disease not known to be of infectious origin (p. 219), 70% of patients are HLA-DR4 compared with 28% of normal people. There is a genetic component in susceptibility to rheumatic fever (pp. 278–9), which perhaps operates via the immune response genes controlling the production of the streptococcal antibodies that cross-react with heart muscle.

Isolated human populations generally show a high mortality after the first encounter with traders, invaders or explorers from Europe or Asia. Although some of this is due to combat, social disruption, etc., most of it appears to be due to the new infectious diseases encountered, such as measles and tuberculosis. One suggestion is that this is based on MHC genetics. It is an advantage for a host population to have a good deal of diversity (polymorphism) in MHC genes controlling immune responses. This makes it more likely that there will be a suitable MHC molecule to accommodate peptides from any novel infectious agent, so that an immune response can be initiated (Ch. 6). We know that many of the populations from isolated islands and continents show less MHC polymorphism than the invading or exploring peoples,† and perhaps this accounts for their great susceptibility to new infections.

There is a good example in mice of the independent control of immune responses to different microbial antigens. Adult mice of most strains infected with LCM virus generally show severe pathological changes as a result of the cell-mediated immune response to the virus (see Ch. 8). Virus multiplies in exactly the same way in mice of the C57BL strain, but they do not develop disease because they have a weak T-cell response to LCM (lymphocytic choriomeningitis) virus antigens. However, mice of the C57BL strain generate a vigorous immune response to ectromelia (mousepox) virus. This virus grows in the liver and is often lethal in many strains of mice, but the vigorous immune response of C57BL mice ensures the early inhibition of virus multiplication in the liver, and allows them to survive. In both these examples, C57BL mice show greater resistance to disease; resistance to LCM virus disease because of a weaker immune response, and resistance to the disease mousepox because of a stronger immune response.

---

* A monoclonal antibody to HLA B27 reacted also with defined antigens on a certain strain of *Klebsiella pneumoniae*, providing an example of molecular mimicry (see pp. 187–9). But there are indications that the bacterial antigens can interact with the B27 molecule and make it immunogenic, and in any case other factors must be involved because not all spondylitics have the B27 antigen.

† Europeans, sub-Saharan Africans and East Asians have 30–40 different MHC (A, B) alleles, whereas native North Americans and natives of Polynesia or Papua New Guinea have only 10–20.

There are obviously other ways in which genetic factors in the host influence susceptibility,* but little is known. For instance, there are 12 genes encoding slightly different α-interferons (see pp. 324–7) in man, although they all bind to the same receptor on cells. Genetically determined receptors control susceptibility to certain bacterial infections (see pp. 14–17). For instance, pigs susceptible to *E. coli* K88 diarrhoea have receptors for the K88 surface component on their intestinal epithelial cells. Receptors are controlled by an autosomal dominant gene.

A fascinating type of parasitism that is at the borderland between infection and heredity should be mentioned here. The vertically transmitted RNA tumour viruses studied in mice and chickens (leukaemia and leukosis viruses) have become integrated as DNA into the genome of the host animal and are transmitted vertically (see Ch. 7). There are interactions between viral and host genomes, and viral functions show all degrees of expression from zero to the production of fully infectious viral progeny in cells. This type of infection can itself properly be regarded as a genetic feature of the host. The spongiform encephalopathies (prion diseases) provide another example (see Ch. 10).

Age of host

There are hardly any infectious agents that cause exactly the same disease in infancy, adult life and old age. Susceptibility is generally greater in the very young and the very old, as for example in Q fever, typhoid, bacillary dysentery, or bacterial pneumonia. A 60-year-old man with community-acquired pneumonia is three times as likely to die as a 30-year-old man. The greater susceptibility is for a number of different reasons. In the first place immune responses are weaker in immature and in ageing individuals. Men over the age of 40 years show a gradual decline in the magnitude of antibody and cell-mediated immune responses to standard antigens. In the elderly, T cell and also phagocyte function tends to be decreased. Similar changes are seen in mice. All types of infections therefore tend to be controlled less successfully at these ages, but at the same time there is less immunopathology. However, the evidence for an overall reduction in resistance to infectious disease in elderly people is not impressive.

Infections in infants sometimes spread rapidly and prove fatal without the evolution of the characteristic clinical and pathological changes seen in adults. The infant's greater freedom from immuno-

---

* *Tinea imbricata* is a fungal infection of the skin that is very common in parts of Papua New Guinea. Susceptibility is inherited as an autosomal recessive trait, and there is evidence that it is determined by the composition of sweat. Sweat from affected individuals (collected in rubber 'sleeves' worn during village football matches) shows defective inhibition of fungal growth.

pathology is illustrated in LCM virus infection of mice. Adult mice die when virus is injected into the brain, as a result of the cell-mediated immune (CMI) response to infection, but infant mice remain well because their response is much weaker. Certain latent infections are kept under control by CMI forces, and in older people with failing CMI these infections are more likely to undergo activation. Thus, older people show an increased incidence of activation of healed pulmonary tuberculosis and of activation of varicella-zoster virus in dorsal root ganglia to cause zoster (see Ch. 10).

This is distinct from age-related differences in the *incidence* of infection. It is not surprising that most infections are commonest in children, especially after first contact with other children at school. Exposure can be an important factor. Sexually transmitted diseases are largely restricted to adolescents and adults, but the occasional case of gonococcal vulvovaginitis in girls before puberty shows that the incidence of such diseases is restricted by exposure rather than by susceptibility.

Immunological immaturity makes newborn animals highly susceptible to viral, bacterial and other infections. Some of the human infections that are also more severe in early infancy are illustrated in Fig. 11.1. They become progressively less severe as the child grows up. The commonest infantile infections are those that cause respiratory illness, and those that give diarrhoea and vomiting. Infections that are prevalent in the community, however, will at some time have infected the mother, and the transfer of maternal antibodies (IgG) to the infant via placenta or milk confers protection during this vulnerable period of life. In the rare human infant who is born without maternal IgG, viruses such as herpes simplex or varicella can cause lethal diseases. Newborn mice also are notoriously highly susceptible to a great variety of experimental virus infections, their mothers not having provided them with antibody. In man, IgM antibodies are not transferred from mother to infant and, in so far as these antibodies are important in resistance to Gram-negative bacterial sepsis, infants are particularly susceptible to coliform sepsis. Maternal antibody, whether to herpes simplex, malaria or streptococci is transferred mainly via the placenta in man, but in other animals transfer through the milk is more important. A newborn foal, for instance, takes a feed of antibody- (especially IgA) rich milk (colostrum) within minutes of birth and thereby acquires a protective umbrella against a great number of infectious agents.

Age-related differences in susceptibility are at times attributable to physical or physiological differences. The increased susceptibility of old people to respiratory infection is partly due to factors such as the loss of elastic tissue round alveoli, weaker respiratory muscles and a poorer cough reflex. Both old people and infants sometimes fail to show the usual signs of infection, such as fever. The lungs of infants are particularly susceptible to whooping cough and other bacterial pneumonias,

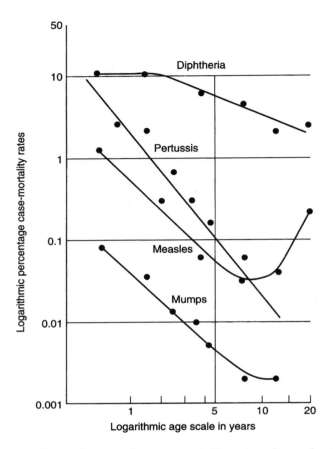

Fig. 11.1 The effects of age on the case-mortality rates of some bacterial and viral infections of man, plotted on logarithmic scales. (Reproduced with permission from Burnet, F. M. (1952). The pattern of disease in childhood. *Australas. Ann. Med.* **1**, 93.)

partly because the airways are narrow and more readily blocked by secretion and exudate. Infants are also the first to suffer the effects of fluid and electrolyte loss, so that infections characterised by fever, vomiting or diarrhoea tend to be more serious at this time of life. Often the reasons for increased susceptibility in infancy are not clear. Respiratory syncytial virus, for instance, often causes serious illness in infancy with croup, bonchiolitis or bronchopneumonia. In adults the virus causes a minor upper respiratory infection, but in early life there is invasion and growth of virus in the lower respiratory tract. It is not known whether this is because respiratory epithelium and alveolar macrophages are more vulnerable to infection than in older individuals, or because host defences are less effective. As an example of age-related susceptibility based on a local physiological difference, the skin of children becomes less susceptible to fungus infections ('ringworm') at puberty, and this is connected with the marked increase in sebaceous secretions at puberty. However, the same increase in sebaceous secretion at puberty leads to greater susceptibility to the skin disease acne, induced by bacteria whose headquarters are sebaceous glands. It is thought that lipases from *Proprionibacterium acnes* hydrolyse triglycerides in sebum to form fatty acids that are responsible for the inflammation in the lesions (and scars in mind and skin).

Certain virus infections are usually milder in childhood, and more likely to be severe when primary infection occurs in adult life. These include varicella-zoster, mumps, poliomyelitis and Epstein–Barr virus infections. Varicella-zoster virus often causes pneumonia in adults, and mumps involves the testicle after puberty, giving a troublesome orchitis. Infections with polioviruses are nearly always asymptomatic in early childhood. When polioviruses first came and caused a 'virgin-soil' epidemic in certain isolated Eskimo communities in the 1940s and in the island of St Helena in 1947, there was a strikingly high incidence of paralysis in adults, but mostly inapparent infections in childhood and old age. Where infection during childhood is the rule, paralytic disease might be expected to be less common. In developed countries, on the other hand (North America, northern Europe, etc.), where there has been a certain amount of interruption of the faecal–oral spread of infection (see Ch. 2), poliovirus infection was often delayed until adolescence or adult life, and as a result paralytic disease had been quite common (until the development of vaccines). Epstein–Barr virus is excreted in saliva, and in developing countries most individuals are infected as young children, undergoing an inapparent infection. In developed countries, where childhood infection is less common, first infections often occur in adolescence or early adult life, following the extensive salivary exchanges that take place during kissing. In this age group, and in these countries, therefore, Epstein–Barr virus causes the more serious disease, glandular fever. It is not known why these infections are more severe in adults, but a more powerful immunological contribution to pathology and disease might be suspected.

The tendency of an acute hepatitis B virus infection to become persistent, and for people in consequence to become carriers, is profoundly affected by the age at infection. Primary infection of adults leads to a persistence rate of about 10%, but when there is infection at birth a persistence rate of over 90% is observed. Susceptibility falls rapidly and by 12 months of age there is a persistence rate of 50%, and between 1 and 4 years a persistence rate of 25%. Neonatal infection is common in the Far East where endemic hepatitis B virus results in maternal transmission, and is important because a proportion of carriers go on to develop primary liver cancer. Fortunately vaccination at birth plus passive protection with immune serum aborts the infection, and the cycle can be interrupted.

Sex of host

There is a slight excess of males over females at birth, but males have a higher mortality, and by old age there is an excess of females. The higher mortality in males is distinct from that due to accidents and wars, and perhaps a slightly increased susceptibility to infectious diseases plays a part. This is also suggested by the observation that, in mice, the difference in mortality between the sexes disappears in the germ-free state.

There are a few examples of sex-related susceptibility to infectious disease, but the differences are usually small and almost nothing is known about mechanisms. For instance, young women show greater mortality from tuberculosis than do men, but later in life the situation is reversed. Mortality and morbidity in whooping cough and infectious hepatitis is also slightly higher in women. A difference in susceptibility, of course, must be distinguished from a difference in exposure. Women have a lower incidence of leptospirosis and jungle yellow fever because they are less often exposed to infection.

It seems likely that hormonal influences are important in sex-related susceptibility, and this is partly because of effects on the immune system. Females generally have higher IgG and IgM levels, develop stronger CMI responses, and are more susceptible to autoimmune diseases. Mechanisms are not clear, but sex hormones (especially oestrogens) influence the differentiation, maturation and migration of lymphocytes, natural killer (NK) cell activity, and phagocytosis by macrophages. Pregnancy certainly affects the severity of many infectious diseases. In pregnant women malaria is more severe, hepatitis A, hepatitis B and hepatitis E are more likely to be lethal, and paralytic poliomyelitis is more common. Hormonal changes during pregnancy are complex. Various new hormones appear, together with substances with known effects on the immune responses such as α-foetoprotein, and there are changes in levels of oestrogens, progesterone and corticosteroids. Increases in corticosteroids would tend to decrease the control of the infectious process (see below) and at the same time inhibit inflammatory responses in the tissues. One should not overemphasise these effects. Pregnancy inevitably brings its own risks, and perhaps some interference with immune responsiveness is unavoidable, but Nature would surely not have tolerated a severe or generalised immunosuppressive handicap. Malnutrition may play a part in the increased susceptibility of pregnant women, especially in developing countries.* In pregnant women, bacterial infections of the

---

* As well as effects on the pregnant mother, malnutrition leads to lower birth weight and reduced survival of offspring. Babies born weighing 4–5 lb are much more likely to die than those of normal weight. In the Gambia, where the average women has 10–12 children, a superbiscuit containing peanuts, dried milk and wheat-soy flour given twice daily has greatly reduced the number of underweight babies, and contributed to a dramatic reduction in child mortality.

bladder are much more likely to ascend to the kidney and cause pyelonephritis. This is partly because in pregnancy the ureter peristalsis that normally 'milks' urine down to the bladder is reduced, making it easier for infection to ascend to the kidney. CMI responses to cytomegalovirus infection weaken during pregnancy, and this virus then reactivates and is shed from the genitourinary tract. Papovaviruses, also, are reactivated and shed in urine during normal pregnancy (p. 350).

The increased susceptibility to infectious disease of pregnant women should be distinguished from the susceptibility of the foetus. The pregnant woman can be regarded as the site of development of a novel set of tissues, including the foetus, placenta, lactating mammary gland, etc., each providing a new and possibly susceptible target for infectious agents. The foetus is exquisitely susceptible to nearly all microorganisms, but access is normally restricted by the placenta. Microorganisms that can infect the placenta, such as syphilis, toxoplasmosis, cytomegalovirus, rubella and smallpox, are then at liberty to infect the foetus (see Table 5.3).

Malnutrition of the host

Malnutrition can interfere with any of the mechanisms that act as barriers to the multiplication or progress of microorganisms through the body. It has been repeatedly demonstrated that severe nutritional deficiencies will interfere with the generation of antibody and CMI responses, with the activity of phagocytes, and with the integrity of skin and mucous membranes. Often, however, nutritional deficiencies are complex, and the identification of the important food factor is difficult. This is reflected in the use of inclusive terms such as 'protein–calorie malnutrition'. Also, at times, it is impossible to disentangle the nutritional effects from socioeconomic factors, such as poor housing, crowding, inadequate hygiene and microbial contamination of the environment. Poverty is a close and ancient companion of infection. Worldwide, the seven most lethal infections in terms of the total numbers killed each year are respiratory infections (general), acquired immunodeficiency syndrome (AIDS), diarrhoeal infections (all forms), tuberculosis, whooping cough, and measles (see Table A.1). The deaths are mostly in developing countries, where communicable diseases still stand as the main public health problem. The period just after weaning is often the most vulnerable, when the nutritional state is poor, and there are many common microorganisms still to be encountered. A study of children in Guatemalan villages provides a good illustration of the synergism between malnutrition and infection. A group of children were studied individually from birth, and their colonisation by various microorganisms and parasites was recorded. There was a loss in body weight after the sixth month, at the time of weaning from breast milk to a deficient diet. Further interruptions in weight gain, and sometimes temporary weight loss, were correlated with measles, various respiratory infections, and infection with *Shigella* and *Entamoeba histolytica*. Body weights at 2 years of age were sometimes little more than a half of those of American children, and there was often almost no gain in weight during the second and third year of life.

Nutrition is affected when there is a bacterial overgrowth in the upper small intestine (see also p. 25). This is common in developing countries and is associated with heavy bacterial contamination of water supplies. The increased numbers of bacteria degrade bile salts to cause malabsorption of fat (steatorrhoea); they impair absorption of carbohydrate, they bind vitamin B_{12}, whose shortage leads to anaemia, and they further interfere with absorption of nutrients when they produce enterotoxins. Children in Jakarta with malnutrition had more than 10^5 bacteria ml^{-1} in the upper jejunum (and as many as 10^7–10^8 ml^{-1}), whereas normal children had less than 10^4 bacteria ml^{-1}.

It seems clear that protein deficiency tends to depress in particular the CMI response, which, together with reduced C3 levels, lowered production of secretory IgA and reduced killing of bacteria by polymorphs, causes an increased susceptibility to many infectious diseases. Children with protein deficiency, the extreme form being represented by the clinical condition called kwashiorkor, are uniquely susceptible to measles. This is a result of their weaker CMI response to the infection, the lowered resistance of mucosal surfaces of the body, and perhaps to the higher contamination of the environment with the microorganisms that cause secondary infections. All the epithelial manifestations of measles are more severe. Life-threatening secondary bacterial infection of the lower respiratory tract is common, as well as otitis media, sinusitis, etc. Conjunctivitis occurs, especially if there is associated vitamin A deficiency, and at times progresses to severe eye damage and blindness. The tiny ulcers in the mouth that constitute Koplik's spots in normally nourished children can enlarge to form massive ulcers or necrosis of the mouth (cancrum oris). Instead of an occasional small focus of infection in the intestine, there is extensive intestinal involvement with severe diarrhoea, which exacerbates the nutritional deficiency.* Even the skin rash is worse, with numerous haemorrhages that give the condition referred to as 'black measles'. The scarcity of good medical care and antibiotic therapy adds to the serious outcome of the illness, and measles is about 300 times as lethal in developing countries as it is in the countries of northern Europe and North

---

* West African children from 6 months to 2 years old, suffering from acute measles enteritis, were found to lose 1.7 g albumin a day in faeces, the ideal daily intake of protein being 9–10 g a day. Thus measles exacerbates malnutrition as well as vice versa, and not only because of diarrhoea, but also because, during systemic febrile infections like measles, there is always a greatly increased breakdown and excretion of body nitrogen.

America. The case fatality approaches 10%, and in severe famines reaches the tragic figure of 50%. The severe form of measles is seen in unvaccinated children in tropical Africa, and it was also seen in children in European cities in the nineteenth century. The virus ('seed') has not altered, but changes in the 'soil' (host) dramatically enhance the severity of the disease. Increased susceptibility to herpes simplex and *P. carinii* (see Glossary) infection, and to Gram-negative septicaemia is also seen in protein deficiency. Because of the effect on CMI, there is greater susceptibility to tuberculosis. Tuberculosis has often been noted to increase in frequency in times of famine, and this has also been observed in the inmates of concentration camps.

On the other hand, it looks as if certain infections are less severe in malnourished individuals. Typhus, for instance, is said to cause a higher mortality in well-fed than in malnourished individuals, and clinical malaria was suppressed in Somali nomads during the 1970s' famines, only to be reactivated 5 days after refeeding. A 4-year study of 100 000 prisoners in the UK in the 1830s showed that those given the most food (costing 3 shillings per week) had a 23% mortality, presumably largely due to infection, whereas those given least food (costing 10 pence per week) had a 3% mortality. It is not known why malnourished individuals are sometimes less susceptible to infection. A decrease in the vigour of host inflammatory and hypersensitivity responses would be expected, and perhaps there are adverse effects on the nutrition of the infectious agent itself, with depressed replication in the malnourished host.

Vitamin A, B and C deficiencies are known to lead to impaired integrity of mucosal surfaces, which in turn causes increased susceptibility to infection, and adds to the complexity of the picture. In developing countries the severity of measles is greatly reduced when children are given vitamin A supplements. There may be a pre-existing vitamin A deficiency, but measles itself causes reduced vitamin A levels. Mortality is lowered and ocular damage, in particular, is less severe. The effect is not only on the integrity of epithelial surfaces. Children given these supplements show less depletion of Th lymphocytes and increased production of measles-specific IgG antibody, compared with untreated children with measles. Children in Papua New Guinea suffered much less from malaria when given vitamin A supplements.

The commonest mineral deficiency is iron and, by affecting certain enzyme systems, this can increase susceptibility to infection. For example, it causes reduced myeloperoxidase activity in phagocytes, with less hydroxyl radical formation, and this means defective killing of bacteria (see Ch. 4). Zinc, selenium, and vitamin E deficiency, especially in the elderly, can reduce immune and phagocytic function. Chronic diarrhoea leads to zinc deficiency, and by giving zinc supplements to children in New Delhi, Brazil and China, the incidence of cough, pneumonia and diarrhoea was reduced.

Hormonal Factors and Stress

Hormones have an important role in maintaining homoeostasis and in regulating many physiological functions in the body. The hormones with a pronounced effect on infectious diseases are the corticosteroids. This is largely because corticosteroids are vital for the bodily response to stress (see Glossary), and infection, like injury or starvation, is a stress (see Fig. 11.2). It has long been known that the adrenal glands are needed for resistance to infection and trauma. Corticosteroids of various types have a complex and wide range of actions; the most important for infectious diseases are the glucocorticosteroids, which inhibit inflammation and depress immune responses. These cortico-

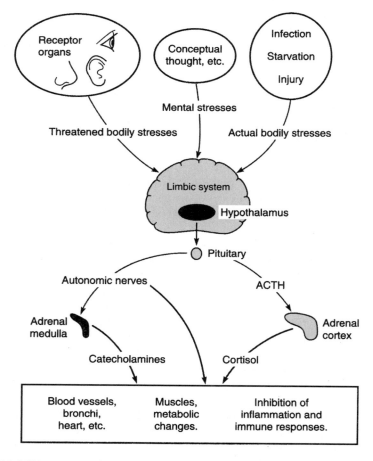

Fig. 11.2 Diagrammatic representation of stress mechanism in man. Various cytokines (IL-1, IL-6 and tumour necrosis factor) act on the hypothalamus, and IL-1 and IL-2 on the pituitary gland.

steroids also stabilise cell membranes and lysosomes, giving cells some protection against damage or destruction. There is, moreover, a great deal of interaction between the neuroendocrine and the immune system which is not included in Fig. 11.2. Not only are immune cells influenced by corticosteroids (see below) and by other mediators generated via the hypothalamic–adrenal axis, but the immune cells themselves (B cells, T cells and macrophages) produce endorphins, adrenocorticotrophic hormone (ACTH), growth hormone and other hormones. Indeed the brain, the endocrine and the immune systems tend to use the same cytokines, peptide hormones and neurotransmitters. For instance, lymphocytes produce growth hormone when cultured *in vitro*. If this is prevented, the lymphocytes stop synthesising DNA, but start again when growth hormone is added. Again, neural cells have receptors for interferons and for interleukin (IL-1), IL-2, IL-3 and IL-6. It seems that discoveries are providing us with more and more mechanisms by which the mind can influence health and disease!

Inflammation makes an important contribution to tissue damage and pathology in infectious disease (see Ch. 8) and injected corticosteroids (or ACTH) have a pronounced anti-inflammatory effect, their therapeutic use in infectious diseases depending on a reduction in the inflammatory pathological components at sites of infection. At the same time they tend to inhibit immune responses. This last action is not completely understood. There is an ill-defined effect on lymphocytes, some of which have receptors for corticosteroids, and inhibition of production and action of immune mediators, such as IL-1 and IL-2. Corticosteroids also prevent the inflammatory expression of the immune response in tissues by blocking the movement of plasma and leucocytes from blood vessels, and this is partly due to inhibition of prostaglandin production.

The inflammatory and immune responses, although on the one hand contributing to pathological changes and disease, are also powerful antimicrobial forces (see Ch. 9). This dual role is reflected in the results of giving corticosteroids in infectious disease. Herpes simplex keratoconjunctivitis or encephalitis, for instance, is temporarily improved by corticosteroids because of the reduction in inflammation, but the simultaneous weakening of antimicrobial forces means that the infection progresses more readily. The net effect is to make the disease worse. For the same reasons a large number of different experimental infections in animals are made more severe by corticosteroid administration.

All the above remarks apply to corticosteroids administered artificially, often in large doses. It is perhaps more relevant to ask what effect the individual's own corticosteroids have on the course of an infectious disease. It is first necessary to say something about the function of the corticosteroid response to stress. Small areas of tissue injury give rise to quite severe but nevertheless locally useful inflammation, mediated by various inflammatory factors. If exactly the same response took place in multiple sites of infection in the body or in response to more extensive tissue injury, the immediate overall result in terms of vasodilation and loss of fluid into tissues would be harmful. An individual who is infected or wounded may need to retain bodily functions for running or fighting, and the effect of multiple unmodified local inflammatory responses might well be incapacitating. When inflammation occurs on a large scale, therefore, it is an advantage to make an overall reduction in its severity, so that the general impact on the host is lessened. This is a teleological way of looking at the function of corticosteroid hormones, which also makes sense of their metabolic function in mobilising energy sources. The response to stress of the autonomic nervous system, involving arenalin-mediated changes in preparation for bodily action (fight or flight) is more obviously interpreted in these terms. During an infection there is an increase in the rate of corticosteroid secretion, just as in response to other bodily stresses such as hunger, injury or exposure to cold. Rises in urinary 17-ketosteroids are seen, for instance, in Q fever and sandfly fever infections in man. There is also an increased rate of utilisation of corticosteroids by tissues. Inflammatory and immune responses thus take place against the dampening and modifying background of increased corticosteroid levels, which ensure that continued bodily function and balance (homoeostatis) is maintained. When the corticosteroid response is depressed, as in Addison's disease (see Glossary), the consequences of infection or tissue injury are very severe, and affected patients therefore have to be given increased doses of corticosteroids during infections.* Bilaterally adrenalectomised animals usually show greatly increased susceptibility to tissue damage and death in experimental infectious diseases.

It can be concluded therefore that increased circulating levels of corticosteroid hormones are necessary for a successful host response to infectious disease. Administering additional amounts of corticosteroids is not necessarily of value unless the host's own corticosteroid response is known to be subnormal, or if it is for the moment more important to reduce inflammation than to control infection. Otherwise, additional corticosteroids tend to promote the infection by decreasing the effectiveness of antimicrobial forces, as discussed above.

When corticosteroids are given, they not only make any infection that happens to occur at the time more severe, but also favour the lighting up of persistent infections that are normally held in check by immune forces. Tuberculosis in man is often activated or made worse by corticosteroid administration. Stress tends to act in the same way,

---

* It may be noted that in Cushing's syndrome there is also a greatly increased susceptibility to infection because of excessive production of corticosteroids from the adrenal cortex. Abnormally high corticosteroid levels promote infection for reasons referred to above, and bacterial infections have been leading causes of death in these patients.

probably because of increased secretion of corticosteroids. One classical example in animals is psittacosis, a chlamydial infection of parrots and budgerigars. These birds normally carry the microorganism as a persistent and harmless infection, localised in the spleen. Following the stress of transport in cages, exposure to strange surroundings or inadequate diet, the infection is activated in the bird, and the microorganism begins to be excreted in the faeces. Human infection can then take place by inhalation of particles of dried droppings from the cage, causing the troublesome disease psittacosis, with pneumonia as a common feature.

In humans, mental stress in the form of anxiety calls into action the same physiological changes which were designed to deal with physical stresses (see Fig. 11.2). For instance, in a university boat race the crew had increases in corticosteroid production that enabled them to sustain the physical stress of the race, but the coxswain was found to have an increase of equal magnitude. The evidence for mental stress acting on infection is not impressive, and for the most part involves immune responses. For example, a study of a symptomatic human immunodeficiency virus (HIV)-infected individuals showed that those judged to be stressed had lower counts of cytotoxic T cells and NK cells. In another study, 48 students were given three doses of hepatitis B vaccine during examination periods over the course of 6 months, and those who had seroconverted after the first injection were less likely to have been stressed and anxious. Standard stress and anxiety assessment scales were used. It is possible but not proven that sustained mental stress, by causing persistent rises in circulating corticosteroids, lowers resistance to persistent infections and other infections that occur during the period of stress (see illness clustering, below). Stress appears to influence the recurrence (reactivation) of oral and genital herpes.

Infections are sometimes more severe when the host animal lives under crowded conditions. Increased transmission as well as stress responses, can play a part. Intestinal coccidiosis in domestic animals is generally asymptomatic, but clinical disease is seen when a heavy parasite load is carried. This is favoured under crowded conditions because of increased transmission. The increased rate of meningococcal and streptococcal disease when people are crowded together is due to increased transmission.

The adrenal cortex itself is not often involved in infectious diseases but, if it is, the infection in the cortex tends to be extensive. Examples include tuberculosis and histoplasmosis in man and various viral, bacterial, fungal and protozoal infections in experimental animals. Infectious agents localising in the adrenal cortex encounter a high concentration of corticosteroid hormones originating from cortical cells. Antimicrobial forces are therefore weakened locally, and the infection is exacerbated. Active adrenal foci of infection are often seen at a time when foci elsewhere in the body are healing.

There is usually a change in susceptibility to infection during pregnancy, as discussed earlier in this chapter, and this is due to hormonal changes. The relative importance of oestrogens, progesterone and corticosteroids is not clear. Oestrogens are necessary for maintaining the resistance of the adult vagina to most bacterial infections, as described on pp. 44–45. The male sex hormones responsible for the changes in the testicle at puberty can be regarded as causing this organ's susceptibility to mumps virus infection. Insulin is also worth mentioning because the metabolic changes in poorly controlled diabetes in some way increase susceptibility to staphylococcal, fungal and tubercular infections (see pp. 50–51). Clearly there are hormones that control the health and well-being of cells and tissues in all parts of the body and, in this sense, serious hormonal disturbances could always affect the course of infectious diseases. It would be surprising for instance if untreated cretins showed a completely normal response to infections. Such effects would scarcely be worth mentioning were it not for the existence of this category called 'hormonal factors'.

Stress proteins

So far, the word stress has been used to describe the response of the infected host. It can also refer to the response of the individual host cell, or to the response of the infecting microorganism or parasite. In other words, infecting bacteria, etc. and host cells have their own stress responses, elicited during infection, heat or other stimuli. Stress proteins are recognised in most living organisms. They help protect cells from harmful effects of stress, and often function by helping with the correct folding, translocation and assembly of other proteins, acting as 'molecular chaperones'. They are present under normal circumstances but are produced in much larger amounts in response to stress. There are families of heat shock proteins (hsp), and hsps 60, 70 and 90 are well-studied examples. *S. aureus* has four classes of heat shock genes, and hsps are often dominant antigens of microbes. For instance, in human mycobacterial infection, up to 40% of the total T-cell response is to bacterial hsp 65. The fact that the human equivalent, also produced in the infection, is hsp 60, and shows considerable sequence similarity with hsp 65, gives opportunities for cross-reactive autoimmune responses by the host. However, the importance of such responses in mycobacterial or other infections has not been established. Viruses do not have their own stress proteins, but there are indications that they make use of those produced in the infected cell. In the adenovirus-infected cell, the adenovirus E1A gene product itself causes increased synthesis of host hsp 70, which seems to have a role in the handling of viral proteins and the assembly of virus particles.

Other Factors

A host of miscellaneous factors influence the course of infectious diseases, and some of them merit particular mention. Certain lung conditions resulting from the inhalation of particles have an important effect on respiratory infection. Silicosis is a disease due to the continued inhalation of fine particles of free silica. It occurs in coal miners and in various industries where sandstone and similar materials are used. There is a great increase in susceptibility to tuberculosis, which is more likely to cause serious or fatal disease. This is because lung macrophages, which play a central role in resistance to respiratory tuberculosis, become damaged or destroyed following the phagocytosis of the free silica particle. When intact macrophages containing nonlethal amounts of silica phagocytose tubercle bacilli, the bacteria grow faster, the cell dies, and the progeny bacteria are released sooner.

Nowadays, most people spend 90% of their lives indoors, and air exchange with the outside world is much less than it used to be, but some exposure to atmospheric pollutants is inevitable. The air is polluted in many towns and cities, especially with substances derived from the combustion of commercial, domestic and automobile fuels. These include SO_2, nitrogen oxides, CO, ozone, benzene, acid aerosols, and also particles. Although these particles form a small proportion of the total mass of particles suspended in air, they are important because they include small (<2 μm diameter) particles, which are stable, penetrate deep into the lungs and may bear acidic gases or contain toxic elements such as lead. In many countries, pollutants (but not CO_2!) have been reduced by clean air laws, catalytic converters, and the use of lead-free fuels. The commonly measured pollutants are SO_2 and particulates (smoke). For both, the upper limit (24 h mean) recommended by the World Health Organisation (WHO) is 100–150 μg m^{-3}, but these values are commonly exceeded. Can atmospheric pollution increase the severity of respiratory infections? It has been reported that people with chronic bronchitis produce larger volumes of morning sputum and note a worsening of symptoms when SO_2 values in air reach 250 μg m^{-3}, and there is an increase in respiratory mortality when levels exceed 750 μg m^{-3}. In the great London smog of 1952, before the Clean Air Bill greatly improved the quality of London air, SO_2 levels reached 8000 μg m^{-3}, and there were 4000 excess respiratory deaths. The morbidity and mortality, however, is seen in the respiratory cripples (chronic bronchitis, etc.), in the very old and in other susceptible individuals.* One feels that atmospheric pollution must also be having a long-term harmful effect on the lungs of normal

---

* The same vulnerable groups also experience increased mortality in influenza epidemics.

people, but what about respiratory infections? Although exacerbations of asthma and of cardiopulmonary disease are well established, there is no convincing evidence that normal people exposed to atmospheric pollution experience an increase in the severity of acute respiratory infections. A careful study of 20 000 children and adults in four geographical areas of the USA has shown that high levels of SO_2 and suspended sulphates are significantly associated with excess acute respiratory disease, much of which can be assumed to be infectious. The effect was most marked after more than 3 years' exposure, and it was independent of cigarette smoking and socioeconomic status, two of the factors that had always been difficult to dissociate from atmospheric pollution in previous studies. Cigarette smoking can be regarded as self-induced atmospheric pollution, and many interesting observations have been made. For instance, cigarette smoke inhibits ciliary activity, the debris-laden alveolar macrophages of smokers show less bactericidal activity, and lung pathogens, such as pneumococci and *Haemophilus influenzae*, attach more readily to pharyngeal cells from smokers. Cigarette smoking is certainly associated with chronic bronchitis, but the evidence linking cigarette smoking with susceptibility to acute respiratory disease in otherwise healthy individuals is conflicting. For instance, one study of 1800 students at a military college in the USA during the Hong Kong 'flu epidemic showed that those who smoked 21 cigarettes a day had a 21% higher incidence of clinical influenza, but other studies have failed to show an effect. On the other hand, a recent (2000) study of 228 smokers and 301 nonsmokers showed that smokers were four times as likely (and passive smokers 2.5 times as likely) to develop pneumococcal disease as nonsmokers. The subjects were immunocompetent, otherwise healthy people aged 18–64 years.

It is a widespread popular belief that people are less resistant to infectious diseases when they are in a poor mental state, and there is in fact some evidence that psychological factors influence susceptibility. This is seen in the phenomenon of illness clustering. In two studies in the USA, the illnesses and significant life events of several thousand people were recorded over a period of about 20 years. It was found that in a given individual, illnesses of all kinds, not only psychosomatic conditions such as peptic ulcers but also bacterial infections and tumours, tended to occur in clusters. There was a significant association of these illness clusters with stressful life situations, such as the death or serious illness of a close relative, personal injury, career crises, etc. It can be difficult to interpret results and, indeed, not all studies have given the same results. Little is known of the mechanism by which such events influence infectious diseases. Presumably it involves the stress response and the known effects of the nervous system on immune responses (see p. 380).

Simple fatigue generally has little effect on susceptibility to infection, but violent exercise in the early stages of poliomyelitis is known to predispose to paralysis in the exercised muscles. The exercise must be done during the preparalytic stages of infection, when virus is spreading from the alimentary canal to the central nervous system. It is associated with dilation of capillary blood vessels supplying the spinal cord neurons that innervate the exercised muscles. Perhaps circulating virus is more likely to invade such regions of the spinal cord. Paralytic poliomyelitis also tends to involve muscles that receive injections during the preparalytic stages of the infection, especially with materials such as pertussis vaccine. In this case too, the injection causes capillary dilation in the appropriate region of the spinal cord.

Exposure to changes in temperature and sitting in draughts are traditionally regarded as influencing infectious diseases. Careful studies with common cold viruses have not provided any evidence for this. Volunteers infected intranasally with a standard dose of virus were exposed to cold, but failed to show detectable changes in the incidence or severity of infection even after standing naked in draughty corridors. The effect of changes in relative humidity has been less carefully studied. Experimentally, ciliary activity in segments of respiratory epithelium is impaired by reductions in the relative humidity of the overlying air. Increases in air temperature in heated buildings lead to substantial reductions in relative humidity unless the air is humidified. The lower respiratory tract would tend to be protected because of humidification of inhaled air by the turbinate mucosa, but the nasal mucosa would be exposed to the dry air, and an effect on ciliary activity and thus on respiratory infection might be expected.

The local concentrations of key elements sometimes determine microbial growth in tissues. For instance, nearly all bacteria require iron, but the body fluids of the host contain iron-binding proteins, such as lactoferrin and transferrin, which limit the amount of free iron available. Hence certain bacteria show greatly increased virulence after administration of iron to the host, and patients with excess iron in the blood may show increased susceptibility to infection. The lethality for mice of *Pseudomonas aeruginosa* is increased 1000-fold by the injection of iron compounds to saturate the iron-binding capacity of serum transferrin. The ability of bacteria to compete with the host for iron can be an important factor, and virulent bacteria such as pathogenic *Neisseria* and enteric bacilli that produce their own iron-binding compounds (collectively called siderophores), are able to circumvent the host restriction on the availability of iron (see also Ch. 8).* Oxygen is a key element for other bacteria. It is essential for many, such as the tubercle bacillus, but *Clostridium perfringens*, for instance, is strictly anaerobic and multiplies best in tissues that are anoxic as a result of

---

* Malaria parasites induce the formation of transferrin receptors on the surface of infected red blood cells.

interruption to their blood supply. Bacterial multiplication is actually inhibited in the presence of oxygen, and patients with gas gangrene are treated by exposure to oxygen in a pressure chamber. *Clostridium tetani* also requires local anoxic conditions in tissues, whether produced by severe wounds or by trivial injuries due to splinters, thorns or rusty nails. It may be noted that some of the most successful invaders of the respiratory tract show optimal growth in the presence of up to 5–10% CO_2 (e.g. tubercle bacilli,* pneumococci). The gases bathing the lower respiratory tract normally contain about 5% CO_2.

Foreign bodies in tissues often act as determinants of local microbial virulence. The term foreign bodies includes foreign particles that are too large to be phagocytosed. Foreign bodies presumably act by interfering with the blood supply and also by serving as a continuous source of multiplying microorganisms, giving them physical protection in nooks and crannies from phagocytes and other antimicrobial forces. Foreign bodies potentiate various clostridial infections (see above) and particularly staphylococci infections. Necrotic bone fragments in chronic osteomyelitis act as foreign bodies, hindering treatment and giving a source of bacteria for flare-up of infection many years later. The ability of staphylococci to cause a local lesion after introduction into the skin is increased about 10 000-fold if the bacteria are implanted on a silk thread. Skin is generally more susceptible to infection when wet, as well as following injury. Wet pastures and minor foot injuries predispose to various types of 'footrot' in cattle, sheep and pigs, due to infection with *Fusiformis* spp. or other bacteria. Patients with plastic devices inserted at the body surface or in deeper tissues show increased susceptibility to commensals such as *S. epidemidis*. The devices act by interrupting the integrity of host defences or by forming a surface for bacteria to grow on, and include catheters in veins, cerebrospinal fluid shunts, prosthetic hips and knees, cardiac pacemakers, etc.

Certain drugs influence resistance to infectious disease and, of the self-administered drugs, alcohol is the commonest. In various studies, intoxicated animals have been found to have impaired ciliary activity, impaired removal of inhaled bacteria, defects in phagocytosis or poor closure of the glottis. Most of these phenomena have not been satisfactorily demonstrated in man, and polymorph function, for instance, appears normal, although there is impaired migration of polymorphs from blood vessels. The position is clearer for chronic alcoholics, many of whom have alcoholic liver disease. These individuals have reduced polymorph counts in the blood and are more likely to develop bacterial (especially pneumococcal) pneumonia. Alcoholics also show increased

---

* Tubercle bacilli commonly cause lesions in the apical regions of the lung, perhaps because oxygen and CO_2 tensions in these regions favour bacterial growth or depress host defences.

susceptibility to pulmonary tuberculosis, but it is not clear how much is due to impaired host defences and how much to the alcoholic lifestyle. Lung infections can also be acquired by inhaling anaerobic bacteria from the mouth while in a drunken stupor.*

Those who inject themselves with narcotics are particularly susceptible to infection. To a large extent this is due to the insanitary techniques used, and it results in skin sepsis or more serious systemic infections, such as endocarditis. In heroin addicts, staphylococcal infections are not so common as might be expected, apparently because street heroin contains quinine which has antistaphylococcal action. Shared syringes may transmit hepatitis B virus infection or HIV and, as with alcohol, susceptibility to pulmonary infection is increased during drug-induced stupor. One disadvantage of regular marijuana smoking is that it lowers stomach acid and thus increases susceptibility to bacterial infection of the intestine (see p. 27). The multiplication of HIV in T cells is enhanced by alcohol, morphine or cocaine, suggesting that these may increase susceptibility by boosting up what would otherwise have been a noninfectious dose of virus. Co-infection with herpes viruses (simplex or cytomegalovirus) also enhances HIV multiplication.

The influence of immunosuppressive drugs on infection is an important feature of hospital medicine at the present time, and is referred to in Ch. 9.

References

Ahmed, S. A., Penhale, W. J. and Talal, N. (1985). Sex hormones, immune responses and autoimmune diseases: mechanisms of sex hormone action. *Am. J. Pathol.* **121**, 531.

Blalock, J. E. (1994). The syntax of immune neuroendocrine communication. *Immunol. Today* **15**, 504.

Cann, A. J. (1997). 'Principles of Molecular Virology', 2nd edn, Academic Press, London.

Chandra, R. K. (1983). Nutrition, immunity and infection: present knowledge and future directions. *Lancet* **i**, 688.

Cohen, S. and Williamson, G. M. (1991). Stress and infectious disease. *Psychol. Bull.* **109**, 5–24.

Coutsoudis, A. *et al.* (1992). Vitamin A supplementation enhances

---

* During sleep, normal individuals often aspirate material from the nasopharynx. This can be demonstrated by introducing 1.0 ml of an ^{111}In-labelled indium chloride solution into the nose every half hour during sleep; a gamma scan carried out after waking reveals the presence of the labelled materials in the lungs. In other words, the lungs are regularly contaminated during sleep with microorganisms from nose and throat, but this does not lead to trouble as long as host defences (see pp. 21–25) are intact.

specific IgG antibody levels and total lymphocytic numbers while improving morbidity in measles. *Pediatr. Infect. Dis.* **11**, 203.

Dorman, C. J. (1994). 'Genetics of Bacterial Virulence'. Blackwell Scientific Publications, Oxford.

Fields, B. N. and Byers, K. (1983). The genetic basis of viral virulence. *Phil. Trans. R. Soc. Lond. B* **303**, 209 (survey of reovirus studies).

French, J. G. *et al.* (1973). The effect of sulfur dioxide and suspended sulfates on acute respiratory disease. *Arch Environ. Health* **27**, 129–133.

Gardner, I. D. (1980). The effect of ageing on susceptibility to infection. *Rev. Infect. Dis.* **2**, 801–810.

Godlee, F. (1991) Air pollution – road traffic and modern industry. *Brit. Med. J.* **303**, 1539–1543.

Gracey, M. S. (1981). Nutrition, bacteria and the gut. *Brit. Med. Bull.* **37**, 71–75.

Huxley, E. J. *et al.* (1978). Pharyngeal aspiration in normal adults and in patients with depressed consciousness. *Am. J. Med.* **64**, 564–568.

Jindal, S. and Malkovsky, M. (1994). Stress response to viral infection. *Trends Microbiol.* **2**, 89.

Joklik, W. K. (1985). Recent progress in reovirus research. *Annu. Rev. Genet.* **19**, 537–575.

Kan, Y. W. and Dozy, A. M. (1980). Evolution of the haemoglobin S and C genes in world populations. *Science* **209**, 388.

Luzatto, L. (1979). Genetics of red cells and susceptibility to malaria. *Blood* **54**, 961–976.

MacKenzie, J. S. (1982). Viral vaccination in the malnourished. *In* 'Viral Diseases in South-East Asia and the Western Pacific' (J. S. MacKenzie, ed.), pp. 166–173. Academic Press, London.

Mata, L. (1982). Malnutrition and concurrent infections. Comparison of two populations with different infection rates. *In* 'Viral Diseases in South-East Asia and the Western Pacific' (J. S. MacKenzie, ed.), pp. 56–76. Academic Press, London.

Mata, L. J. (1975). Malnutrition – infection interactions in the tropics. *Am. J. Trop. Med. Hyg.* **24**, 564–574.

Meyer, C. G., May, J. and Stark, K. (1998). Human leukocyte antigens in tuberculosis and leprosy. *Trends Microbiol.* **6**, 148–154.

Murray, J. and Murray, A. (1977). Suppression of infection by famine and its activation by refeeding – a paradox? *Perspect. Biol. Med.* **20**, 471–484.

Ogasawara, M., Kono, D. H. and Yu, D. T. Y. (1986). Mimicry of human histocompatibility HLA-B27 antigens by *Klebsiella pneumoniae*. *Infect. Immun.* **51**, 901.

Peterson, P. K., Chao, C. C. *et al.* (1991). Stress and the pathogenesis of infectious disease. *Rev. Infect. Dis.* **13**, 710–712.

Pincus, S. H., Rosa, P. A., Spangrude, G. J. and Heinemann, J. A. (1992). The interplay of microbes and their hosts. *Immunol. Today* **13**, 471–473.

Report of Study Group (1979). Acne. *J. Invest. Dermatol.* **73**, 434–442.

Weinberg, E. D. (1978). Iron and infection. *Microbiol. Rev.* **42**, 45–66.

Zinkernagel, R. M. (1979). Association between major histocompatibility antigens and susceptibility to disease. *Annu. Rev. Microbiol.* **33**, 201–213.

12

Vaccines and How they Work

Introduction

Probably the greatest achievement in medicine in the twentieth century has been the great reduction in the incidence of infectious disease. Smallpox has been eliminated, and most of the old scourges such as tuberculosis, cholera, diphtheria and typhoid have been brought under control, at least in the developed countries of northern America, northern Europe, Australia, etc. giving us the opportunity to die of other things later in life. This revolution in infectious diseases was in the first place the result of dramatic improvements in sanitation and public health, which provided clean water supplies, adequate disposal of sewage and better housing. The downward trends in many infectious diseases were in progress early in the twentieth century, well before antibiotics and vaccines had been invented.

Improvements in water supplies and sewage disposal obviously have a great impact on enteric diseases such as cholera and typhoid. Better housing and nutrition have had an important influence on other diseases. Tuberculosis, referred to as the Great White Plague in the cities of nineteenth-century Europe, and notoriously promoted by crowding and poverty, has been steadily declining as a cause of death as standards of housing and nutrition have improved. Infectious diseases like typhus and plague have receded as people and their dwellings have become free from the lice, fleas and rats that were necessary for the spread of these diseases. But all these infections are still present in the world and the people of developed countries are protected from them only so long as they continue to be protected from lice, fleas, rats, poverty, crowding and contaminated food and water. A general breakdown in the organisation and structure of modern society would lead to food shortages and allow the lice, fleas, rats and contaminated water to return, together with many of the old diseases. This is what happens on a limited scale during wars and in natural disasters such as earthquakes. War, famine and pestilence traditionally ride together. Once an infectious agent has been totally eradicated on a global scale, it cannot of course return. This is difficult to achieve with infections such as malaria, plague and yellow fever because they have vectors and animal reservoirs (see Glossary), but infections restricted to man and involving no other host can be totally eradicated if all human infection is prevented. Smallpox came into this last category, and it was totally eradicated from the world by a relentless vaccination programme carried out by the World Health Organisation (WHO). Smallpox eradication was also made easier because the virus does not persist in the body and therefore cannot reactivate. Polio virus is another example, where it is hoped to bring about complete eradication of disease within the next few years. Other infections restricted to man and which do not cause persistent infection and reactivation include whooping cough, bacillary dysentery, and measles. Although many infectious diseases were already declining following general improvements in public health, the decline was greatly accelerated by the development of vaccines to prevent diseases and antibiotics to treat infections. Vaccines, used on a large scale, have been a major antimicrobial force in the community. As each microbial agent has been isolated and identified, its cultivation under artificial conditions has generally led within a short time to the development of a vaccine. Many infections, especially virus infections such as measles and poliomyelitis, have receded wherever effective vaccines have been used. Some vaccines are better than others. Yellow fever has proved to be one of the best vaccines, while vaccines against typhoid and cholera have so far remained comparatively unsatisfactory. There has been complete failure to develop effective vaccines for many important human diseases such as trachoma, human immunodeficiency virus (HIV), malaria, syphilis and gonorrhoea.

Infectious diseases remain the greatest health problem for most people (see Table A.1) and for most animals in the world. More and more, the practitioners of medicine or veterinary science will be concerned with the prevention rather than with the cure of diseases. Advances in immunology and microbiology are leading to the development of many new vaccines, and many better vaccines, and it is the purpose of this chapter to survey some of the principles governing the development and use of vaccines.

What is a vaccine?

A vaccine* is a material originating from a microorganism or other parasite that induces an immunologically mediated resistance to

---

* The word vaccine (Latin, *vacca* = cow) derives from the vaccinia (cowpox) virus inoculated to protect against smallpox.

disease.* Material with similar structure and activity can also be produced artificially rather than obtained from the actual micro-organism or parasite.

What do we ask of an ideal vaccine?

1. That it promotes effective resistance to the disease, but not necessarily to the infection.
2. That resistance lasts as long as possible.
3. That vaccination is safe, with minimal and acceptable side effects. Standards have changed for human vaccines, and today we are more safety conscious than we used to be. The smallpox vaccine, which remained more or less unchanged for more than a 100 years, would never have been licensed if introduced a few years ago. Rather lower safety standards are acceptable for most veterinary vaccines. A vaccine, even if not completely safe, should be safer than exposure to the disease, assuming that the risk of exposure is significant. Attitudes to a given vaccine's safety depend on whether the safety of the individual or the protection of the community is under consideration. When a vaccine gives protection to the community, the community owes a debt to any individual damaged by the vaccine.
4. That the vaccine is stable, and will remain potent during storage and shipping. The fact that yellow fever virus can be freeze-dried and transported unrefrigerated in the tropics has been a great asset favouring the success of this vaccine. Poliovirus cannot be successfully freeze-dried, but vials containing the live (oral, Sabin) vaccine show a colour change when overheated.
5. That the vaccine is reasonably cheap, if it is for large-scale use, or for use in developing countries.

General Principles

Effective resistance to infection or disease depends on the vaccine having certain properties, and there are often different requirements for different types of infection. Some important general principles are listed below.

---

* Vaccines are distinct from specific antibodies which are given to confer passive immunity. Immunoglobulin pooled from normal adults generally contains enough antibody to hepatitis A virus to confer protection for a few months. But for other viruses (hepatitis B, rabies, mumps, varicella-zoster) immunoglobulin from known immune donors is necessary.

1. The vaccine should induce the right type of immune resistance

The relative importance of antibody and T cells in resistance to disease has been discussed in Chs 6 and 9. Vaccines should induce the type of immunity that is relevant for the particular microorganism. Resistance to tuberculosis or typhoid seems to require effective T-cell-mediated immunity, whereas resistance to yellow fever or poliomyelitis requires a good antibody response. If the wrong type of response is induced, protection is inadequate, and once or twice the disease when it occurred has even been made more serious. This is highlighted in the use of formalin-inactivated respiratory syncytial virus (RSV) in clinical trials in the 1960s resulting in vaccinees succumbing to severe lung disease following a natural infection with the virus. This condition was reproduced in mice where Th2 responses dominated the normally protective Th1 response, resulting in the accumulation of eosinophils and lung pathology. Eosinophilia was a condition identified in the blood of the vaccinees. In the natural infection, immunity to the F protein of RSV is linked to protective immune responses. In the inactivated RSV vaccine, the F protein is damaged and the immune response is targeted to the G protein, favouring the induction of Th2 responses.

2. The vaccine should induce an immune response in the right place

For resistance to infections of epithelial surfaces, it is more appropriate to induce secretory IgA antibodies than circulating IgG or IgM antibodies. Thus, secretory IgA antibodies might give valuable protection against influenza or cholera, but not against rabies or yellow fever which by-pass epithelial surfaces and enter the body through bite wounds (see Ch. 2). Even the secretory antibody response must be in the right place; antibodies in the intestine will not protect the nose or throat. Unfortunately, in spite of attention to these principles, live polio vaccine (Sabin) remains almost the only one that induces a good IgA-mediated immunity.

3. The vaccine should induce an immune response to the right antigens

A given microorganism contains many different antigens, as discussed in Ch. 6 and as illustrated by the number of genes (Table 12.1). There are many hundreds or thousands of antigens in the case of protozoa, fungi* and bacteria, and in virus infections from as little as three (poly-

---

* *Candida albicans*, for instance, contains at least 78 water-extractable antigens, and from *E. coli* 1100 proteins have been resolved by two-dimensional gel electrophoresis.

Table 12.1. Sizes of genome of microorganisms

Microorganism		No. of genes[a]
Viruses	Polyomavirus	6
	Poliovirus	5
	Influenza virus	10
	Adenovirus	30
	Herpes virus	160
	Poxvirus (vaccinia)	300
Chlamydias	Trachoma	800
Mycoplasmas	*Mycoplasma* spp.	900
Rickettsias	*Rickettsia prowazeki* (typhus)	1000
Bacteria	*Neisseria gonorrhoeae*	1000
	E. coli	3000
Protozoa	Malaria	12 000

[a] Known, or calculated (as number of medium-sized proteins that can be coded for) from molecular weight of nucleic acid.

omavirus) to more than 100 (herpes and poxviruses) are produced. Immune responses to many of these antigens develop during infection. Resistance to infection, however, depends principally on immune responses to the smaller number of antigens on the surface of the microorganism. The relevant surface antigens have been isolated and characterised for certain viruses, but much less is known of the surface antigens that induce resistance to chlamydia, bacteria, fungi and protozoa. Vaccines consisting of killed whole bacteria, for instance, inevitably induce a very large number of irrelevant immune responses.

4. Resistance to some infectious diseases does not depend on immunity to the infectious agent

In certain infections such as tetanus and diphtheria, disease is entirely due to the actions of toxins as discussed in Ch. 8. Immunity to the disease requires only an effective antibody to the toxin. For the production of vaccine, therefore, a toxin is modified by chemical or physical treatment (alcohol, phenol, ultraviolet irradiation) so that it is no longer toxic, but maintains its antigenic character. The resulting toxoid is a very effective vaccine when combined with an adjuvant (see below).

5. There are important differences in principle between killed and live vaccines

The primary response to an antigen is classically distinguished from the secondary response. After the first injection of an antigen, the immune response begins and at the same time the antigen itself is generally degraded and disposed of in the body. The second injection of antigen now induces a greatly enhanced response, and subsequent injections give further boosts (see Fig. 12.1). Each killed vaccine must therefore be given in repeated doses if an adequate immune response and resistance is to be induced. The microorganisms in live vaccines, on the other hand, multiply in the host after administration. The antigenic mass contained in the vaccine itself is small but it is increased many thousand times following growth of the microorganism in the body. The effective dose is greatly amplified in this way, and the primary merges into the secondary immune response, giving a high level of immunity (Fig. 12.1). Only one dose of vaccine is therefore needed to produce satisfactory immunity. Nearly all the successful viral vaccines, both medical and veterinary, consist of living attenuated virus. Examples of different types of vaccine are given in Table 12.2, and differences between live and killed vaccines are summarised in Table 12.3.

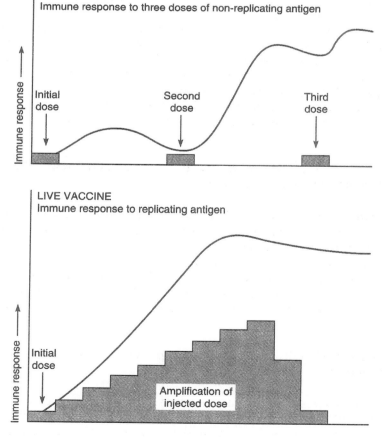

Fig. 12.1 Comparison of immune responses to live and to killed vaccines.

Table 12.2.   Types of vaccines

Vaccine	Live vaccines	Killed vaccines
Viral	Smallpox[a] Rubella[a] Measles[a] Poliomyelitis (Sabin) [a] Yellow fever[a] Mumps[a] Varicella-zoster Also 10–20 commonly used veterinary vaccines (dog, cat, cattle, horse, chicken, pig, sheep)	Poliomyelitis (Salk)[a] Influenza Rabies (human diploid cell)[a] Hepatitis A Hepatitis B[b]
Bacterial	BCG[h] Brucella (veterinary use)	Cholera Typhoid Whooping cough[f]
Bacterial polysaccharide vaccines		Pneumococcus[c] – 23 anti- genically distinct polysaccharides Meningococcus – serogroups A and C[d] *Haemophilus influenzae* b[c]
Rickettsial		Typhus
Bacterial toxoid vaccines	Diphtheria,[a] tetanus[a] *Clostridium perfringens* (veterinary use)	
Helminths	Cattle lung worm No effective vaccines for human helminth infestations	
Important diseases with no effective vaccine available	Trachoma, chlamydial urethritis,[e] malaria Syphilis, gonorrhoea, trypanosomiasis, leprosy,[f] schistosomiasis Human herpes viruses (herpes simplex,[f] Epstein–Barr virus,[f] cytomegalovirus),[f] rotavirus,[f] respiratory syncytial virus,[f] HIV Rheumatic fever[g]	

[a] Highly effective.

[b] The first vaccine produced by recombinant DNA technology after the cloning of the surface antigen (HBSag).

[c] There are 84 pneumococcal serotypes but most serious illnesses are due to the 23 more common types. Children less than 2 years old generally give poor antibody responses to polysaccharide vaccines. Capsular polysaccharides are being used to produce vaccines to other bacteria, such as *Haemophilus influenzae*. Polysaccharides are T-independent antigens and their immunogenicity in infants can be enhanced when they are converted to T-dependent antigens by conjugation with protein 'carriers'.

[d] Unfortunately, most cases of meningitis in the UK and USA are due to serogroup b, but this particular polysaccharide is poorly immunogenic in man.

[e] The chlamydia responsible for trachoma, urethritis, salpingitis, conjunctivitis and lymphogranuloma inguinale (see Table 2.3) exist as at least 15 serotypes. In diseases such as trachoma, promoted by flies, crowding and shortage of water, and where re-infection is probably important, improvements in hygiene may in the end be as important as vaccines. In one study in Mexico, a daily face wash reduced the incidence of trachoma in children from 48% to 10%.

[f] New vaccines undergoing clinical trials; oral vaccines for rotaviruses, respiratory syncytial virus, *H. pylori* and influenza will be available before long.

[g] Still an important disease in three-quarters of the world's population. A vaccine would induce immune responses to M proteins from the relevant streptococcal types (see p. 279), but there is the possibility that these responses could lead to heart damage, as in the disease itself. The M proteins, however, have type-specific determinants localised to 20 amino acid residues, which are highly immunogenic when coupled to polylysine.

[h] BCG is derived from a strain of *M. bovis* isolated in 1908 from a cow with mastitis; after 230 subcultures over the course of about 15 years, the bacteria had lost their virulence. Highly effective in children, but less so in prevention of pulmonary tuberculosis in adults. We await a new twenty-first century vaccine for tuberculosis.

Table 12.3. Comparison of live and killed vaccines

Live	Killed
Must be attenuated by passage in cell culture or bacteriological media	Can be produced from fully virulent microorganisms, e.g. poliovirus (Salk), typhoid vaccines
Given as a single dose[a]	Given in multiple doses
Smaller number of microorganisms needed	Large number of microorganisms needed
Tend to be less stable	Tend to be more stable
Adjuvant not required	Adjuvant often required
Can often be given by a natural route	Generally given by injection
Induces antibody and T-cell responses	Induces antibody but poor T-cell responses
Possibility of spread of infection to unvaccinated individuals	Spread is not possible

[a] Live polio (Sabin) vaccine is an exception. Each dose contains the three types of poliovirus that interfere with each other's replication in the intestine, and it must be given on three occasions to ensure an adequate response to each type.

6. The disease should be serious enough to justify vaccination

Rubella, for instance, is a very mild disease, and vaccination would not be worthwhile were it not for the fact that infection during pregnancy can lead to serious damage to the foetus. This incidentally is the only vaccine that is used to protect an as yet nonexistent individual. Many coxsackie and echo virus infections cause little or no illness and vaccines are not therefore required. Vaccines are sometimes given particularly to certain groups of individuals. This is generally because they are susceptible to some complication of the infection, as in the case of rubella infection of the foetus in pregnant women. Similarly, older people and those with chronic respiratory diseases are often given influenza virus vaccines because they are susceptible to influenza pneumonia. Children with leukaemia have been given a live varicella-zoster vaccine because varicella is often fatal in these children. Again, the 23-valent pneumococcal polysaccharide vaccine can be given to children with sickle cell disease, who are very susceptible to pneumococcal infection. Restricted vaccination is sometimes based on the likelihood of exposure to the disease as with vets or workers in quarantine kennels who receive rabies virus vaccine, and dentists at risk from infected blood who receive hepatitis B vaccine.

7. Factors determining the duration of resistance

Clearly the longer protection lasts the better; no vaccine would prove popular if an injection were required every 6 months throughout life. The duration of resistance to disease depends to some extent on the type of infection. In the case of systemic infections with an incubation period of a week or two, a low residual level of immunity gives resistance to disease, because even if re-infection does occur, the immune response is boosted during the incubation period and the infection is terminated before the onset of disease. Repeated subclinical booster infections may be important in maintaining immunity to diseases such as measles and rubella. Infections of the body surfaces, in contrast, have incubation periods of only a few days, and if there is a low residual level of immunity* re-infection can occur and cause disease before there has been time for the immune response to be boosted and the infectious process controlled. Thus it is difficult to induce long-lasting immunity to parainfluenza virus infections or to gonorrhoea, but easier for measles or mumps.

We have little understanding of the factors responsible for the longlasting immune responses to microorganisms that are seen in the absence of persistent infection or re-infection. Immunity to live yellow fever virus vaccine, for instance, is probably lifelong, although the infection is not a persistent one, and viable virus is apparently completely eliminated from the body. Perhaps small amounts of viral antigen remain sequestered in some site in the body (see p. 163). Live vaccines give longer lasting protection than killed ones, if the infectious agent persists in the body and produces antigens, to give continuous stimulation of immune responses (BCG, Marek's disease vaccine for poultry).

8. The concept of attenuation

It would seem ridiculous to use the naturally occurring disease agent as the vaccine, because it would tend to cause the disease that one wishes to prevent. In the early days of smallpox vaccination, however, living virus from the scabs of smallpox patients was used as a vaccine. Lady Mary Wortley Montagu, wife of the British Ambassador to Turkey, brought this type of vaccination ('variolation') to England over 250 years ago. It was effective, but could be fatal, and was made illegal in 1840† when Jenner developed his calf lymph vaccine. Usually it is necessary to reduce the pathogenicity of the micro-

---

* Secretory IgA responses tend to be short-lived compared with IgG responses. Accelerated secondary IgA responses are seen, but are weaker than with IgG (see pp. 163–164).

† Variolation had a mortality of about 1%, as compared with a mortality of 15–20% for smallpox itself (the milder form of smallpox, variola minor, did not arise until around 1900). Variolation was carried out a good deal in England, especially in the 1760s, someone called Daniel Sutton having variolated 300 000 people, and Jenner himself was variolated as a boy in 1756.

organism by growth in artificial media such as cell culture (measles, Sabin polio) or bacterial growth media (BCG). This is an empirical procedure, depending on the fact that prolonged passage of a micro-organism in an artificial system tends to select mutants better suited to growth in that system than in the original host. Unless the micro-organism can be conveniently cultivated artificially such attenuation is impossible, and this is why on most occasions live vaccine for a given virus soon follows its successful cultivation *in vitro*. Attenuation has usually been a 'blind' procedure, and the microorganism has to be tested for virulence during its continued cultivation in the laboratory. The yellow fever vaccine strain of virus (17D) arose in this way, and only arose once, by what amounts to sheer good fortune. Nowadays attenuation can sometimes be carried out more rationally. For instance, influenza and respiratory syncytial virus mutants have been produced that grow poorly at 37°C, the temperature of the lower respiratory tract, but well at the temperature of the nose, 33°C. These temperature-sensitive (ts) mutants multiply after instillation into the nose and induce immunity, but are unable to spread to the lower respiratory tract.

Other approaches to attenuation rely on identifying 'virulence' genes in microorganisms and either removing or modifying these genes by genetic manipulation (see p. 410). A novel approach to attenuation has been achieved with the herpes viruses. This involves deleting the glyco-protein H gene from the virus genome. This glycoprotein is essential for virus maturation and in its absence only an abortive infection is possible. Propagation of the infectious virus *in vitro* is only possible if a source of gH is provided by a cell line into which the gH gene has been transfected. Such attenuated viruses are unable to spread from cell to cell and hence to a new host. Nevertheless, this is an efficient means of stimulating immunity.

The process of attenuation must be taken far enough so that the vaccine does not cause disease. An early live measles vaccine (Edmonston strain) caused fever and a rash, and human gammaglob-ulin was administered at the same time to decrease the severity of the vaccine disease. Attenuation, however, must not be taken too far, because the microorganism may then fail to replicate fully enough to induce a good immune response.

9. The concept of monotypic microbes

Certain microorganisms are antigenically much the same, wherever and whenever they occur, so that resistance to disease, once estab-lished, is secure. This is so for polio, measles, yellow fever or tubercu-losis. Sometimes a given disease is caused by a number of microorganisms which differ antigenically, and resistance to only one of them will not provide resistance to the disease. There are dozens of antigenically distinct types of streptococci, for instance, and resistance to streptococcal infection is not complete until there have been immune responses to them all. The same is true for the common cold, which can be caused by more than 100 antigenically distinct viruses belonging to at least five different groups. Some microorganisms are undergoing repeated antigenic changes during the course of their circulation in the community. Respiratory viruses in particular are evolving rapidly in this way. Vaccination against today's strains may give no protection against tomorrow's variant. This is true of influenza viruses in man, which always tend to be one step ahead of the vaccinators, and of foot and mouth disease virus in animals.

10. Adjuvants

Adjuvants are materials that increase the immune response to a given antigen without being antigenically related to it. Aluminium salts act in this way, and in diphtheria and tetanus vaccines the toxoids are combined with aluminium hydroxide or phosphate. The aluminium salt converts the soluble toxoid into a particulate precipitate and thus increases immunogenicity. Killed *Bordetella pertussis* bacteria, as used in the current pertussis vaccine, have a slight adjuvant action, and increase the immune response to other vaccines that are given at the same time.

Various oils are effective as adjuvants. The vaccine material is generally administered with the oil as a water-in-oil emulsion, and the mechanism of action is partly because of the very low breakdown of the mass of oil and consequently slow release of antigen. However, mineral oils, at least, are potentially carcinogenic, and oils tend to cause local sterile abscesses after injection. Finally, mycobacterial products act as adjuvants, and Freund's original complete adjuvant consists of killed, dried mycobacteria (usually *Mycobacterium tuberculosis*) suspended in mineral oil. Mycobacterial adjuvants can cause granulomas and are not acceptable for human or animal use. Alternatives to whole bacteria are subunits of the bacterial cell wall or bacterial toxins. Muramyl dipeptide, for instance, a synthetic product that is also a component of the cell wall of various bacteria and responsible for the adjuvant activity of mycobacteria shows great promise. Cholera toxin and the heat-labile toxin of *E. coli* are potent mucosal adjuvants and immunogens associated with overt disease of the gut. However, by manipulating the gene of these toxins using targeted mutagenesis, it is possible to engineer toxins deficient in disease production, but still capable of providing adjuvant activity.

A microbial product that has attracted a lot of interest as an adjuvant is bacterial DNA. Bacterial DNA provides a potent stimulation of the immune system due to CpG motifs containing unmethylated CpG dinucleotides.* The adjuvant activity of the CpG ODN motifs (see footnote) is linked to entry into lymphocytes and antigen presenting cells causing an upregulation of key cytokines (interleukin-12 (IL-12), tumour necrosis factor-α (TNF-α), interferon-γ (IFN-γ)) which in turn trigger a cascade of immune responses. When these structures are delivered with a vaccine, they rapidly augment the host response leading to the protective immunity even against fast-acting and dangerous pathogens such as Ebola virus and anthrax.

A number of alternative adjuvants and delivery systems are being explored (see Table 12.4), and some of them are in veterinary use. These include the use of microspheres composed of lactic and glycolic acids which encapsulate the vaccine. The complexes are biodegradable causing a gradual release of the antigens to the immune system. A similar approach is to enclose vaccines in synthetic lipid vesicles (liposomes). The antigens are not only released slowly, but can also fuse with cell membranes, delivering the antigen intracellularly where it becomes processed and presented via MHC class I molecules.

11. Interference

The ultimate objective of vaccination programmes might be to administer all vaccines at the same time, thus giving a once and for all protection against everything that matters. There is evidence, however, that when too many antigens are combined, they sometimes interfere with each other (antigenic competition), so that the immune response to each is not so great as if they had been given separately. More importantly, live viruses occasionally interfere with each other. Live measles virus vaccine, for instance, could inhibit the growth of other live virus vaccines given at the same time, perhaps by inducing interferon. Live poliovirus vaccine (Sabin) given at the same time as other vaccines is unlikely to interfere with them because it grows in a different part of the body, establishing an exclusively intestinal infection. Sabin vaccine, however, contains the three distinct strains of poliovirus, and these tend to interfere with each other during their multiplication in the intestine. They can also be interfered with by any naturally acquired enteroviruses that happen to be multiplying in the intestine at the same time. After the first dose there is often a response to only one of the strains. The same Sabin vaccine is therefore given three times, to ensure that a satisfactory response to each of the virus strains takes place. This is the reason for this apparent exception to the

---

*A CpG motif contains an unmethylated CpG dinucleotide flanked by two 5′ purines and two 3′ pyrimidines. These motifs, present in bacterial DNA or as short oligodeoxynucleotides (ODN), cause activation of dendritic cells, macrophages and natural killer (NK) cells, and polyclonal stimulation of B cells. CpG motifs are also a feature of insect DNA, but not mammalian DNA.

principle about single doses of live vaccines being adequate (see above). However, the theoretical problem of interference seems less important in practice and there are immense advantages (cost, organisation) in using combined vaccines. For instance, there is widespread use of MMR vaccine (a mixture of live measles, mumps and rubella virus vaccines) given to children aged 1 year or older, sometimes together with live oral poliovirus vaccine, and the live varicella-zoster virus vaccine will be added before long.

12. The age at which vaccines should be given

Human infants are born with a supply of maternal IgG antibody derived from the placental route, and they thus acquire resistance to all infections to which the mother had antibody-mediated immunity. They also receive secretory IgA antibodies in colostrum and milk, and these provide some protection against intestinal infections. Live polio-vaccine (Sabin) is less likely to immunise in the first few months of life because secretory IgA antibodies from maternal milk inhibit the growth of the vaccine virus.

The first encounter with many microorganisms thus takes place under an umbrella of maternal immunity, and when infection takes place it is likely to be mild yet at the same time significant enough to generate some immunity in the infant. Maternal antibody persists for up to 6 months after birth and vaccines, particularly live vaccines, are likely to be less effective if given before this time. Diphtheria, tetanus and whooping cough vaccines are not given until the infant is 3–6 months old. Certain infections, however, such as measles and whooping cough are particularly severe in infants and very young children (two-thirds of deaths from whooping cough occur during the first year of life), and there is a need to give protection as soon as possible after maternal immunity has faded.* Pertussis vaccination, therefore, is commenced at 3 months to give protection during the first year of life, and live measles vaccine is given at 1 year.† The complications of diph-theria and pertussis vaccination are commoner in older individuals, and this is another reason for giving these vaccines early in life. Rubella is designed to protect the foetus, and in the UK at present live

---

* In developed countries, measles vaccine is not generally given before the age of 12 months. In developing countries, however, where measles still accounts for about a million deaths a year, children are infected during the first year of life, and there is a need for a vaccine that is effective ideally at 2–6 months, when maternal immunity is fading.

† Each year more than 100 000 infants die of tetanus, usually acquired from an infected umbilical stump. Vaccination of mothers is an effective method of prevention, as it would also be for group A streptococcal or respiratory syncytial virus infection.

rubella vaccine is given to girls aged 10–13 years, who are old enough to develop a good immune response, but presumably too young to be pregnant. Finally, it may be worthwhile vaccinating elderly people against infections to which they are particularly susceptible, such as influenza or pneumococcal pneumonia.

In the poultry industry, the use of *in ovo* technology (vaccination of eggs) has been a major breakthrough in the large-scale vaccination of chickens against troublesome pathogens, such as Mareks disease virus, infectious bursal disease virus and Newcastle disease virus. Combination vaccines against several viruses are possible without compromising survival at hatching.

13. Problems of testing

All vaccines have to be tested, and their safety and effectiveness evaluated. This may take many years. Human vaccines must be tested on human beings. Preliminary trials in experimental animals are useful, but do not always give results applicable to man. The effectiveness of a vaccine cannot be reliably assessed merely by measuring the immune responses induced because it is not known which of the many responses induced is the one which gives protection; it must be tested in those exposed to natural infection. Trials of this sort are difficult when the disease is an uncommon or serious one (e.g. rabies). In the case of hepatitis B, the vaccine was tested by being given to people (male homosexuals) whose lifestyle made it likely that they would soon become infected. Trials must, of course, be large enough and well planned; large numbers may be needed for the proper evaluation of safety. Many vaccines are still of uncertain status because trials have been too small, or because of differences in the potency or composition of a given type of vaccine. About 200 new vaccines are still in the research and development stage.

14. Problems of vaccine production

Even when a vaccine of known potency has been developed, its large-scale use depends on efficient methods of mass production. For instance, there would be problems with a leprosy vaccine derived from whole bacteria because these bacteria cannot be grown in culture but must be obtained from artificially infected armadillos. There are not enough armadillos in the world to give an adequate supply of bacteria. A similar problem would have arisen with the hepatitis B virus vaccine derived from the blood of human carriers. Genetic engineering (see below) has solved this problem, and doubtless the key immunogen(s) for protection against leprosy will be produced in a similar way.

15. Problems with immunological unresponsiveness

Failure to give an adequate response to a vaccine may be due not only to immaturity (see p. 404) or immunosuppression, but also to the following factors.

1. Malnutrition (see Ch. 11).
2. Interference. Very frequent enterovirus infections in children in developing countries can interfere with the 'take' of live poliovirus vaccine. There are suggestions that failures of BCG vaccination in India are due to previous infection with related mycobacteria, which induced immune tolerance to cross-reactive but protective antigens in the BCG vaccine.
3. Unknown factors. For instance, infants develop poor antibody responses to the polysaccharides of type b meningococci, and it is precisely this type that causes serious infection in this age group. The cause for this is not known, but one solution is to couple covalently a protein (e.g. outer membrane protein of the meningococcus) to the polysaccharide, which then induces a good (and now T-dependent) response. Again, there are reports that hepatitis B vaccine fails to induce protective antibodies in certain individuals. It is not known why, but there is the possibility that the unresponsiveness is genetically controlled.

Complications and Side Effects of Vaccines

Virulent infectious material in vaccine

Virulent microorganisms are inactivated to make killed vaccine. If inactivation has been incompletely carried out, vaccination introduces infection. On one occasion, incomplete inactivation of virulent poliomyelitis virus by formaldehyde in the killed (Salk) vaccine caused paralytic poliomyelitis in large numbers of vaccinated children. There have also been injurious effects because of the presence of additional unsuspected microorganisms in a vaccine. Live yellow fever virus vaccine was at one time stabilised by the addition of human serum, and during the last war thousands of US servicemen vaccinated against yellow fever became jaundiced because of the presence of hepatitis B virus in the 'normal' serum. Poliomyelitis virus for the killed (Salk) vaccine was produced in normal monkey kidney cells and then inactivated by formaldehyde. Subsequently, however, the normal monkey kidney cell cultures were shown to contain a papovavirus, SV40 (simian virus 40), which was present in the vaccine and was not inactivated by the formaldehyde treatment. This virus transforms cells and causes tumours experimentally. The thousands of children who had been injected with live SV40 virus were therefore followed with some anxiety, but fortunately there were no harmful effects.

Allergic effects of vaccines

First, nonmicrobial antigens in vaccines can cause allergic responses, especially in vaccines that are given more than once. Vaccines containing penicillin and egg proteins, for instance, have given trouble. Even the microbial components in a vaccine may sometimes give allergic responses (local swelling, rash, etc.) in hypersensitive individuals. Second, certain vaccines induce autoimmune-type responses in the host. The post-vaccinial encephalitis, or the peripheral neuritis (Guillain–Barré syndrome) that very occasionally occurs a week or two after administration of various killed or live virus vaccines (as well as after natural infection) appears to arise in this way. For instance 10 per million Americans developed the Guillain–Barré syndrome after being given inactivated influenza virus vaccine in 1976 in a nationwide attempt to protect against swine influenza. Possible autoimmune side effects must be considered, especially with vaccines for infectious diseases where there is a significant amount of immunopathology. Microbial components in the vaccine might induce cross-reactive auto-immune responses that could be harmful (e.g. trachoma, rheumatic fever).

Toxicity of vaccines

Large numbers of killed salmonellas are present in the vaccine for typhoid, and this vaccine therefore contains large amounts of endo-toxin (see Ch. 8). Accordingly, fever and malaise are not uncommon sequels to vaccination, although these side effects are reduced by injection of the vaccine by the intradermal route.*

Influenza virus vaccines containing inactivated whole virus particles often give troublesome febrile and local reactions in children. This is not seen with vaccines containing disrupted virus material, nor of course with live virus vaccines. The current pertussis (whooping cough) vaccines cause a variable incidence of neurological sequelae. Up to 1 in 5000 vaccinated children may be affected with more serious effects in 1 in 100 000. This whole cell vaccine contains at least 49 different antigens,

---

* Typhoid remains a world-wide problem, with an estimated 12.5 million cases per year, and carriers are common in affected areas (for instance, nearly 700 carriers per 100 000 people in Santiago, Chile). The standard killed whole cell vaccine gives up to 70% protection for about 3 years. Recent alternatives (licensed) are: (1) injection of purified Vi polysaccharide antigen; (2) oral dosing with 2×10^9 organisms of a live attenuated strain (Ty 21a) of *S. typhi*. These give fewer side effects and probably better protection.

and the main toxic component is the lymphocytosis promoting factor, which also causes hypoglycaemia and histamine sensitisation. It contributes to the illness in the natural infection. New pertussis vaccines, already in use in several industrialized countries, consist of purified antigens rather than whole bacteria. They contain the lymphocytosis promoting factor (converted into a toxoid), together with other antigens that protect against bacterial colonisation of the respiratory tract.

Harmful effects on the foetus

After vaccination against smallpox, or during infection with the disease, the foetus was often infected and killed. Live rubella virus vaccine can infect the foetus but it is not clear whether it causes damage. Live hog cholera and bluetongue vaccines can certainly cause foetal infection and malformations in domestic animals. Because maternal ill health or infection is a risk to the foetus, it is a good general rule not to give vaccines to pregnant women, especially live vaccines, and especially during the first trimester of pregnancy. On the other hand, immunisation during pregnancy can give specific protection to the newborn (see pp. 164–165). Tetanus vaccine is often given to pregnant women in developing countries to prevent neonatal tetanus and certain veterinary vaccines can be administered during pregnancy to protect the offspring.

Effects on immunodeficient host

Live vaccines often cause illness in children with immunodeficiencies. Children with agammaglobulinaemia may develop paralytic disease following vaccination with live polio (Sabin) vaccine. Deficiencies in cell-mediated immunity are the most important, and the consequences of giving measles or BCG vaccine to children with thymic aplasia have been described in Ch. 9. The infection caused by the live virus or bacterium is readily controlled in the normal individual and induces a good immune response; it is not controlled and may give rise to serious or lethal infection in the immunodeficient child. Live vaccines can also cause disease in those that are immunodeficient as a result of malignant disease of the lymphoreticular system.

There has been widespread concern about possible harmful effects of live vaccines in HIV-positive individuals. The immune response, moreover, may be poor. But because of the risk of severe disease, measles, mumps, rubella and polio* vaccines are given as well as the inactivated

---

* Live (Sabin) polio vaccine may be excreted in faeces for long periods (type 2 was excreted for 15 years in one case) and inactivated (Salk) vaccine is often given.

vaccines, whether or not there are symptoms of AIDS-related complex (ARC) or AIDS. BCG, however, is not generally given to HIV-positive individuals.*

The Development of New Vaccines

Problems

There are four important problems in vaccine development.

1. A failure to grow satisfactory quantities of the microorganism in the laboratory. Leprosy, human papilloma virus, hepatitis C virus and hepatitis B virus come into this category.
2. When crude preparations of killed microorganisms are used as a vaccine, they often give poor protection against disease. Only a small number of the microbial antigens that are present induce a protective immune response, and in most cases (gonorrhoea, herpes viruses, typhoid, trachoma, etc.) the key antigens have not been identified. As often as not, there is ignorance as to the type of immune response that needs to be induced. When we do know this, how can we induce Th1 as opposed to Th2 responses? (see p. 153.)
3. Live vaccines often give effective protection, but the virus vaccines that are most needed are for the herpes virus group (see Table 12.2). Live vaccines for these viruses may be difficult to achieve in man because of the possibility of virus reactivation later in life (see pp. 342–346) and because of the remote possibility that they might induce cancer (e.g. Epstein–Barr virus, see p. 358).
4. Some of the most successful microorganisms induce ineffective immune responses in the host or actually interfere with the development of effective immune responses (see Ch. 7). Examples include gonorrhoea, syphilis, HIV and the herpes virus infections. This is likely to be a problem with live vaccines but, if a particular microbial component is responsible for this activity, it could be eliminated from a killed vaccine.

Many vaccines contain large numbers of irrelevant antigens, derived either from the microorganism itself or from the culture system used to produce it. It would be better to replace these crude soups with cocktails of defined polypeptides. Sometimes it does not matter if the relevant protective antigen or antigens are not known, as long as relatively

---

* Tuberculosis is a major complication of AIDS. World-wide, an estimated four million HIV-infected individuals suffer also from tuberculosis. Less pathogenic alternatives to BCG are being sought, such as *Mycobacterium vacca*, an environmental saprophyte (isolated from cow-dung) that induces protection without the damaging responses.

clean preparations of virus are available, as in the case of the inactivated rabies vaccine produced from human diploid cells. However, for many bacterial and protozoal infections we need to know more about the role of microbial surface components in pathogenesis.

There are a few bacterial diseases where the relevant antigen has been identified. Capsular polysaccharides can be used to induce protection against pneumococcal, meningococcal or *Haemophilus influenzae* infections (see Table 12.2). Capsular materials are readily obtained by growing bacteria in the laboratory, but it may be noted that many of the genes that code for the enzymes that synthesise capsule have been cloned. The K88 and K99 adhesins (p. 15) that are responsible for the attachment of *E. coli* to the gut wall of piglets and calves can be used to immunise the pregnant mother whose antibodies will then protect the newborn animal against *E. coli* diarrhoea. Also, the diseases due to the action of a toxin can be approached by developing a toxoid vaccine, and promising work is in progress with a cholera toxoid for oral immunisation. Capsular polysaccharides that promote T-cell-independent immune responses are notoriously poor immunogens in infants. However, when such polysaccharide antigens are coupled to a carrier protein, then strong T-cell-dependent immune responses are induced with IgG antibodies and long-lived memory responses. This approach revolutionised vaccination against *Haemophilus influenzae* b, a cause of pneumonia and central nervous system disorders in infants, and has now opened new avenues for producing meningococcal and pneumococcal vaccines in the future.

The problem of growing the microorganisms can be solved by genetic engineering. If the DNA that codes for the relevant antigen (e.g. hepatitis B surface antigen) can be obtained, it is incorporated into a plasmid which is introduced into a bacterium (*E. coli*) or a yeast (*Saccharomyces cerevisiae*). The antigen can then be bulk-produced from cultures. All hepatitis B vaccine is now produced in this way.

It is also possible to develop live avirulent vaccines by removing or inactivating the genes that confer virulence. For instance, the gene for the cholera toxin has been cloned in *E. coli*, altered by mutation and then re-introduced into virulent *Vibrio cholerae*. This gives a strain of bacteria that multiplies without producing the toxin when given orally, and induces immunity to cholera.

Transgenic plants are an exciting addition to the mass production of vaccines with the potential to be delivered as part of the normal diet. To produce vaccine-transgenic plants, the gene for a protective antigen is introduced under the control of plant-specific DNA regulatory sequences and integrated into the genome of the plant. This vaccine antigen can then be transferred from generation to generation in the normal way. Recent successes include the introduction into potato tubers of the capsid protein of Norwalk virus (a virus associated with acute gastroenteritis), the LT fusion protein of *E. coli* and CT-B subunit of cholera toxin. A clinical trial involving eating raw potatoes con-

taining the LT fusion protein demonstrated specific IgG and IgA antibodies sufficient to overcome a high dose of virulent *E. coli*. If these antigens can still retain immunogenicity after they are cooked, then this
will make a tremendous impact on delivering vaccines on a large scale
to developing countries, where enteric diseases are a major problem.

An exciting new development in vaccination is the use of nucleic acid
vaccines. In this situation nucleic acid encoding a foreign antigen
(vaccine) is directly introduced into tissue, resulting in the transfection
of host cells and the expression of the foreign protein. DNA vaccines
are composed of a bacterial plasmid, with a strong viral promoter, the
gene of interest and polyadenylation termination signals. The plasmid
can also act as an adjuvant by containing CpG motifs. Delivery of DNA
vaccines is by intramuscular injection or by gene-gun, a process
involving the plasmid DNA being coated on gold particles which are
then shot into skin by a high-pressure gas jet. This process has been
shown to promote antigen presentation by both the MHC class I and
class II pathways. An early example of this technique was the introduction of the gene encoding influenza haemagglutinin into the muscle
of mice, resulting in the induction of both T-cell and antibody immunity
which protected the animals against an influenza A virus challenge.
DNA vaccines have been used widely to protect animals against a
variety of pathogens, including those against dangerous pathogens and
against pathogens where the antigens have proved difficult to isolate
in any quantity, particularly those of complex parasites. A summary of
vaccine delivery systems and adjuvants is given in Table 12.4.

Synthetic peptides as vaccines

For a given protein only a small proportion of the molecule is important
as an immunogen. This consists of a short sequence of amino acids (a
continuous antigenic site) or a bunch of amino acids which are not
immediately linked to each other but are brought together by the
conformation of the protein (a discontinuous antigenic site). Once the
protein is denatured the latter ceases to exist. Examples of the former
include a 14-amino-acid peptide of diphtheria toxin, and similar-sized
peptides from hepatitis B virus surface antigen or from the VP1
polypeptide of foot-and-mouth disease virus are active, in so far as they
can induce protective antibodies when covalently linked to a carrier
protein. Small peptides like this can be chemically synthesised, by-
passing the need to produce the entire polypeptide in *E. coli* or in
yeasts. However, even a continuous peptide has to have the same
conformation as in the whole protein, and most antibodies are made to
irrelevant conformations of it and are not protective. Rules governing
the conformation of proteins are not understood at present, and guess-
work is a poor substitute. There are also problems with immunogenicity (the need for carriers, adjuvants, etc.), and there is no progress

Table 12.4.   Summary of vaccine delivery systems and adjuvants

Aluminium salts	Aluminium hydroxide or phosphate (alum). Precipitate soluble antigen making the complexes more immunogenic
Emulsions	Water-in-mineral oil, e.g. Freund's incomplete adjuvant. Killed M. tuberculosis or muramyl di- or tripeptides added to stimulate strong T-cell responses – Freund's complete adjuvant
Microspheres	Lactic and glycolic acids encapsulate antigens. Biodegradable, causing slow release of vaccine
ISCOMS	Immune stimulating complexes. Composed of glycosides in Quil A (adjuvant), cholesterol, phospholipids and antigens. Forms a capsule 30–40 nm diameter
Liposomes	Lipid vesicles which encapsulate antigens. Glycoproteins can be inserted into liposome membranes
Natural mediators	The vaccine contains, or expresses, IL-1, IL-2 or IFN-γ
Nucleic acid vaccines	Genes encoding antigens can be introduced directly into cells using liposomes or 'gene-gun' leading to protein expression and immunisation
Recombinant viruses	Genes encoding foreign antigens introduced into viral genome leading to expression of protein following infection of cells, e.g. vaccinia, adenovirus, plant viruses
Recombinant bacteria	Foreign genes can be introduced into chromosome or plasmids. Include M. tuberculosis BCG, attenuated (aromutants) S. typhimurium and S. typhi
Transgenic plants	Foreign genes expressed in leaves or in potato tubers induce protective immunity when delivered orally

at all on the problem of discontinuous antigenic sites. A major limitation to the use of peptides as vaccines is the MHC and finding peptide combinations that will cover the various MHC haplotypes. These problems suggest that the use of synthetic peptides as vaccines is still some way off.

Attenuated viruses and bacteria as carriers

Finally, it is possible to introduce the gene for any given viral protein into the genome of an avirulent virus that can then be administered as a live vaccine. The foreign viral protein is produced in infected cells, and induces an immune response. This has been done mostly with vaccinia virus, into which genes from viruses such as hepatitis B, influenza, rabies, herpes simplex, HIV and foot-and-mouth disease have been introduced. This has the supreme advantage that genes from up to ten or more different infectious agents could be introduced into the same strain of vaccinia virus, a single inoculation of which would simultaneously immunise against a wide range of infections. Unfortunately, smallpox-vaccinated people cannot be immunised in this way, and vaccinia virus is not considered safe enough for people by modern standards (see above). Such concerns have resulted in the genetic engineering of highly attenuated vaccinia virus and the use of canary poxvirus as vaccine vehicles. The latter is unable to grow fully in mammalian cells, but is capable of initiating protective immune responses to antigens by the virus. Other virus vectors are being investigated, notably adenovirus which has been used to target antigens to initiate mucosal immune responses. A major advantage of recombinant viruses is their ability to elicit a wide range of immune responses, in particular cytotoxic T cells. The 17D yellow fever virus vaccine is highly successful, yet the related viruses, dengue and Japanese encephalitis, are still important public health problems. One promising approach to vaccination against the latter viruses is to insert their envelope glycoprotein genes in place of the glycoprotein gene in the yellow fever virus vaccine.

Bacteria have also been used to deliver antigens to the immune system. Here foreign genes introduced into the chromosome or into plasmids of attenuated *Salmonella typhimurium* and mycobacteria BCG have been used successfully to initiate immune responses at mucosal surfaces. In the case of intestinal pathogens, the avirulent bacteria, given orally, colonise and multiply, the polypeptides from the pathogens are produced, and gut immunity develops against the intestinal pathogen.

New vaccines will certainly be produced and some of them will have a major impact on the health of people in developed and quite possibly in developing countries. The difficulty of this task, in what is regrettably still largely an empirical science, is no better illustrated than in the enormous and so far abortive efforts to find a vaccine against HIV. But it is unfortunately true that much human illness and death is due to diseases that can be prevented by the effective use of existing vaccines. This is largely a matter of money and organisation, of politics, economics and inequalities in health care. Millions of deaths have already been prevented as a result of WHO immunisation programmes. It looks as if poliomyelitis has been eradicated from the Americas and the WHO aims at global eradication. But diphtheria, pertussis, tetanus, measles, polio and tuberculosis still kill more than two million children each year. What is needed are a heat-stable oral polio vaccine, a single-dose tetanus toxoid vaccine, a measles vaccine that can be given earlier in life, and vaccine 'cocktails', in which a single dose confers immunity to many different infections. Also there is a great need for vaccines that do not require injections. At present 24 of the 27 standard vaccines have to be injected, and the number is rising. It matters in developed countries, where a child receives about ten infections by the age of 18 months, as well as in developing countries, and obviously oral vaccines and combination vaccines will solve the problem. One promising approach, once the oral route is decided on, is to produce edible vaccines. As mentioned above there is rapid progress in this area and it will not be long before safe vaccines are available for consumption; always assuming that the Western world does not completely abandon the idea of genetically modified foods. This could lead to the exciting prospect of banana 'supervaccines'. Plantations of bananas will surely be much cheaper than the products from production plants in pharmaceutical companies.

Successful new vaccines will come from encouraging laboratory advances in immunology and biotechnology and at the same time attending to practical problems such as vaccine supply, quality testing, licensing and financing. Then it is a matter of organisation and enthusiasm. December 5th 1996 was a mass vaccination day in the Indian subcontinent and on that day no less than 118 million children were vaccinated against poliomyelitis. Everyone hopes that the WHO vaccine initiatives will 'serve the children of the world by having the fruits of high science speaking to their immune system, rather than allowing children to reach specific immunity via the cruel lottery of the disease process' (Sir Gustav Nossal, Kyoto, November 1993).

References

Ada, G. L. (1993). Towards phase III trials for candidate vaccines. *Nature (Lond.)* **364**, 489–490.

ASM News (1993). Vaccine linked to rare adverse health effects. ASM News **59**, 547–548.

Berzofsky, J. A., Ahkers, J. D., Derby, M. A., Pendleton, C. D., Arichi, T. and Belyakov, I. M. (1999). Approaches to improve engineered vaccines for human immunodeficiency virus and other viruses that cause chronic infections. *Immunol. Rev.* **170**, 151–172.

Cardenas, L. and Clements, J. D. (1993). Stability, immunogenicity and expression of foreign antigens in bacterial vaccine vectors. *Vaccine* **11**, 126–135.

Cox, W. I., Tartaglia, J. and Paoletti, E. (1993). Induction of cytotoxic T lymphocytes by recombinant Canary pox (ALVAC) and attenuated vaccinia (NYVAC) viruses expressing the HIV-1 envelope glycoprotein. *Virology* **195**, 845–850.

Cox, J. C. and Coulter, A. R. (1997). Adjuvants – a classification and review of their modes of action. *Vaccine* **15**, 248–256.

Crowle, A. J. (1988). Immunization against tuberculosis; what kind of vaccine? *Infect. Immun.* **56**, 2769–2773.

Donnelly, J. J., Ulmer, J. B., Shiver, J. W. and Liu, M. A. (1997). DNA vaccines. *Ann. Rev. Immunol.* **15**, 617–648.

Fenner, F. (1982). A successful eradication campaign. Global eradication of smallpox. *Rev. Infect. Dis.* **4**, 916.

Haq, T. A., Mason, H. S., Clements, J. D. and Arntzen, C. J. (1995). Oral immunization with a recombinant bacterial antigen produced in transgenic plants. *Science* **268**, 714–716.

Heilman, C. A. and Baltimore, D. (1998). HIV vaccines – where are we going? *Nature Med. Vaccine Suppl.* **4**, 532–534

Kaper, J. B., Lockman, H., Baldini, M. M. and Levine, M. M. (1984). Recombinant nontoxigenic Vibrio cholerae strains as attenuated cholera vaccine candidates. *Nature* **308**, 655–658.

Katz, S. L. (1997). Future vaccines and a global perspective. *Lancet* **350**, 1767.

Klinman, D. M., Verthelyi, D., Takeshita, F. and Ishii, K. J. (1999). Immune recognition of foreign DNA: A cure for bioterrorism. *Immunity* **11**, 123–129.

Liu, M. A. (1998). Vaccine developments. *Nature Med. Vaccine Suppl.* **4**, 515–519.

McGhee, J. R. and Mestecky, J. (1990). In defence of mucosal surfaces. Development of novel vaccines for IgA responses at the portals of entry of microbial pathogens. *Infect. Dis. Clinics North Am.* **4**, 315–341.

Miller, L. H. and Hoffman, S. L. (1998). Research toward vaccines against malaria. *Nature Med. Vaccine Suppl.* **4**, 520–524.

Minor, P. D. (1992). The molecular biology of poliovaccines. *J. Gen. Virol.* **73**, 3065–3077.

Mor, T. S., Gomez-Lim, M. A. and Palmer, K. E. (1998). Perspective on edible vaccines – a coming of age. *Trends Microbiol.* **6**, 449–453.

Pardoll, D. M. (1998). Cancer vaccines. *Nature Med. Vaccine Suppl.* **4**, 525–531.

Rappuoli, R., Pizza, M., Douce, G. and Dougan, G. (1999). Structure and mucosal adjuvanticity of cholera and *Escherichia coli* heat-labile enterotoxins. *Immunol. Today* **20**, 493–500.

Sirard, J-C., Niedergang, F. and Kraehenbuhl, J-P. (1999). Live attenuated *Salmonella*: a paradigm of mucosal vaccines. *Immunol. Rev.* **171**, 5–26.

Appendix

Table A.1. Top ten infectious disease killers, world-wide

Disease	Deaths (millions) in 1998[a]	Comments
Respiratory infections	3.5	Mostly children
AIDS (HIV)	2.3	[b]
Diarrhoeal infections (viruses, bacteria, helminths)	2.2[c]	Mostly children
Tuberculosis	1.5[d,e]	Sinister synergism with HIV, forming a lethal combination
Malaria	1.1	Nearly all in Africa, and nearly all children under age 5
Measles	0.9	[d]
Tetanus of newborn	0.4	[d] Bacteria from soil contaminate cut surface of umbilical cord
Whooping cough	0.35	[d]
Syphilis	0.16	
Meningitis	0.14	

AIDS, acquired immunodeficiency syndrome; HIV, human immunodeficiency virus.
[a] World Health Organisation Report 1999. By sheer numbers, infectious and parasitic diseases are the biggest killers (for comparison, tobacco kills an estimated 3.5 million people each year). Most of the deaths are in developing countries, and directly or indirectly due to poverty. In the past 50 years more have died of malaria, tuberculosis and AIDS than in all the wars combined. Note that typhus, typhoid, Lassa and Ebola fevers and schistosomiasis, come further down the list.
[b] Nearly all heterosexually acquired. Mainly adults (leaving orphans), although children can be infected *in utero*. More than 14 million deaths since the epidemic began.
[c] This works out at nearly 7000 deaths a day, mostly in parts of Africa, Asia, and Latin America, where the average 2–3-year-old has diarrhoea for a total of 2–3 months each year.
[d] Preventable by standard vaccines.
[e] The number infected with the related bacteria (*M. leprae*) causing leprosy, a disease once feared throughout the world, has fallen dramatically from 10–15 million to less than 1 million in the past 10 years, thanks to multidrug therapy.

Table A.2. Bacteria of human importance

Organism	Diseases	Other features
Gram-positive cocci		
Staphylococcus aureus	Boils, septicaemia, food poisoning	Common skin commensal. Phage typing identifies virulent strains. Enterotoxin causes food poisoning
Streptococcus pyogenes	Tonsillitis, scarlet fever, erysipelas, septicaemia	Also causes glomerulonephritis, and rheumatic fever, with immunopathological basis
Streptococcus viridans group (*Streptococcus sanguis*, etc.)	Infective endocarditis	Oral commensals settle on abnormal heart valves during bacteraemia
Streptococcus mutans	Dental caries	Regular inhabitant of mouth; initiates plaque on tooth surface
Streptococcus pneumoniae	Pneumonia, otitis, meningitis	Normal upper respiratory tract commensal, can spread to infected or damaged lungs
Gram-negative cocci		
Neisseria gonorrhoeae	Gonorrhoea	Obligate human parasite
Neisseria meningitidis	Meningitis	Obligate human parasite; increased upper respiratory carriage in epidemics
Gram-positive bacilli		
Corynebacterium diphtheriae	Diphtheria	Natural host man. Noninvasive disease due to toxin
Bacillus anthracis	Anthrax	Pathogen of herbivorous animals who ingest spores. Occasional human infection
Clostridium spp.	Tetanus, gas gangrene, botulism	Widely distributed in soil and intestines
Gram-negative bacilli		
E. coli	Urinary tract infections, infantile gastroenteritis	Normally inhabit the intestine (man and animals). Many antigenic types
Salmonella spp.	Enteric fever; food poisoning	*Salmonella typhi* – natural host man; invasive. Other *Salmonella* – 1000 species, mainly animal pathogens
Shigella spp.	Bacillary dysentery	Obligate parasite of man. Local invasion only
Proteus spp.	Urinary tract and wound infection	Common in soil, faeces. Occasionally pathogenic
Klebsiella spp.	Urinary tract and wound infection, otitis, meningitis, pneumonia	Present in vegetation, soil, sometimes faeces. Pathogenic when host resistance lowered
Pseudomonas aeruginosa	Urinary tract and wound infection	Common human intestinal bacteria. Resists many antibiotics
Haemophilus influenzae	Pneumonia, meningitis	Human commensal. Invades damaged lung
Bordetella pertussis	Whooping cough	Specialised human respiratory parasite

Organism	Diseases	Other features
Yersinia pestis	Plague	Flea-borne pathogen of rodents. Transfer to man as greatest infection in human history
Brucella spp.	Undulant fever	Pathogens of goats, cattle and pigs with secondary human infection
Legionella pneumophila	Legionnaire's disease	Respiratory pathogen of man, often acquired from contaminated air-conditioning units
Vibrio cholerae	Cholera	Obligate parasite of man. Noninvasive intestinal infection
Helicobacter pylori	Peptic ulcer	Infection widespread. Pathogensis unclear, but cytotoxin is important and urease helps wth residence in acid environment
Acid-fast bacilli		
Mycobacterium tuberculosis	Tuberculosis	Chronic respiratory infection in man; killed 1.5 million in 1998. Enteric infection with bovine type via milk
Mycobacterium leprae	Leprosy	Obligate parasite of man. Attacks skin, nasal mucosa and nerves. About 1 million lepers in the world
Miscellaneous		
Treponema pallidum	Syphilis	Obligate human parasite. Sexual transmission. Related nonvenereal human bacteria
Actinomyces israeli	Actinomycosis	Normal inhabitant human mouth
Leptospira spp.	Leptospirosis (Weil's disease, etc.)	Mostly pathogens of animals. Human infection from urine of rats, etc.
Mycoplasma spp.	Pneumonia, urethritis	Airborne transmission of *M. pneumoniae*. Lack true cell wall – hence no staining with Gram stain and no sensitivity to beta-lactams
Rickettsia spp.	Typhus, spotted fevers, Q fever (*C. burnetii*)	Obligate intracellular parasites. Gram negative. Acquired from animal reservoir via biting arthropods or aerosol (*C. burnetii*)
Chlamydia spp.	Trachoma, urethritis, pneumonia	Obligate intracellular parasites. Acquired by direct human contact, or from infected bird (psittacosis)

Table A.3. Characteristics of different bacteria and comparison with viruses

	Regular bacteria	Mycoplasmas	Rickettsias	Chlamydias	Viruses
DNA and RNA	+	+	+	+	–
Cell wall muramic acid	+	–	+	+	–
Binary fission	+	+	+	+	–
Growth in nonliving media	+	+	–	–	–
Examples of microbe	Staphylococci *Mycobacterium tuberculosis* *Treponema pallidum*	*Mycoplasma pneumoniae*	*Rickettsia prowazeki* *Coxiella burnetii*	*Chlamydia psittaci* *Chlamydia trachomatis*	Herpes simplex Rhinovirus
Human diseases produced by the above microbes	Abscesses Tuberculosis Syphilis	Atypical pneumonia	Typhus Q fever	Psittacosis Trachoma	Cold sore Common cold

Resolving power of the naked eye = 100 μm
Resolving power of light microscope = 0.2 μm (= 200 nm)

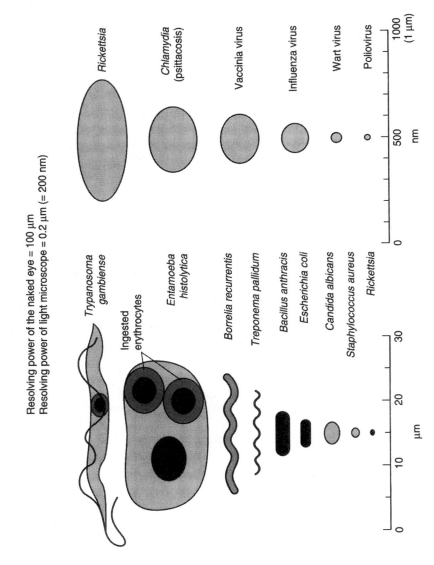

Fig. A.1 Relative size of microorganisms.

Table A.4. Fungi and protozoa of human importance

Organism	Diseases	Other features
Fungi		
Candida albicans	Thrush, dermatitis, etc.	Normally present on body surfaces; occasionally pathogenic
Dermatophytes (*Trichophyton* spp. *Epidermophyton* spp. *Microsporum* spp.)	Ringworm	Infection of skin, hair, nails (e.g. *Tinea pedis* – athlete's foot). Some species acquired from animals
Cryptococcus neoformans	Meningitis	Occurs in bird droppings, soil; causes skin, lung, CNS disease in immunocompromised
Blastomyces spp.	Blastomycosis	Soil fungi in the Americas; systemic infection in man
Histoplasma capsulatum	Histoplasmosis	Soil fungus in the Americas; can give lung lesions and systemic illness in man
Pneumocystis carinii	Pneumonia	Worldwide infection, causes disease in immunocompromised
Protozoa		
Plasmodia (four species)	Malaria	Mosquito transmitted; persistent infection in man
Toxoplasma gondii	Toxoplasmosis	Widely distributed in animals and birds; transplacental infection in man
Trichomonas vaginalis (flagellate)	Trichomoniasis (urethritis, vulvovaginitis)	Genito-urinary infection in both sexes, often asymptomatic
Giardia lamblia (flagellate)	Low-grade intestinal disease	Water-borne outbreaks occur
Trypanosoma spp. (flagellate)	Trypanosomiasis	Transmission by biting insects (animal reservoir); three species pathogenic for man
Leishmania spp. (flagellate)	Leishmaniasis (kala-azar, oriental sore, etc.)	Transmission from animal host to man via sandflies
Entamoeba histolytica	Amoebic dysentery	Invasion of intestinal mucosa; may spread to liver
Balantidium coli (ciliate)	Dysentery	Infection of man from pigs

Table A.5. Viruses of human importance

Nucleic acid	Virus group[a]	Example	Envelope from infected cell	Miscellaneous properties
ss DNA	Parvovirus	Human parvovirus	−	Mild disease in children; infects blood-forming cells in bone marrow
ds DNA	Poxvirus	Vaccinia Myxoma, moll. contagiosum	−[b]	Large complex viruses
	Herpesvirus (*herpes* = creeping)	Herpes simplex (HHV1, HHV2) Varicella–zoster (HHV3) Epstein–Barr virus (HHV4), cytomegalovirus (HHV5), HHV6, HHV7, HHV8	+	Latency and oncogenicity
	Adenovirus (*adeno* = gland)	Adenovirus (types 1–33 in man)	−[b]	May give latent infections of lymphoid tissue
	Papovavirus (<u>papi</u>lloma <u>polyo</u>ma <u>va</u>cuolating virus)	Wart virus, polyoma (JC and BK) viruses, SV40	−[b]	Oncogenic in experimental animals, persistent infection common
	Hepadnavirus	Hepatitis B virus	+	Not readily grown *in vitro*
ss RNA	Picornavirus (*pico* = small + RNA)	(a) Enterovirus 3 polioviruses 33 echoviruses 30 coxsackie viruses Hepatitis A virus (enterovirus 72)	−	Acid stable (pH 3) Infect alimentary tract ± heart, liver, CNS. Coxsackie viruses pathogenic for baby mice
		(b) Rhinoviruses (100 types)	−	Common cold viruses, acid labile (pH 3)
	Togavirus (*toga* = cloak)	Alphaviruses – Semliki Forest virus, Sindbis virus, Ross River virus, etc.	+	Multiply in arthropods, over 100 viruses
		Rubivirus	+	Rubella in man
	Flavivirus (*flavus* = yellow)	Flaviviruses – yellow fever virus, dengue virus, etc.	+	Multiply in arthropods, over 100 viruses
		Pestivirus	+	Bovine viral diarrhoea; hog cholera
		Hepatitis C virus	+	Hepatitis in man; not grown *in vitro*

Family	Viruses		Notes
Orthomyxovirus (*myxo* = mucin)	Influenza types A, B and C	+	Respiratory infections (intestinal in birds)
Paramyxovirus	Parainfluenza types 1–4, Respiratory syncytial virus, Mumps, measles viruses	+	Respiratory ± generalised infection
Coronavirus (*corona* = crown)	Common cold agents	+	Group includes mouse hepatitis virus and the toroviruses (Table 8.7)
Retrovirus	HIV; leukaemia viruses in man (HTLV1), mouse, cat, etc; mammary tumour viruses	+	World-wide 34 million people infected with HIV in 1998, 2/3 of them in sub-Saharan Africa
Bunyavirus	Rift Valley fever, sandfly fever, hantaviruses	+	Human infection from mosquitoes (RVF, La Crosse), sandflies, or directly from rodents (hantaviruses). Cause febrile, sometimes haemorrhagic illness
Rhabdovirus (*rhabdo* = bullet)	Rabies, vesicular stomatitis of horses	+	
Arenavirus (*arena* = sand)	Lymphocytic choriomeningitis, S. American haemorrhagic fevers, Lassa fever	+	Normally infect rodents; may give serious disease in man
Calicivirus (*calix* = cup)	Vesicular exanthem virus	–	Vesicular lesions in swine, dolphins, calves, dogs, chimpanzees, etc.
	Norwalk virus	–	Diarrhoea in humans
Filovirus	Marburg and Ebola viruses	+	Cause severe disease in occasional nosocomial outbreaks in Africa; natural reservoir unknown
ds RNA			
Reovirus (**r**espiratory **e**nteric **o**rphan)	Reoviruses types 1–3	–	Harmless in man
	Colorado tick fever virus Rotaviruses (*rota* = wheel)	–	Infantile gastroenteritis

[a] Prions (see pp. 351–3) are therefore not included in this table.

[b] Do not have a conventional envelope, but proteolytically processed viral antigens appear on surface of infected cells which are then susceptible to immune lysis.

Conclusions

One of the most important conclusions from this survey of microbial infection and pathogenicity is that various microorganisms have developed many of the theoretically possible devices that enable them to overcome or by-pass host defences. Microorganisms evolve rapidly compared with their vertebrate host species, and can generally be expected to be one step ahead. They are also quick to take advantage of changes in the host's way of life, and the comparatively recent increases in human density, for instance, have been exploited, especially by the respiratory viruses. For similar reasons the venereal route of infection has become more and more promising from the microorganism's point of view. Modern syphilis appears to have originated from an ancestral nonvenereal spirochaete similar to yaws, which infected the skin in warm countries and was spread by contact. The venereal form arose in the towns and cities of temperate countries where skin-to-skin contact was reduced because people wore clothes, and mucosal spread offered greater possibilities. Contemporary venery has led to a flowering of chlamydial and other infections of the genital tract.

Unless an infectious agent is transmitted effectively it will not survive. Transmission is an essential requirement (Table 1.1). It is a striking feature of human infections acquired from arthropods and from animal reservoirs that almost none of them are transmitted sequentially from person to person. If human to human transmission occurs at all, it fails to be maintained for more than a few transfers (Lassa fever, monkeypox). Infections of this type depend on human proximity to infected arthropods, birds, or mammals. They are often restricted geographically and they are eliminated when the source of infection is eliminated (e.g. rabies, malaria, yellow fever). In contrast, nearly all the infections that spread directly from person to person, maintain themselves independently of arthropods and animals (Fig. 2.12).

Transfer of microorganisms via urine, faeces, and food has greatly decreased as a result of public health reforms, at least in developed countries, but aerosol and mucosal (kissing, venereal) transfer occurs with ever-increasing efficiency (Fig. 2.12). The mucosal route of infection is more difficult to control. Mucosal contacts are part of loving and caring, at the core of man's humanity, and as long as people are people, microorganisms will have the opportunity to spread in this way.

Control of the spread of respiratory infections poses even greater problems. For the foreseeable future they will predominate in our crowded communities. One person soon infects scores of others and new infections,* such as pandemic strains of influenza, now spread throughout the world with formidable speed. This is the type of infection that could present a major threat to the human species. It might be prudent to study more thoroughly the aspects of pathogenesis that are concerned with respiratory transmission, and to identify the genetic determinants that would confer on a sexually or faecally transmitted microorganism the capacity to be transferred by aerosol. Also, because most of our new infections come from arthropods or animals, we need to know more about the pathogenic determinants that enable them to infect us and be transmitted between us. Since it first appeared in Africa some time before the 1960s, human immunodeficiency virus (HIV) has already changed and become more readily transmissible by the sexual route. Perhaps there is an unsurmountable pathogenic block to respiratory transmission, but in acquired immunodeficiency syndrome (AIDS) patients the virus is known to be present in alveolar macrophages and broncho-epithelial cells. A further short step into respiratory epithelium could convert HIV into a true major threat to our species.

Because microorganisms can evolve so rapidly, there is a real possibility that a particularly unpleasant one could emerge at any time. For instance, there have been no influenza A pandemics of very great significance since the 1918 outbreak. If a new strain appeared that spread with great facility and was at the same time highly lethal, say by invading cardiac muscle to produce myocarditis, the human population could be decimated before vaccines controlled it or a more stable type of virus–host balance emerged (see Ch. 1).

On the other hand, we are learning much more about infectious agents and infectious diseases, as outlined in this book. Vaccines have been of immense importance in the past and hold great promise for the future. The evolution of a microorganism can be decisively terminated by the proper application of knowledge. Smallpox, the most widespread and fatal disease in England in the eighteenth century and a major cause of blindness, has been totally eradicated from the earth.

It is important to contrast the incidence of infectious disease in different parts of the world, as discussed in Ch. 2, and to remember the

---

* Important respiratory infections are not necessarily new ones. Tuberculosis remains as a major problem worldwide, number four in the top ten list of infectious killers (Table A.1), and about a third of the world's population is thought to be infected. A total of 98% of the deaths are in poorer, developing countries. *M. tuberculosis* is still evolving and has not yet finished with the human race.

differences between the 'haves' and the 'have nots'. In developing countries, there are millions of unnecessary deaths from infection each year (Table A.1), and these are due, directly or indirectly, to poverty. About 1300 million people still live in grinding poverty, on less than US$1 a day, in spite of the overall growth of the world's economy. In 1995, the richest fifth of the world's population had 82 times the income of the poorest fifth. In developing countries, many of the old infections have been eliminated, but various latent, persistent, and opportunistic infections remain, especially in those kept alive by modern medicine and who have serious defects in antimicrobial resistance.

We need greater understanding of disease processes and pathogenicity, not only for its basic biological interest, but also because it helps with the development of vaccines, with the control of persistent and latent infections, and with our ability to deal with any strange new infections that arise and threaten us.

References

Hackett, C. J. (1963). On the origin of the human treponematoses. *Bull. W.H.O.* **29**, 7–41.

Zumla, A. and Grange, J. M. (1998). Tuberculosis. *Brit. Med. J.* **316**, 1962–1964.

Glossary

Active immunity   Immunity acquired actively following infection or immunisation by vaccines.

Addison's disease   Disease resulting from destruction of adrenal glands, characterised by weakness, debility and very great susceptibility to the stress of infection, trauma, etc. Other features include spontaneous hypoglycaemia and pigmentation.

Adjuvant   A material that enhances the immune response to an antigen.

Agglutination   Clumping together of proteins (in antigen–antibody reactions) or microorganisms, or red blood cells (haemagglutination).

Agonist   Something that stimulates a biological response by occupying cell receptors.

Aleutian disease virus   This virus infects mink and causes a fatal immunopathological disease in the type of mink that are homozygous for a recessive gene conferring the Aleutian coat colour.

Anamnestic response   Secondary immune response (see Primed).

Anterior horn cells   The main motor neurons in the anterior horn (as seen in cross-section) of the spinal cord, supplying striated muscle.

Antigen presentation   Display on the cell surface of processed peptides in combination with MHC proteins.

Antigen processing   Proteolytic digestion of a protein to form peptides which combine with MHC proteins to be presented on the cell surface to specifically reactive T cells.

Antigenic determinant (= epitope)   The small site on the antigen to which antibody attaches. Large antigens such as proteins carry several different antigenic determinants on the molecule, against which several different antibodies are formed.

Antigenic site   A cluster of epitopes/antigenic determinants (q.v.).

Apoptosis   Death of a cell in which it rounds up and is phagocytosed. It is not lysed. Involved in the natural regulation of cell numbers in tissues and occurs as the result of signals transmitted by adjoining cells. Is triggered by some viruses and inhibited by others (see Ch. 8).

427

Arthus response   Inflammatory reaction formed at the site where antigen is given to an animal possessing precipitating antibody to that antigen. Characteristically, oedema, haemorrhage and necrosis appear after a few hours ('immediate hypersensitivity'), and complement, polymorphs and platelets are involved in the reaction.

Attenuated   Reduced in virulence for a given host, often as a result of continued growth of a microorganism in an artificial host or culture system.

Autoimmunity   Immunity (humoral or cell mediated) to antigens of the body's own tissues. Can cause tissue damage and disease, but also occurs as a harmless consequence of tissue damage.

Avidity   Refers to the strength of binding of $\geq$ 2-valent antibodies to multivalent antigens. (Affinity is a more precisely used term referring to the strength of binding of one antibody combining site to a monovalent antigen.)

Babesia   Intracellular protozoan parasites of deer, cattle, rodents, humans, causing the disease babesiosis. Transmitted by ticks. Occur in Africa, North America, Asia, Europe; more than 70 species. Similar to Plasmodia (malaria parasites) and multiply in red blood cells.

Bacterial cell wall   Constitutes up to 20% dry weight of cell. Basically peptidoglycan (= mucopeptide = polymer of aminosugars cross-linked by peptide chains) containing components unique to microorganisms (e.g. muramic acid). Peptidoglycan may constitute nearly all of wall (certain Gram-positive bacteria), sometimes with additional polysaccharides and teichoic acids. Gram-negative bacterial cell walls are mostly lipopolysaccharides and lipoproteins, with little mucopeptide (p. 94).

Bacteriocin   Complex bacteriocidal substance released by certain bacteria, active against related bacteria, e.g. colicins produced by *E. coli*; pyocins produced by *Pseudomonas aeruginosa*.

Basement membrane   A sheet of material up to 0.2 μm thick lying immediately below epithelial (and endothelial) cells and supporting them. Contains glycoproteins and collagen and to some extent acts as a diffusion barrier for microorganisms. Thickness and structure varies in different parts of the body.

B-cells   Population of lymphoid cells derived from bone marrow developing without the need for the thymus. Differentiate to form antibody-producing cells. Compose 10–20% circulating lymphocytes in man.

Capsid   Protein coat enclosing the nucleic acid core of a virus.

CD (cluster differentiation) antigens   Antigens on cell surfaces, serving various functions and used to identify different cell types (e.g. CD4 on helper T-cells). More than 100 different CDs are described.

Cell-mediated immunity (CMI)   Specific immunity mediated by and transferrable to other individuals by cells (T cells), not by serum.

Challenge Administration of antigen or pathogen to provoke an immune reaction, usually in a primed individual.

Chemokines Small molecular weight molecules acting as chemoattractants and activators of lymphocytes and macrophages. There are four families defined by the position of the first two cysteines in their sequence: CC (27 members), CXC (15 members), CX3C (1 member), and C (2 members). They bind specifically to a seven-transmembrane G-protein-coupled receptor, for which there exists an equally diverse family of chemokine receptors.

Coccus Spherical or ovoid bacterium.

Colicins See Bacteriocin.

Commensal ('table-companion') Associated with a host, often deriving nourishment from host, but neither beneficial nor harmful.

Complement An enzymic system of serum proteins, made up of nine components (C1–C9) that are sequentially activated in many antigen–antibody reactions. It is a unique cascade system in which a small triggering event is amplified into a large response. Complement can also be activated directly, without an initial antigen–antibody reaction; this is the alternate complement pathway. Complement is involved in immune lysis of bacteria, and of some viruses and other microorganisms. It plays a part in phagocytosis, opsonisation, chemotaxis and the inflammatory response.

Connective tissue Forms an all-pervading matrix, connecting and supporting muscles, nerves, blood vessels, etc. Consists of a muco-polysaccharide 'ground substance' containing cells (fibroblasts, histiocytes, etc.), collagen and elastic fibres.

C-reactive protein A protein with subunits of M_r 24 300 that happens to react with the C carbohydrate of the pneumococcus. It is synthesised in the liver and is detectable in the serum when inflammation or tissue necrosis has taken place. It binds to substances from microorganisms and damaged tissues, activating the complement system.

Cryptococcus neoformans A yeast-like fungus found universally in soil, occasionally causing local or generalised infection in man.

Cushing's syndrome A disease resulting from excessive secretion of hormones from the adrenal cortex. Patients show wasting of muscle and bone, fat deposits on face, neck and back, and small blood vessels are easily ruptured.

Cytokines A group of at least 20 proteins, including interferons and interleukins. They are the hormones of the immune system, mediating interactions between immune cells, and having pathological as well as protective actions on infectious diseases.

Defective virus replication Incomplete virus replication, with production only of viral nucleic acid, proteins or noninfectious virus particles.

Defensins Peptides present in tears and in phagocytes that act against and destroy many microbes (bacteria, viruses, fungi) by punching holes in outer membranes.

Delayed-type hypersensitivity (DTH) Hypersensitivity reaction visible 1–2 days after introduction (usually intradermally) of antigen into a sensitised individual. An expression of cell-mediated immunity (cf. Arthus reaction).

Dendritic cell A large, specialized antigen-presenting cell with long tree-like (dendritic) processes, present in lymphoid tissues; not phagocytic and does not bear Fc receptors. Similar to Langerhan's cell in skin.

Dorsal root ganglia A series of ganglia lying dorsal to the spinal cord (as seen in cross-section). Contain cell bodies of principal sensory neurons, each receiving impulses along fibres from skin, etc., and sending impulses along shorter fibres to spinal cord.

ELAM-1 Endothelial cell leucocyte adhesion molecule-1. Inflammatory mediator, binds to polymorphs, enabling them to stick to endothelial cells lining capillaries and venules.

Enanthem Lesions of mucosae (e.g. mouth, intestines) in virus infections (cf. Exanthem).

Endocytosis The uptake of material by the cell into membrane-lined vesicles in the cytoplasm. The term includes pinocytosis (uptake of fluids) and phagocytosis (uptake of particles).

Endogenous pyrogen (= Interleukin-1) Substance released from leucocytes (in man) acting on hypothalamus to produce fever. Endotoxin (q.v.) causes fever by liberating endogenous pyrogen.

Endotoxin Toxic component associated with cell wall or microorganism. Generally refers to lipopolysaccharide of Gram-negative bacilli, the toxic activity being due to lipid A (see Fig. 4.4 and Fig. 8.15).

Enterotoxin Toxin acting on intestinal tract.

Envelope Limiting membrane of virus derived from infected host-cell membrane.

Exanthem Skin rash in virus infections (cf. Enanthem).

Exotoxin Toxin actively secreted (e.g. cholera toxin) or released by autolysis from microorganism (e.g. tetanus toxin).

Fimbriae (pili) Thread-like processes (not flagella) attached to cell walls of certain bacteria, often mediating attachment to host epithelial cell.

Fomites Comprehensive word for patients' bedding, clothes, towels, and other personal possessions that may transmit infections.

Germinal centre A rounded aggregation of lymphocytes, lymphoblasts, dendritic cells and macrophages. Germinal centres develop in primary nodules (follicles) of lymphoid tissue in response to antigenic stimuli.

Gram-negative Losing the primary violet or blue during decolorisation in Gram's staining method. The method, developed by Hans

Gram, a Danish physician, in 1884, gives a simple and convenient distinction between groups of bacteria. The staining reaction reflects differences in cell wall composition (see Fig. 4.4), but the mechanism is not clear.

Gram-positive Retaining the primary violet or blue stain in Gram's method.

Granuloma A local accumulation of densely packed macrophages, often fusing to form giant cells, together with lymphocytes and plasma cells. Seen in chronic infections such as tuberculosis and syphilis.

Haemolysis Destruction of red blood cells. Caused by bacterial toxins, or by the action of complement on red cells coated with specific antibody.

Hapten A small molecule which is antigenic (combines with antibody) but is not immunogenic, i.e. does not induce an immune response *in vivo* unless attached to a larger ('carrier') molecule.

Heat shock proteins (hsp) A family of proteins that control the correct folding of other proteins, acting as 'molecular chaperones'. They are induced in both microbe and phagocyte during the stress of infection (e.g. raised body temperature); and also have immunological roles (e.g. in antigen processing).

Heterophile antibody Antibody to heterophile antigens which are present on the surface of cells of many different animal species.

HLA (see MHC).

Horizontal transmission The transmission of infection from individual to individual in a population rather than from parent to offspring.

Humoral immunity Specific immunity mediated by antibodies.

ICAM-1 Intercellular adhesion molecule-1.

Immune complex A complex of antigen with its specific antibody. Immune complexes may be soluble or insoluble, and may be formed in antibody excess, antigen excess, or with equivalent proportions of antibody and antigen. They may contain complement components.

Immune tolerance An immunologically specific reduction in immune responsiveness to a given antigen.

Immunopathology Pathological changes partly or completely caused by the immune response.

Infarction Obstruction of blood supply to a tissue or organ.

Integrin A family of at least 18 cell adhesion receptors (e.g. fibrinogen receptor, laminin receptor) expressed on many cell types, mediating adhesion of cells to each other or to extracellular components. After interaction of integrin with ligand, vital signals affecting differentiation, proliferation, etc. are transmitted to the cell interior.

Interleukins Cytokines; a group of 18 different proteins, all of them cloned and sequenced, that carry vital signals between different immune cells.

Interleukin-1 (= endogenous pyrogen) Produced by macrophages, promotes activation and mitosis of T and B cells. Causes fever as well as a variety of effects on muscle cells, fibroblasts and osteoblasts.

Interleukin-2 Produced by T (especially Th) cells; essential for the continued proliferation (clonal expansion) of activated T cells.

Interleukin-3 Multicolony stimulating factor; stimulates precursor cells (e.g. in bone marrow) to divide and form colonies of polymorphs, monocytes, etc.

In vitro 'In glass', that is to say not in a living animal or person.

In vivo In a living animal or person.

Kinins Low molecular weight peptides generated from precursors in plasma or tissues and functioning as important mediators of inflammatory responses. C2 kinin is derived from complement, and other kinins from α_2-globulins.

Lactate dehydrogenase elevating virus A virus that commonly infects mice, and multiplies only in macrophages. The macrophages fail to remove certain endogenous enzymes from the blood and an infected mouse is identified because there is a rise in the level of plasma lactate dehydrogenase. Infection is lifelong, and there are no pathological lesions or harmful effects.

Latency Stage of persistent infection in which a microorganism causes no disease, but remains capable of activation and disease production.

LD$_{50}$ (lethal dose 50) Dose that kills 50% of test animals/cells. A direct measure of virulence.

Legionellosis Infection with *Legionella pneumophila*. The bacteria colonise cooling towers, creeks, showerheads, air conditioning units, etc., and are inhaled after becoming airborne. Some patients develop pneumonia.

Leishmaniasis Disease caused by protozoa of genus *Leishmania*, e.g. cutaneous leishmaniasis (Delhi boil, etc.) or generalised leishmaniasis (kala-azar).

Leucocytes Circulating white blood cells. There are about 9000 mm^{-3} in human blood, divided into granulocytes (polymorphs 68–70%, eosinophils 3%, basophils 0.5%) and mononuclear cells (monocytes 4%, lymphocytes 23–25%).

LCM Lymphocytic choriomeningitis virus. Naturally occurring virus infection of mice displaying many phenomena of great biological interest, e.g. vertical transmission, immunopathology, noncytopathic infection of cells.

Lymphokine A cytokine released by primed lymphocyte on contact with specific antigen. Involved in signalling between immune cells. Important in CMI; cf. monokine (e.g. IL-1) produced by monocytes and macrophages.

Lysosome Cytoplasmic sac present in many cells, bounded by a lipoprotein membrane and containing various enzymes. Plays an important part in intracellular digestion.

Lysozyme An enzyme present in the granules of polymorphs, in macrophages, in tears, mucus, saliva and semen. It lyses certain bacteria, especially Gram-positive cocci, splitting the muramic acid-β-(1→4)-*N*-acetylglucosamine linkage in the bacterial cell wall. It potentiates the action of complement on these bacteria. Presumably lysozyme is not exclusively an antibacterial substance because large amounts are present in cartilage. It is present in glandular cells in the small intestine, especially in the Brazilian ant bear, where its chitinase-like activity may help with the digestion of insect skeletons.

Marek's disease virus A herpes virus, commonly infecting chickens, and causing lymphocyte infiltration of nerves with demyelination and paralysis, and lymphoid tumours. Infectious virus present in oral secretions and feather follicles. Controlled successfully by a live virus vaccine.

Memory cells Sensitised cells generated during an immune response, and surviving in large enough numbers to give an accelerated immune response on challenge.

MHC (major histocompatibility complex) A region of the genome coding for immunologically important molecules.

Class I MHC molecules are HLA (human leucocyte antigen A, B, C) in man and H2 (K, D, L) in mice. They are associated with β_2 microglobulin and expressed on the surface of nearly all cells. They confer uniqueness on the cells of each individual and ideally the class I characteristics of donor and recipient should be matched for successful organ transplantation.

Class II MHC molecules (HLA-DP, DQ, DR in man; H-2 IA, IE in mice) are present on antigen-presenting cells (some macrophages, dendritic cells, Langerhans cells).

Monoclonal antibody A given B cell makes antibody of a certain class, avidity and specificity. Serum antibody consists of the separate contributions from tens of thousands of B cells. Dr Caesar Milstein discovered how to induce an individual B cell to divide and form a large enough population (clone) of cells to give bulk quantities of the unique antibody. This is a monoclonal antibody.

Natural antibodies Antibodies present in normal serum, reacting with a wide range of organisms. To a large extent they reflect specific responses to previous subclinical infections, e.g. normal sera lyse many Gram-negative bacteria because of antibodies induced by the normal intestinal flora.

Nosocomial infection An infection acquired in hospital.

Nucleocapsid Viral nucleic acid enclosed in a capsid consisting of repeating protein subunits.

Opsonin (Greek *opson*, a seasoning or sauce). Serum component that combines with antigen or the surface of a microorganism and promotes its phagocytosis by polymorphs or macrophages.

Otitis media Infection and inflammation of the middle ear.

Passive immunity Transfer of preformed antibodies to nonimmune individual by means of blood, serum components, etc., e.g. maternal antibodies transferred to foetus via placenta or milk, or immunoglobulins injected to prevent or modify infections.

Pathogenic Producing disease or pathological changes.

Persistent infection An infection in which the microorganism persists in the body, not necessarily in a fully infectious form, but often for long periods or throughout life.

Phage typing Different strains of *Salmonella typhi, Staphylococcus aureus*, or *Mycobacterium tuberculosis* can be distinguished on the basis of their different susceptibility to a battery of bacteriophages.

Pili (see Fimbriae).

Plaque forming cells (p.f.c.) Refers to lymphocytes that form areas of lysis in a layer of erythrocytes to which the lymphocytes are immunologically sensitised.

Plaque forming units (p.f.u.) Refers to virus that kills cells and forms plaques (holes) in cell sheets.

Plasma cell B cell which has differentiated to form rough surfaced (ribosome studded) endoplasmic reticulum, with basophilic cytoplasm. It is the major antibody-producing cell.

Plasmid A small extrachromosomal piece of genetic material in bacterium, replicating autonomously in the cytoplasm. It may carry 50–100 genes. Plasmids are common in Gram-negative bacilli, and also occur in staphylococci.

Pleural and peritoneal cavities Potential cavities surrounding organs of thorax and abdomen. Lined by 'mesothelial' membrane and containing macrophages and other cells.

Pneumocystis carinii Exceedingly common fungal parasite of respiratory tract of man and various animals; normally of zero pathogenicity. Little is known of its structure, life cycle or epidemiology. It attaches to host cells *in vitro* by means of a tubular projection but does not enter the cell except when phagocytosed, e.g. by an alveolar macrophage. It causes pneumonia in immunocompromised individuals, either by reinfecting them or by being reactivated from a persistent state.

Polyclonal activator Something that activates many clones of lymphocytes. Infections that activate B cells in this way cause the formation of large amounts of circulating antibody directed against unknown antigens as well as against the infectious agent, and often against host tissue antigens.

Primary infection The first infection with a given microorganism.

Primed Exposed to antigen for the first time to give a primary immune response. Further contact with the same antigen leads to a secondary immune response.

Prion Infectious particle containing neither DNA nor RNA, consisting of host protein (Pr^c) that has been converted into a self-replicating form (Pr^{sc}). Thought to be the infectious particle causing scrapie, kuru, Creuztfeld–Jakob disease (CJD), bovine spongiform encephalopathy (BSE), etc.

Properdin system Consists of Factor A (a serum protein). Factor B (a β-glycoprotein) and properdin. Not completely defined and role not understood, but may have antibacterial and antiviral action. It is an alternative pathway for the activation of complement, in which C1, C2 and C4 are short-circuited.

Pyogenic Causing production of pus.

Pyrogen A substance causing fever.

Reservoir Animal (bird, mammal, mosquito, etc.) or animals in which a microorganism maintains itself independently of human infection.

Reticulocytosis Presence in blood of increased numbers of an early form of red cell (reticulocyte), due to increased rate of production in bone marrow.

Reticuloendothelial system A system of cells that take up particles and certain dyes injected into the body. Comprises Kupffer cells of liver, tissue histiocytes, monocytes, and the lymph node, splenic, alveolar, peritoneal and pleural macrophages.

Schistosomiasis (= bilharzia) A disease with urinary symptoms common in many parts of Africa. Caused by the fluke (trematode) *Schistosoma haematobium*; larvae from infected snails enter water and penetrate human skin.

Shedding The liberation of microorganisms from the infected host.

SSPE (subacute sclerosing panencephalitis) A rare complication of infection with measles virus, occurring in about 1 per 100 000 cases. The incubation period of about 10 years classifies it as a 'slow' virus infection. Noninfectious mutant virus slowly spreads through the brain causing deterioration of brain function and death.

Streptococci Classified into groups A–H by antigenic properties of carbohydrate extracted from cell wall. Important human pathogens belong mostly to Group A (= *Streptococcus pyogenes*), which is divided into 47 types according to antigenic properties of M protein present on outermost surface of bacteria.

Streptolysin O Exotoxin produced by *Streptococcus pyogenes*. Oxygen labile, haemolytic and a powerful antigen.

Streptolysin S Exotoxin produced by *Streptococcus pyogenes*. Oxygen-stable, causing β haemolysis on blood agar plates, but not demonstrably antigenic.

Stress Physical or mental disturbance severe enough to initiate a coordinated response originating in the cortex and hypothalamus, and involving either the autonomic nervous system or pituitary–adrenal axis. Catecholamines and corticosteroids are released in an attempt to counter the harmful systemic effects of the disturbance (or often the threatened disturbance in the case of mental stress).

Symbiotic Living in a mutually beneficial association with the host.

Systemic infection Infection that spreads throughout the body.

T cells (T lymphocytes)   Population of lymphoid cells whose development depends on the presence of the thymus. Responsible for cell-mediated immunity. Compose 75% circulating lymphocytes in man. Distinguished by having on their surface CD4 proteins or CD8 proteins which define their reactivity with cells bearing MHC I or II proteins, respectively. All helper T cells are CD4.

Teleology   Doctrine that biological phenomena generally have a purpose, serving some function.

T-independent antigen   Antigen that directly stimulates a B cell to form antibody without the need for a helper T cell. These antigens (e.g. polysaccharides) have repeated determinants that cross-link Ig receptors on B cells. The antibodies formed are mostly IgM.

Titre   (1) A measure of units of antibody per unit volume of serum, usually quoted as a reciprocal of the last serum dilution giving antibody-mediated reaction, e.g. 120. (2) Measure of units of virus per unit volume of fluid or tissue. Usually given in log_{10} units per ml or g, e.g. $10^{5.5}$ p.f.u. ml^{-1}.

TNF (tumour necrosis factor)   A cytokine, first recognised as a product of activated macrophages (see Table 6.1). Plays a role in disease production as well as in host defence.

Toxoid   Toxin rendered harmless but still capable of acting as antigen.

Toxoplasma gondii   A protozoan parasite of the intestine of cats, which also infects mice, humans, sheep and other animals. Humans ingest oocysts, originating from cat faeces or cysts from infected meat, and about half of the inhabitants of the UK eventually develop antibodies. It is generally asymptomatic, but disease (toxoplasmosis) sometimes occurs, and infection during pregnancy can result in congenital abnormalities involving the brain and eyes.

Transfer factor   A preparation derived from disrupted human leucocytes which on transfer to other individuals can supply certain missing CMI responses. The active constituent is unidentified, but it has been successfully used to treat chronic mucocutaneous candidiasis.

Transformation   A change in the behaviour of a cell, for instance after infection with an oncogenic virus, so that it acquires the properties of a cancer cell. Transformed cells undergo continued mitosis so that the cells in a monolayer are not inhibited from growth by contact with neighbouring cells, and continue to multiply and form a heap of cells. The word also refers to changes in a lymphocyte associated with onset of division.

Tuberculin test   A skin test for delayed hypersensitivity to antigens from *Mycobacterium tuberculosis*. In man the antigen is introduced into the skin by intradermal injections (Mantoux test) or by multiple puncture (Heaf test and tine test).

Vector   As used in this book the word refers to an arthropod that carries and transfers an infectious agent. Quite separately, a vector means a replicating genetic unit such as a virus or a plasmid, which will carry and replicate a segment of foreign DNA that has been introduced into it.

Vertical transmission   The transmission of infection directly from parent to offspring. This can take place *in utero* via egg, sperm, placenta, during birth (contact with infected birth canal), or post-natally via milk, blood, contact.

Viraemia   Presence of virus in the bloodstream. Virus may be associated with leucocytes (leucocyte viraemia), or free in the plasma (plasma viraemia), or occasionally associated with erythrocytes or platelets.

Virion   The complete virus particle.

Index

Note:
Page numbers followed by the letter n indicate a footnote; **bold** page numbers refer to the Appendix